钟扬文选

钟　扬 ◎ 著
本书编委会 ◎ 编

复旦大学出版社

《钟扬文选》编委会

主　任：焦　扬　许宁生

主　编：袁正宏　刘承功　金　力

副主编：陈玉刚　张人禾　陈浩明

编　委：（按姓氏笔画排序）

　　　　王卫东　包晓明　严　峰　李子凡

　　　　陈　洁　陈家宽　周　鹏　赵佳媛

　　　　南　蓬　曾　艺　楚永全

本书由复旦大学出版资助基金资助

序

许宁生

社会需要好老师，钟扬教授就是一名好老师。

对于一名教师而言，他可以投身科研，也可以专注教学，还可以将无限的热忱投入到服务社会中去。当这些选择汇集在一个人身上的时候，便呈现出一种缤纷的轨迹。在整理《钟扬文选》的过程中，我们看到了他闪光的人生，看到了作为一名教师扎根祖国大地的家国情怀。

钟扬生前是复旦大学生命科学学院教授、博士生导师。他常说，教师是他最看重的身份。他认为，"每个学生都是一粒宝贵种子，全心浇灌就会开出希望之花"。他坚持教书育人、立德树人，坚持有教无类、因材施教，始终爱生如子、甘为人梯，始终为人师表、以身作则。他不忘初心、牢记使命，以坚定信念和高远志向，至诚报国的奉献精神，把人生理想和国家发展结合起来，把科研方向和社会需求结合起来，把"顶天"与"立地"有机紧密结合起来。他牢牢把握学科前沿领域的重大科学问题，潜心钻研、孜孜奋斗，获得一系列重要原创成果；同时直接对接国家战略需求和重大需求，数十年如一日奋战在生态文明建设一线，为建设美丽中国做出了贡献。

在他的努力下，西藏大学申请到了第一个理学博士点，为藏族培养了第一个植物学博士，带出了西藏第一个生物学教育部创新团队，带领西藏大学生态学科入选国家"双一流"学科建设名单，填补了西藏高等教育的系列空白；克服重重困难，坚持9年在上海引种红树林，尝试为上海海滨增加生态屏障。他热心社会公益事业，连续17年参与科普志愿服务，成绩卓著，是深受欢迎的"科普明星"。他把科学研究的种子播撒在雪域高原和上海海滨，为祖国的生态文明建设和科学教育事业做出了杰出的贡献。在他身上，集中体现了新时代科技工作者的一系列优秀精神品质。

钟扬教授特别注重积累跨学科知识和培养勇于开拓的创新精神，始终保持一颗善于发现问题、解决问题的心，不断攀登科学研究高峰。他提到，科学研究的本质和创新活动的基本范式就是不断尝试错误、改正错误，以此无限接近真理。钟扬教授说过，"只要国家需要、人类需要，再艰苦的科研也要去做"，在他的日常科研活动中，有曲折，甚至也有失败，但他不畏艰难、勇于探索，纵使经历各种苦厄，也不忘科研初心，越挫越勇，直至成功。

为了开展科研工作，不管路途多偏远、环境多险恶、条件多艰苦，克服重重困难，长期超负荷工作甚至冒着生命危险。凭着执着进取的敬业精神，只要对研究有帮助，钟扬教授都会亲身投入，义无反顾地去探索。他的一生都在与时间争分夺秒，践行着矢志不渝的追梦精神。他身上所体现出的高尚师德、优良师风、崇高师道，值得我们每一个人学习。

钟扬教授常说，当一个物种要拓展其疆域而必须迎接恶劣环境挑战的时候，总是需要一些先锋者牺牲个体的优势，以换取整个群体乃至物种新的生存空间和发展机遇。先锋者为成功者奠定了基础，他自身就是一个先锋者，倾其所有智慧和力量，带领团队协同创新、攻坚克难，为人类存储下绵延后世的基因宝藏、智慧和希望的种子。

当今世界，新一轮科技革命蓄势待发，要抢占未来发展的制高点，要实现中华民族的伟大复兴，需要广大科技工作者不忘初心、牢记使命、不懈奋斗。钟扬教授是我们身边的榜样，在全面建成小康社会、夺取新时代中国特色社会主义伟大胜利的奋进道路上，我们当以科学精神引领科技发展潮流，夯实创新研究之路，坚持把报效祖国作为自己的人生追求，把奉献社会作为自身的价值体现，把服务人民作为自己的肩头责任，用一流的科学研究和科研成果贡献自己的智慧和力量，回报祖国和人民。

（作者为复旦大学校长）

目 录

序 　　　　　　　　　　　　　　　　　　　　　　许宁生 / 1

上编
　要做时间的主人　　　　　　　　　　　　　　　　　　/ 3
　第一次……　　　　　　　　　　　　　　　　　　　　/ 7
　引用率低说明了什么?　　　　　　　　　　　　　　　/ 10
　道德底线干细胞研究试探中　　　　　　　　　　　　/ 12
　我国生物信息学教育的发展与挑战　　　　　　　　　/ 14
　宏观生物学与交叉学科人才培养的理念与实践　　　　/ 20
　生命的节拍　　　　　　　　　　　　　　　　　　　/ 25
　倘若生命失去节律　　　　　　　　　　　　　　　　/ 31
　从战争中学习战争　　　　　　　　　　　　　　　　/ 33
　美丽实用的分子进化　　　　　　　　　　　　　　　/ 35
　人类与病毒共舞　　　　　　　　　　　　　　　　　/ 39
　分子进化　　　　　　　　　　　　　　　　　　　　/ 41
　番木瓜的故事　　　　　　　　　　　　　　　　　　/ 53
　学会"与流感共舞"　　　　　　　　　　　　　　　/ 57
　西藏，已不再遥远　　　　　　　　　　　　　　　　/ 62
　进化论与进化生物学的发展　　　　　　　　　　　　/ 64
　达尔文进化论的科学本质与贡献　　　　　　　　　　/ 71
　进化论的分子时代　　　　　　　　　　　　　　　　/ 76
　世界之巅上的断思　　　　　　　　　　　　　　　　/ 84
　生物信息学专业规划的理念与实践　　　　　　　　　/ 87
　从艰苦中提取欢乐——访植物学家钟扬教授　　　　　/ 95

生命之树常青 / 103
"影响因子"与科研创新无关 / 107
生命的高度 / 110
中国适应性生存与可持续发展 / 112
获得,还是失去,这是个进化问题 / 121
播种未来(钟扬院长文字配音稿) / 125
创新文化是一种"试错"文化 / 127
怎样真正提升研究生的能力 / 130
原创的,就是世界的——钟扬教授专访 / 137
在我失联的日子里 / 144
复旦博学文库(第一辑)总序 / 150
研究生培养质量提升的解决之道 / 152
弘扬劳模精神 培育创新人才 / 158
一个招办主任儿子的高考 / 161
复旦博学文库(第二辑)总序 / 166
钟扬老师给家长的一封信 / 168
达尔文进化论过时了吗? / 170
基因密码——生命与哲学 / 175
教育最重要的是释放学生的学习力 / 183
我和科学队长 / 190
世界屋脊上的种子收集者 / 192

下编

钟扬教授各学术领域代表性论文 / 201
 Ⅰ. 生物数据模型与信息系统 / 201
  荷花品种的数量分类研究 / 203

Data model and comparison and query methods for interacting classifications in a taxonomic database / 217

HICLAS: a taxonomic database system for displaying and comparing biological classification and phylogenetic trees（摘要） / 244

PlantQTL-GE: a database system for identifying candidate genes in rice and *Arabidopsis* by gene expression and QTL information（摘要） / 245

PBmice: an integrated database system of *piggyBac* (PB) insertional mutations and their characterizations in mice（摘要） / 246

MPSS: an integrated database system for surveying a set of proteins（摘要） / 247

Ⅱ. 分子进化分析方法及应用 / 248

Testing hybridization hypotheses based on incongruent gene trees / 250

Detecting evolutionary rate heterogeneity among mangroves and their close terrestrial relatives / 278

The *Schistosoma japonicum* genome reveals features of host-parasite interplay（摘要） / 293

Solution structure of Urm1 and its implications for the origin of protein modifiers（摘要） / 294

Molecular evolution of the SARS coronavirus during the course of the SARS epidemic in China（摘要） / 295

Photosynthetic metabolism of C-3 plants shows highly cooperative regulation under changing environments: A systems biological analysis（摘要） / 296

Ⅲ. 生物多样性与植物基因组分析 / 297

Discovery of a high-altitude ecotype and ancient lineage of *Arabidopsis thaliana* from Tibet / 299

The genome and transcriptome of *Trichormus* sp. NMC-1: insights into adaptation to extreme environments on the Qinghai-Tibet Plateau / 308

Testing the effect of the Himalayan mountains as a physical barrier to gene flow in *Hippophae tibetana* Schlect. (Elaeagnaceae)（摘要） / 336

Genetic diversity and population structure of *Lamiophlomis rotata* (Lamiaceae), an endemic species of Qinghai-Tibet Plateau（摘要） / 337

Fine- and landscape-scale spatial genetic structure of cushion rockjasmine, *Androsace tapete* (Primulaceae), across southern Qinghai-Tibetan Plateau （摘要） / 338

Genetic diversity in *Primula obconica* from Central and South-west China as revealed by ISSR markers（摘要） / 339

附：钟扬教授学术论文总目 / 340

钟扬教授著作选摘 / 372

 Ⅰ. 专业著作 / 372

  《数量分类的方法与程序》前言 / 372

  《分支分类的理论与方法》前言 / 374

  《简明生物信息学》前言 / 376

 Ⅱ. 专业译著 / 378

  《水生植被研究的理论与方法》编译前言 / 378

  《延续生命》译者的话 / 380

 Ⅲ. 科普著作 / 382

  《基因计算》后记 / 382

 Ⅳ. 科普译著 / 386

  《林肯的DNA》译后记 / 386

  《大流感》译后记 / 388

  《DNA博士》译后记 / 391

 Ⅴ. 科普译著审校 / 395

  《造就适者》导读：证据的力量 / 395

 附：钟扬教授著作总目 / 398

上编

要做时间的主人

时间，对每一个人来说，都是宝贵的，对我们青少年尤其宝贵。虽然如此，可不是人人都理解它的价值。

记得有一年除夕的晚上，我们一群要好的小伙伴在外面噼噼啪啪放了一通爆竹之后，都跑到我家来做猜谜语游戏。我爸爸是很喜欢孩子的，他也参加进来与我们同乐。他给我们出了一个题："世界上最宝贵的东西是什么？"我们有的回答是手表，有的回答是金子，还有的回答是电视机……大家说了许多答案，可爸爸老是皱着眉毛不点头。他看我们张张小脸上都快急出汗来了，才解答说："世界上最宝贵的东西是时间。其他东西若失掉了，还可以买回来，时间失掉了，无论用多少金钱也买不回来。"

这是我四岁那年的事。后来我回忆起那次有趣的游戏，越来越觉得它有意义。在我入学以后，那次游戏在我学习中起到了料想不到的作用，成了鞭策我前进的力量。

我的家庭条件还是比较好的。爸爸妈妈都受过高等教育。来我家串门儿的叔叔阿姨们，都是知识分子，可以说是"往来无白丁"。他们在我家惊奇地发现，我爸爸妈妈教育孩子的方法很特别，从不见他们督促孩子学习。叔叔和阿姨们问是怎么回事儿，爸爸对他们说："一个孩子，如果光靠大人督促着去学习，他绝不会有大出息的！关键是看他能不能自觉地学习。"

爸爸这话虽然是对客人说的，却给我敲了警钟。关键在自己，这话一点不错。老师是无私的，他在讲课的时候，总希望每个同学都能消化他讲授的知识，可是有的人学得多，有的人学得却很少。我在听课的时候，思想非常集中，老师讲的话，一句也不让它漏掉，一堂课的时间，一分钟也没浪费。我看到有些同学在课堂上根本就不用心听，不是在做小动作，就是两眼不停地朝外张望。我觉得他们表面上看来是坐在那儿听课，实际上是在消磨时间。

在小学读书的时候，功课不太重，有些作业，在课堂上就做完了。有的孩子放学回家，草草完成课堂上未完成的作业，就跑出去嬉闹。我呢，做完作业之后，就看课外书，练书法，背诗词，学画画，有时还学点音乐知识。尽管我的爱好这么广泛，但有个中心，那就是课本。课本不钻透，什么事也不干。课本吃透后，想学什么就学什么。课本以外的知识，虽然不完全与课本有直接关系，但它可以帮助活跃自己的脑神经，促使自己多思考一些问题，增强自己的思维能力，这样，它又反过来促进了对课本的学习。但是有一条，我从来不在那些毫无意义的事情上花费半分钟的时间。

时间是宝贵的，它对每个人也是公平的，从不偏爱谁，但有的人得的多，有的人却得的少。这种多或少，全由你珍惜或者浪费来决定的。知识跟着时间走，多抓住一分钟，就多一份收获。基于这种认识，我的学习一向抓得很紧，每次测验，成绩都是前一二名。有些叔叔阿姨们来我家"取经"，看看我是怎么学习的，回去好教育自己的孩子。可当他们走进我家一看，都大失所望。他们有时看到我在画画，有时看到我在读诗歌，有时还看到我在"吱啦吱啦"地拉二胡。他们平时就反对自家的孩子搞这些玩意，能把我这"经"拿回去传播吗？我听到他们惋惜地对我爸爸说："你让孩子搞这些玩意，不分散他的学习精力吗？"爸爸笑着回答："谁管他！他愿怎么干就怎么干呗！"

我知道爸爸说的并不是真心话。我若真的糟蹋时间，他一定会干涉的。他知道我从来就不浪费时间，课余养成的多种爱好还有助于学习。

记得我上小学三年级的时候，有一天上午，老师布置了一道作文题，叫大家当场作好。我的作文写好后，交给了老师。他看后赞不绝口，说我这篇作文结构严谨，人与物、情与景写得也很得当。他认为一定是我爸爸教的。其实，我爸爸妈妈都没给我讲过怎样写作文。这篇作文写得比较好，与我业余时间学画画很有关系。平常画人物素描写生，讲形体结构，讲素描关系，讲明暗调子等。写文章与画画有些地方是相通的，让我给"套"上了。此后这篇作文在报纸上发表了，有人问我是怎么写出来的，我告诉他是我在画画中"画"出来的。他们听罢很不理解。

　　课余有多种爱好是不会影响课本学习的，这点我是深有体会的。就像人吃菜一样，老是吃肉也会感到腻口，有时吃点豆芽儿或青菜什么的，就会感到异常可口。会学习的人，对时间的支配也是这样。真正浪费时间的人有几种：一种是到处嬉闹，干些毫无意义的事；另一种就是在大人严格监视下，除课本外不准看任何书、干任何事，规规矩矩坐在那里，貌似在啃课本，实际上思想早已开小差了。这两种人虽然表现形式不一样，其结果却都是一样的。前者是当了抛撒时间的浪荡公子，后者是做了时间的奴隶。

　　时间有时也很狡猾，要珍惜它，就得学会驾驭它。你若稍有疏忽，它就会悄悄地从你身边溜走。记得是在我上小学三年级那年秋天，上边号召学生"反潮流"，一时弄得学校大乱。在课堂上学生可以骂老师，老师哪还敢管学生？这正是时间容易溜走的机会。就在这时，我佯装有病，向学校请了病假，回到家里自学，不会的就找爸爸妈妈问，或者趁饭后时间到老师家里请教。爸爸妈妈对我装病不去上学，不但不反感，还夸我这样做得对。时间最公正，它不会亏待勤奋学习的孩子，也不会便宜不干正事的懒汉，学校秩序好一些之后，我把小学三、四年级的课程全部学完了，一些"反潮流"的英雄们，却把原先学的知识忘光了。我朝前跳到五年级读书，他们却朝后"退"到二年级"热剩饭"吃。

　　我上初中之后，"四人帮"被粉碎了。校风大变，同学们都是争先恐后地学习，你追我赶，谁也不甘落后。这时，我感到时间不够用，必须快马加鞭，跑在时间的前面。有件事使我感触很深，逼着我在学习的征途上不得不快上加快。

　　那是一个明媚的早晨，我拿着一本英语书跑到城外山上早读。一会儿来了一个农村的小伙子，担着一担土粪往山坡地里送，跑得很快，累得满头大汗。撒完了粪，他用羡慕的目光看了我一眼，挑起空担子，又慢慢下坡了，看样子他在边走边想心思。看着他，我心里突然翻起了浪潮——中国的农民还这样苦，劳动方式还这样落后，多么需要用科学文化来改变这种现状啊！这件事对我触动很大，后来我以此素材写了一首诗发在《中岳》杂志上。

时间，对于懒惰者像一只蜗牛，背着你慢吞吞地爬行；对于勤奋者却像滔滔的江水，驮着你日夜向前！我在初中阶段，把高中的课程全部自学完毕。1979年我破格参加高考，以较优异的成绩被录取到中国科技大学少年班。

有不少同学问我学习到底有什么诀窍，我实在不好回答。如果说真有诀窍，我体会也只有这么七个字：要做时间的主人。

本文发表于《少年大学生谈学习》（安徽少年儿童出版社，1984年12月），署名为中国科技大学少年班79级钟扬，收入本书时略有修改。

第一次……

首先，我得申明一下，这确实是我第一次自己发帖子。我喜欢看复旦自己的BBS，有时实在想说点什么，就写在纸上，让学生代发。由于每次换名字，经常有人对钟老师的真实性产生怀疑。其实，到目前为止，还未发现一例仿冒事件。即使有，只要不是恶性的、重大的、需要学校领导来谈话的，我不会追究。我自己不发帖子的一个主要原因是打字太慢，按一学生的话说："再努力三年（刚好是他读博士的时间），钟老师就能达到菜鸟的水平了。"看他说话的神态，我觉得就像我鼓励他咬紧牙关，三年拿下博士，其实我心里觉得恐怕至少要五年才行的样子。的确，在我这个年龄段，要找一个既不会五笔，拼音还不好，只能是"一指禅"打字，又得以写作为生的人似乎还不太容易。我挺羡慕你们活跃在各网站的写手们，如不是亲眼所见，我以为各位都有三名以上的女秘书呢。此时此刻，我正在异国他乡，如今有汉字窗口就觉得挺不错了，哪里还指望得上在上海欢乐的学生们，只好一字字敲着。只要计算机不"挂"，一天时间还怕弄不出一张帖子！

其次，我要感谢你们这个网站，帮了我的大忙。事情是这样的：我从广州回来后没多久，就到东京来了。出门比较急，一些该带的文献没有带，该联系的朋友没有地址。心想没多久时间，只好请各位原谅了。不料，上周武汉大学一位很出色的青年学者黄双全（正在日本做博士后）又是发E-MAIL又是打电话，愣给联系上了。一下子就按"系统树"的方式重建了本人在日本的关系。双全说，他读了一个叫"静生"的人在一个叫"义妹"的网站发的一篇关于我的文章,得知近况,一定要与我切磋。在感谢他寄来了崭新的论文同时,我当然想看看这个网站、这名静生、这篇文章到底如何。——顺便提一句,我和我的学生(Vivian)刚刚翻译了一本沃森的自传《基因·女郎·伽莫夫》(有兴趣的话,我可以另贴些片断),交上海科技教育出版社出版,以纪念双螺旋发现50周年。彼得·鲍林在这本书的序中写道:"作为受害者之一,我认为这本书中的故事不全是真实的。"撇开这些句子的广告效果不谈,也从一个侧面说明要回忆一些难以对证的故事,描写一个说不清的人有多么困难。如沃森者也未能幸免。——所以,我就不对这篇文章本身说什么了。同时,在这个网站见到那么多熟悉的名字,让我大吃一惊,觉得脱离科学院、脱离这个圈子太久了。至于帖子嘛,如果没有那么多火药味该有多好啊!

  再次,在孩子出生前,我就想好了名字,都用红树植物命名。我最近主要与施苏华教授等合作,研究红树植物。比如说,"海桑"如何?即使不是学植物的,也会想成"沧海桑田"的意思,整个一书卷气。无奈,孩子他姥爷的名字中有一"海"字,不能用了。哪位有兴趣的,不妨拿去,丫头小子兼宜(似乎桑涛的女儿就叫桑出,浑然大成),本人分文不收,也无须致谢。话说孩子出生后,我的学生朱彬得我真传地在院里贴了条告示:"钟扬教授和张晓艳博士的遗传学实验取得巨大成功。结果为两新种:钟云杉、钟云实。"一时争相讨论。由于两个名字都只用拉丁名和英文名,还真出了一些趣事。认为分类学重要,又急切盼望论文毕业的同学真的在讨论复旦一下子出两新种是否可以发一篇点数高的文章,比如 *Nature* 如何。也有同学说似乎命名法上有些问题,至少应该是两人共同发表,定名人只有钟老师一人似乎有那什么之嫌。我只是将帖子转给了一些人,并未专门说明。桑涛很感慨地回信道:"我好长时间都没有发新种了。"话里面似乎又有什么打算。而且,他很理解一位新父亲在时间上的困境,再不跟我谈合作的事了。其实可能是我以此作幌子,让他产生误解。红雅告诉我,虽然你钟扬有本事做什么像什么,但转移方向去发新种好像不太对劲吧(原话)。红雅啊红雅,你为什么总是那么忙,认真看一下信的时间都没有呢。顺便说一句:我认为只要有可能应当都用植物给孩子命名。一来花花草草那么多,植物志那么厚,要想重名都难；二是不用动脑，就有一外国人能发音的英文名和拉丁名，今后国际交往多方便；三是如果蔚然成风，会给分类学在社会上带来多大的影响啊。最不济，就算我辈吵吵闹闹，恩恩怨怨，低头不见，抬头也只当没看见，到了孩子们那一辈，一看名字就像找到了组织，就能聚在一起说说这个的爹，那个的娘，该有多好。关于这两个植物学下一代，静生是有说法的：大凡真正的植物分类学家都生女儿。他举了不少例子，包括他自己，各位也可以对号入座。我只是觉得奇怪，这样的"负选择"有显著性差异吗？为什么？可以想象，他"恶狠狠"地说："你小子口口声声热爱植物分类，这不就是明证吗，还一次来俩！"话说到这份上，我热爱植物分类学之心，算是只有明月可鉴。

最后，想借此机会给所有认识的老师、同学和朋友拜个早年。也许太早了点，不过国外都不过春节的。希望网站好人做到底，转达我的祝愿。我工作的地方在东京的文部科学省数理统计研究所（ISM），E-mail：zhong@ism.ac.jp，什么时候这个信箱不通了，就说明我已回到复旦了。

本文来自网络，发帖时间为2002年12月19日1:00—3:00pm，收入本书时略有修改。

引用率低说明了什么？

任何科研论文一经公开发表，就有可能被国内外学者阅读和引用。目前，被引用次数已成为国际学术界判断科研论文内在价值的一项通用指标。一般认为，一篇论文被引用次数越多，其学术质量就越高，创新性也越强。例如，有人统计过获诺贝尔奖的论文平均被引用次数在1 000次以上。

为了评价学术刊物的质量，位于美国费城的科学信息研究所（ISI）专门建立了一套基于引用率（被引用次数除以论文篇数）的影响因子（Impact Factor）指标。一份ISI收录刊物（即通常所称的SCI源刊物）的影响因子是用过去两年发表的论文总数来计算的。以生命科学研究领域最顶尖的杂志之一《细胞》(*Cell*)为例，其2003年和2004年论文被引用总次数为16 746次，而发表论文总数为569篇，因而该刊物2005年的影响因子为29.431。其他国际著名学术刊物如*Science*和*Nature* 2005年的影响因子则分别为30.927和29.273。ISI还将论文引用率作为评价一个国家科学竞争力的一项重要指标予以公布。

近年来，我国科研人员在国际学术刊物上发表论文的势头有增无减，已在151个国家和地区中名列前茅（前10名），但平均引用率较低（排在100名之后）的问题也引起了广泛的关注。中国科学院文献情报中心的报告指出，我国学者在国际上发表论文的80%左右在低被引用区甚至零被引用区。这一低引用率现象究竟说明什么呢？

（1）原始创新能力不足。通常，某一个研究方向的"开山之作"的引用率会大大高于后续论文。近20年来，我国科学研究的投入和产出均达到空前的规模，但"敢为人先"的科研创新精神决不是在短时间内可以培养出来的。

（2）方法学上新意不足。新方法和新技术的发展对现代科学研究的作用不言而喻。许多高引用率论文常常以新方法见长，而习惯沿用前人方法则是我国科学论文的通病。

（3）关注学科主流不够。高引用率论文往往表明其在一段时间内所占据的学科主导地位，而我国一些科学家在某个学科"支流"上的工作较多，缺乏正面竞争意识。

究其原因，不难发现计算论文数量的简单化管理方式必然导致追求数量胜于质量的后果，这已成为我国目前科研评价体系中的"软肋"。此外，为增加论文数量而将一篇学术论文拆分为两篇论文的现象屡见不鲜；有意或无意地不引用他人（特别是国内同行）文献的现象也依然存在，这些问题均是降低我国科研论文引用率的部分因素。

诚然，一个学术期刊的平均引用率并不能完全代表其中每一篇文章的引用率，而一个国家的平均引用率更不能代表一个学科的发展水平和一名学者实际工作的业绩。我们应当看到，一方面任何国家科学研究的发展都经历过从量变到质变的过程；另一方面，一个国家科研发达并不意味着一个一流的科研单位一定在所有的方向上都是一流的，当然也不是一个二流的单位在所有的方向上一定都居二流。毋庸置疑，我国科学研究近年来取得的长足进步与引进SCI等国际通用计量标准直接相关，全面客观地看待论文引用率指标，正确引导科研人员（特别是青年科学工作者和研究生）探索科学前沿，对加强我国的整体科研创新能力具有积极的意义。

本文发表于《中国高等教育评估》2006年第4期，被社会科学报收入《大学怎么了》（"复兴之路　智库建设"丛书，博芬、王雪编，上海社会科学出版社，2015年12月出版），署名钟扬、张彦、吕红，收入本书时略有修改。

道德底线干细胞研究试探中

美国总统布什于2006年7月19日否决了美国参众两院通过的一项有关人类胚胎干细胞研究的法案，引起了世界各国的高度关注，在国际学术界（尤其是生物医学科学研究领域）产生了强烈的反响。布什否决法案的理由将又一次引发人们对人类胚胎干细胞研究所涉及的技术与伦理问题展开争论。

否决理由的关键是"摧毁人类胚胎等于谋杀"。的确，现阶段获取胚胎干细胞都不可避免地要毁坏胚胎本身。然而，国际上通常接受的英国Warnock委员会的建议是，研究用胚胎的发育时间不得超过14天。美国国家科学院2005年出台的《人类胚胎干细胞研究指导原则》中建议：将第5天的胚胎（囊胚）用于提取干细胞之前，必须事先征得胚胎捐赠者的许可，而用于干细胞研究的人体胚胎不得超过14天的生长期限等。我国科技部和卫生部2003年颁发的《人胚胎干细胞研究伦理指导原则》中也明确表示："利用体外受精、体细胞核移植、单性复制技术或遗传修饰获得的囊胚，其体外培养期限自受精或核移植开始不得超过14天。"这是因为在此期间的人类胚胎既没有器官，也没有神经系统，不能感觉痛苦，不构成道德主体。胚胎学认为，14天前主要形成胚胎外部组织，尚未出现"原胚条"；但14天后，胚胎具有完全不同的特性，出现了"原胚条"，意味着胚胎细胞开始向多个组织和器官发育分化。因而，采用14天前的早期胚胎是规范人类胚胎干细胞研究的关键所在，这在技术上也是可行的。

严格控制这一点，就没有理由将损毁早期胚胎与谋杀生命相等同。从科学技术层面上看，人类胚胎干细胞研究本身争议不多，但在伦理道德层面上的确还存在一些值得商榷的问题。布什支持"人类胚胎干细胞研究超越了人类社会应当尊重的道德底线"的说法，但究竟什么是这一领域的伦理道德底线，并没有一个统一的标准答案。在一些伦理指导原则中，除了上述的囊胚期和14天限制外，还包括禁止将人体细胞用于"人兽混种"动物试验，规范人类胚胎干细胞的获取途径，贯彻受试者的知情同意与知情选择原则以及隐私权保护，禁止买卖人类配子、受精卵、胚胎或胎儿组织等一系列伦理道德条款，并要求各级科学伦理委员会进行审查、咨询和监督。

尽管某些研究机构在恪守这些伦理道德规范方面可能存有一些不足之处，但我们认为，科学社会发展的方向和主流不容否定。我们也认为，现阶段的生命科学研究伦理并非尽善尽美，必要的争论将有助于人们全面了解人类胚胎干细胞研究的重大意义，权衡利弊所在，并最终调整完善现行道德规范，而不是因噎废食，裹足不前。

诚然，布什的决定有其更为复杂的背景，其中不可忽略的是宗教势力的影响。许多美国民众笃信人的生命源于受孕的一瞬间。20多年来，自人类体外受精技术取得成功之后，涉及人类胚胎的侵权官司络绎不绝，几乎每个案件的争执焦点都是人们所持科学道德观与宗教道德观相冲突的结果。此次事件所传达的信号是，人类利用自身进行科学研究的道路依然充满坎坷。

美国政府对人类胚胎干细胞研究不予支持的态度，使其生物医学研究领域的发展蒙上了一层阴影，也使世界范围内众多的科学工作者深感失望。

本文发表于《第一财经日报》2006年7月27日"纪念达尔文专题"栏目第5期，收入本书时略有修改。

我国生物信息学教育的发展与挑战

十多年来，生命科学和技术迅猛发展，特别是人类基因组计划等科学研究计划取得了举世瞩目的成就，也随之产生出海量的生物学数据。人们已经充分认识到，如果不能及时分析和有效利用这些信息，那么这些耗费巨资所获得的数据将变成一堆垃圾。所幸的是，生命科学与信息科学及计算机技术相结合的产物——生物信息学应运而生，并已逐渐成为当代生命科学研究的一个前沿领域和重要的方法论工具。目前，世界上越来越多的政府部门、教育机构和企业都呼吁加快培养各类生物信息学人才。然而，当前的生物信息学教育从经费支持到师资队伍等诸多方面都还面临着不少困难与挑战。

作为唯一一个参加人类基因组计划的发展中国家，我国成功地完成了人类全基因组1%的测序任务。随后，从水稻基因组到家蚕基因组测序，从功能基因组分析到蛋白质组计划实施，众多研究项目推动了生物信息学在我国的蓬勃发展。同时，许多大学和研究所也敏锐地把握到这一发展良机，积极开展生物信息学相关的教育与人才培养工作。2003年，笔者应邀在英国出版的《生物教育杂志》(*Journal of Biological Education*)上撰文介绍了我国生物信息学教育快速发展的状况与趋势，着重比较了一些重点大学的相关本科生课程与研究生培养计划。时隔三年，我们进一步对这一领域的发展作了简要的回顾，并指出了我国现阶段生物信息学教育中存在的若干问题，以期引起广泛的注意与深入的讨论。

上 编

一、本科生教育

自20世纪90年代后期起，许多大学都为生命科学领域的本科生开设了"计算机在生物学中的应用"一类的课程。近年来，相当大一部分学校直接将其转为"生物信息学"课程。目前，我国生物学本科教育主要围绕两个专业——生物科学和生物技术进行，而生物信息学相关课程通常作为这两个专业高年级学生的选修课，且要求学生们已修完大部分专业必修课以及一些计算机课程，如"C++语言"等。教学实践表明，这一安排基本上符合国内本科生教育的实际情况，有利于本科生们掌握生物信息学的基本知识和工作原理，激发他们今后深入研究的兴趣。在复旦大学，"计算机在生物学中的应用——生物信息学基础"这门课程是大学三年级下学期的选修课，至今已有近十年的历史，前五年选修人数逐年递增，至2002年达到了规定的最大选课人数（约140人），现在每年稳定在110人左右。从2003年起，该课程也向复旦大学数学系和计算机科学与工程系的本科生开放。课程内容包括：生物信息学的基本概念、原理和方法（占课程总学时的50%），邀请国内外生物信息学家专题报告（占课程总学时的25%），以及计算机软件练习（如美国国家生物信息技术研究中心提供的BLAST服务工具和华盛顿大学的PHYLIP软件包等，占课程总学时的25%）。问卷调查表明，学生们认为大多数邀请专题报告信息量大，对他们了解生物信息学的全貌很有帮助，而一些基本概念则可以简化或在其他课程（如"分子生物学"）中予以介绍。

从2002年起，我国一些高等院校开始向教育部申请设立生物信息学本科专业，如浙江大学、同济大学、哈尔滨工业大学、华中科技大学、上海交通大学、西南交通大学、郑州大学等校先后获得批准。有些学校（如哈尔滨医科大学和上海交通大学等）已新建了生物信息学系，旨在培养跨生命科学、计算科学、医学、化学、数理科学等不同领域、具备"大科学"素质的复合型人才。

二、研究生教育

2003年之前,生物信息学硕士和博士研究生教育只能以挂靠在不同的二级学科下面或由不同的学科联合组成校内交叉项目的形式来实施。例如,北京大学于2000年在国内率先建立了第一个生物信息学研究生项目。在这个交叉学科项目中,导师来自9个院系(如生命科学学院、数学学院和化学系等),不同院系的在读研究生可以选择在硕士或博士第二年加入该项目。2001年,浙江大学和北京基因组研究所开始组建了一个生物信息学联合研究生项目。其他一些大学则在已有的研究生专业中设立与生物信息学相关的研究方向以及相应的研究生课程,如复旦大学在"遗传学"专业中设置"功能基因组学"方向,在"生态学"专业中设置"生物多样性信息学"方向等,以满足过渡时期内培养生物信息学人才之所需。

自2004年起,生物信息学成为了自主设置博士点试点专业之一,当年即有13个一级学科博士授予点获批自主设置生物信息学二级学科博士点。2005年,生物信息学博士点有了更大发展。据不完全统计,目前有60余所高校和研究所招收生物信息学专业研究生,其所属一级学科包括生物学、医学、农学、数学(统计学)、信息科学以及化学(化工)等。此外,一些学校(如同济大学等)已开始尝试利用专业学位(如工程硕士)来加快培养社会急需的生物信息学应用型人才。

三、网络与服务器条件

生物信息学教育对网络与服务器条件的要求较高。国内目前主要是通过中国教育和科研网(CERNET)来开展相关教学与科研活动的。CERNET由国家投资建设,教育部负责管理,清华大学等高等学校承担建设和管理运行的全国性学术计算机互联网络。CERNET分四级管理:(1)全国网络中心,设在清华大学(北京),负责全国主干网的运行和管理;(2)地区网络中心,设在清华大学、上海交通大学和西安交通大学等10所著名大学;

（3）省教育科研网，分布于全国范围内的36个城市中的38所大学；（4）校园网。国内绝大多数教育机构已被授权使用CERNET。目前，其主干网网宽已提升到2.5Gbps，地区网的传输速率已达155Mbps。CERNET已有12条国际和地区性信道，与美国、加拿大、英国、德国、日本和中国香港特区联网，总带宽在100Mbps以上。

北京大学在CERNET和校园网的基础上于1996年建立了中国第一个生物信息中心，该中心也是EMBnet（欧洲分子生物学网）和APBioNet（亚洲-太平洋生物信息学网）的中国节点。北京大学生物信息中心可向全国提供生物信息学网络服务，并有许多用于学术和教育资源的镜像数据，如分子数据库和软件等。CERNET用户可通过匿名文件传输协议（FTP）免费获取。该FTP服务器每天平均被点击5 000次以上。在复旦大学，与生物信息学相关的本科生和研究生教育及实践活动均在校园网和CERNET基础上进行，如用BLAST工具进行数据库搜索和用PHYLIP软件包进行系统发育分析等。

四、教　　材

最初，Baxevanis和Ouellette的 *Bioinformatics — A Practical Guide to the Analysis of Genes and Proteins*（1998）以及Attwood和Parry-Smith的 *Introduction to Bioinformatics*（1999）两本书被翻译成中文后用作教材。郝柏林和张淑誉出版了一本《生物信息学手册》（2000），也被选为教学阅读材料。2001年，高等教育出版社出版了笔者编著的教材——《简明生物信息学》。在正式出版前，该教材已在复旦大学高年级本科生和研究生教学中采用。随后，在高等教育出版社的支持下，我们又分别于2002年和2003年翻译了两本生物信息学相关的专业书籍——Nei和Kumar的 *Molecular Evolution and Phylogenetics*（2000）以及Mount的 *Bioinformatics：Sequence and Genome Analysis*（2001），后者的英文版被美国许多高校选为生物信息学教材。近来，生物信息学方面的书籍呈快速增长的趋势，已不下百种，其中授权影印国外原版教科书和翻译书籍仍占主导地位。值得指出的是，生物信息学教材编写工作较为困难，主要原因是学科本身发展太快，新的知识不断涌现，相关领域拓展迅速，编写工作常常跟不上内容变动。例如，我们在教学实践中已深感《简明生物信息学》一书已不能满足实际教学工作的需要，但由于种种原因，修订版迟迟未能完成。此外，系统性也是目前生物信息学教材中普遍存在的一个问题。

除常规教材外，网络业已成为一个重要的教育资源。中国生物信息网（http://www.biosino.org/）、中国医学生物信息网（http://cmbi.bjmu.edu.cn/）、上海生物信息技术研究中心（http://www.scbit.org/）以及北京大学生物信息中心（http://www.cbi.pku.edu.cn/）等均提供生物信息学教育的公共在线课程和阅读材料，其中的中文辅导材料为国内研究人员和学生学习生物信息学提供了方便。

五、问题与挑战

目前，我国生物信息学教育发展呈迅速增长趋势，必将为我国生命科学研究赶上国际先进水平提供新的机遇，但其自身也面临着种种困难与挑战。现阶段存在的主要问题有：

（1）缺乏合格的生物信息学师资，教师队伍的整体数量和质量与我国生物信息学教育快速发展的规模极不相称。目前，大多数开设生物信息学专业高校的师资队伍靠新近毕业的博士补充。由于欧美等发达国家的生物信息学正处于方兴未艾之际，薪资待遇等多方面的条件较为优越，回国从事生物信息学教学和科研的人才不多，而国内自行培养的博士生有些刚刚毕业，大多数还在学习阶段，这样一来势必出现一些学校的生物信息学专业匆忙上马，而具有良好专业背景的教师匮乏的现象。有些学校问题还相当严重，合格教师寥寥无几，这将对生物信息学在我国持续健康地发展造成不利的影响，学校本身也难以通过严格的教学评估。

（2）对生物信息学专业人才培养的认识各异，造成课程设置不合理。事实上，国外在生物信息学专业的课程设置方面也缺乏成功的经验，围绕"哪些是生物信息学专业的必修课程"和"生物信息学专业的研究生需要哪些基本学科背景"之类的问题争议颇多。我国高等教育的传统模式在创新性人才和交叉学科人才的培养方面本身就存在不少薄弱环节，如何通过生物信息学专业课程教学与实践加强学生的研究能力，从而加快培养不同专业背景的"复合型"人才是摆在我们面前的一项艰巨任务。

（3）生物信息学教育与其他专业的合作还有待加强。尽管生物信息学是一门新型学科，但与其他学科专业之间存在不少联系。举例来说，统计学（包括生物统计学和医学统计学）专业在我国已有较长的历史，也具有一定优势。生物信息学与统计学的关系极为密切，如能整合统计学教学资源，势必提升生物信息学教育水准。现阶段的问题是不少学校将统计学专业划归管理学院等，不同学院的教师之间又缺乏交流与合作，难以满足生物信息学教育的需求。

有鉴于此，笔者认为解决我国目前生物信息学教育问题的关键，在于适当控制发展规模，调整师资队伍，在不同专业内、不同学校间乃至高校和研究所之间共享专家资源。此外，加强国际合作，引进国外智力，借鉴先进国家的成功经验也将促进我国生物信息学教育的发展。

本文发表于《计算机教育》2006年9月，署名钟扬、王莉、李作峰，收入本书时略有修改。

宏观生物学与交叉学科人才培养的理念与实践

近年来，关于宏观生物学和交叉学科人才培养的问题已有许多论述，也引发了不少争论。这里，我想结合自己所从事的生物信息学与分子进化生物学教学与研究体会，谈谈基于宏观生物学的交叉学科人才培养理念以及案例分析的实践意义。

生物学交叉学科人才培养的首要问题是：我们为什么要培养这样的人才？我觉得关键是生物知识发展的需求所致。长期以来，我们已经明确科学教育应当在向学生传授知识的同时指明学科发展的不同方向，从而培养他们的探索精神与创新能力，然而这样的理念在生物学教育实践中有时只能缓慢前行。宏观生物学研究中延续数百年的"传统"问题和知识积累方式与日新月异的分子生物学为代表的微观生物学发展似乎很难协调，更遑论培养兼具两方面素质的人才了。其实，宏观生物学与微观生物学所面临的都是生物学知识不断发展且日益复杂的需求。2002年的《科学》上曾经刊登了一篇文章介绍"系统生物学（Systems Biology）"，其中类似于太极图的系统生物学思想图解给人留下深刻的印象，应当可以扩展到其他学科领域。在这幅图上，生物学研究的起点是目前存有争议的问题，生物信息学先帮我们做"干实验"——利用各种数据库和前人所提出的假说建模，然后进行计算机模拟。在此基础上，进行系统分析并预测结果。接下来，要用生物学实验来检验这个结果，这里的生物学实验可称为"湿实验"。最后，我们可以根据验证结果提出新的假说，看似回到了研究的起点，实则深化了研究的思想，或者说为下一轮循环提供了新的起点。在上述过程中，宏观问题和微观实验必须紧密结合，即使是宏观实验也可以借助微观的技术手段，由此实现学科交叉。值得一提的是，对实验数据进行分析和诠释，需要生物统计学和整合生物学的思想与训练，而这些学科本身的交叉性就很强。研究成果则体现在是否丰富了现有的生物学知识并提出了更有争议的问题。

现阶段交叉学科人才培养工作中应当明确的理念包括：

1. 学习动机

以生物信息学为例，招生属生物科学还是信息科学并不重要，重要的是了解学生的真实动机和兴趣所在。当今某些生物学科并不很景气，导致许多同学对生物学本身丧失了兴趣，但由此产生的对交叉学科的兴趣是否长久需要进一步考察。此外，即使学生声称对交叉学科有浓厚的兴趣，也要告之光有兴趣是不够的，还要看过去的训练与相关知识方面是否存在明显的优势和劣势。

2. 实际需求

培养交叉学科人才一条有效的经验就是在学习和研究过程中发现和解决科学问题。如果所研究的问题涉及范围较小，只用少数已知的途径就可以解决，那就不需要交叉人才。换言之，交叉人才应满足研究工作的实际需求，会用多种途径来解决纷繁复杂的科学问题。

3. 博士后

国内外培养交叉学科人才相对成功的经验是在博士后阶段。交叉学科的问题往往很复杂，大多数本科生难以理解，一般不具备相关的知识和能力，而在博士后阶段这些问题可能就变得相对容易了。

4. 风险性

应当强调的是，交叉学科人才培养的成功概率很低。培养人才难，而培养交叉学科人才更难，其中蕴含了巨大的风险。可以说，交叉学科在某种程度上其实是利用了生物学中杂交育种的原理：杂交能够产生杂种优势，但同样可能导致杂种不育。

通过多年的生物信息学和进化生物学教育实践，我们逐渐认识到在高年级本科生和研究生交叉学科课程中增加案例分析的重要性，这在很大程度上可以弥补交叉学科人才培养缺乏系统性的问题。原理上讲，这些课程中均可采用两类案例：

1. 数据驱动型

如果你面对的生物学问题中涉及大量数据，但找不到突破口，可能就需要关联分析（association analysis）。以生物信息学为例，其"数据挖掘"的基本内容就是通过关联分析找出原始数据之间并不明晰的关系。关联分析并不同于统计学中的相关分析，因为相关分析经常锁定了若干因子来看它们之间的相关性，而关联分析中可能最初根本不知道要关联的少数因子在何处。显然，这样的训练对后基因组时代的生物学家是必不可少的。

2. 假说驱动型

用微观生物学方法或证据去检验宏观生物学现象及其复杂假说，或者通过微观生物学研究发现一些现象，用宏观生物学原理来进行综合与诠释时，需要假设驱动型方法。它强调多方面证据，必须将宏观和微观两方面有机地结合在一起。

我们先看一个实例。我们实验室一位博士生的论文涉及麻黄碱类化合物的进化起源研究。他的本科和硕士完全是医学和药学背景，因而挑选一个麻黄碱相关的化学分类学研究课题挑战性不大，但他希望从分子生物学水平对这个问题做一些更深入的探讨，试图将宏观生物学和微观生物学中均有的比较生物学方法结合起来。原理上，麻黄碱是通过多巴胺途径对人体产生作用。人体中有一个叫做多巴胺转移蛋白（DAT）的"阀门"，破坏了这个阀门就会产生药物成瘾。通过比较麻黄碱和甲基安非他明以及安非他明、多巴胺和血清胺等几种物质的结构相似性，可以考虑它们在不同物种间的受体相似性。我们尝试用生物信息学方法分析麻黄碱的人类受体——肾上腺素受体（AR）和DAT。将AR序列输入GenBank数据库，用BLAST搜索工具并没有找到相似序列；而将人类DAT序列输入数据库后发现，除了人和小鼠外与其最接近的竟是一种真菌（镰孢菌）。镰孢菌的全基因组已经测序，其编号为FG07634的蛋白跟人类DAT序列相似。依照这一全新的线索，我们进一步研究了麻黄碱的进化假说：

1）麻黄碱是麻黄的一种防御物质，其作用是抗镰孢菌的侵染。前者在植物化学课程中可能涉及，即次生代谢物质是植物的防御物质已成常识，但防御什么呢？需要与具体目标相关联。

2）初步猜想麻黄碱对镰孢菌的防御与镰孢菌中的FG07634蛋白有关，这需要在镰孢菌全基因组范围检查。

3）FG07634蛋白很可能与DAT同源。进一步地，若两种蛋白同源，则功能也相近，所以我们推断FG07634蛋白也是转运蛋白。如果它是转运蛋白，它所转运的物质会不会跟多巴胺相似呢？沿着这个大胆的设想，我们初步发现镰孢菌酸与多巴胺的相似性，在比较不同物种在多巴胺代谢途径上的异同的基础上,对一系列有趣的生物学问题进行了探索。这些问题包括：植物中卡西酮到麻黄碱的分布；为什么美洲大陆的麻黄不存在麻黄碱而旧世界麻黄中存在；相关基因的克隆和调控网络分析；能否利用麻黄碱与真菌的相互关系来提高麻黄碱产量和质量，等等。由此可见，不同学科知识融汇于一点可能令交叉学科领域取得突破性进展。

进化生物学的例子就更多了。我们可以将进化生物学看作是宏观生物学和微观生物学的良好界面。达尔文进化论有两个关键词：一是"变异"，宏观生物学注重形态的变异，微观生物学则看分子水平的变异；二是"进化树"，达尔文的手稿上已经画出了树状结构。如今，我们经常碰到的问题是："为什么要用分子信息研究进化问题呢？"达尔文时代只考虑宏观进化，今天我们有分子数据，就可以用全新的角度来研究进化：

1）DNA序列由四个碱基组成的，比形态变异类型少，便于把握；

2）统计学家已经发展了各种DNA进化模型；

3）可以利用海量的基因组数据；

4）可以检验分子钟假说并估计进化时间。

然而，需要明确的是，仅凭分子数据开展进化研究有时难免偏颇，只有将宏观和微观生物学思想与方法紧密结合，才有可能得出真正有价值的结果。我们可以剖析《科学》和《自然》上发表的三个研究工作。

一个例子是对渡渡鸟的研究。这本应属于保护生物学问题，但这种鸟已经灭绝多时，所以用宏观生物学方法进行研究已经行不通，有人很自然的就想到分子生物学手段。牛津大学的 Alan Cooper 教授测定了大约 1 000bp 的序列，与40多种鸟类进行比较，绘出进化树并在树上标示不同颜色以区分物种的濒危程度，使其成为保护生物学的行动图。我们看到，实验基础是微观水平的，但贯穿其中的却是宏观生物学理念。

另一个例子是关于与人类 HIV 有关的黑猩猩 SIV 的起源。哈佛大学的科学家们搜索了分子证据，一个证据认为黑猩猩 SIV 病毒来源于花鼻猴 SIV 病毒；另一个证据则表明黑猩猩 SIV 来自于红盖猴 SIV。研究人员经过细致的进化分析发现两个证据都是正确的，即来自花鼻猴和红盖猴的 SIV 病毒发生了重组。因此分子生物学构建了一个病毒重组假说，但还需动物学家的帮助。动物生态学研究表明，这两种猴和黑猩猩在中非洲西部具有重叠活动范围，动物行为学研究则发现了黑猩猩是如何捕食这两种猴子的，这才保证了此项研究的最终成功。换言之，只拥有分子生物学技术是无法完成这类研究的。

最后一个例子颇有象征意义。达尔文最重要的工作之一就是研究加拉帕戈斯群岛雀类（又称"达尔文雀"）喙的变异。有一对夫妇是普林斯顿大学的进化生物学家，他们40年如一日，重复达尔文的足迹，坚持观测达尔文雀的变异与进化。最近，他们与分子生物学家合作，以四种喙形态极为不同的达尔文雀作为研究对象，应用基因芯片方法，从上万个基因中找出与雀喙形态强烈相关的候选基因。他们用已测序的鸡基因组作为模板，采用生物信息学途径将基因组中与喙相关的基因都注释出来，筛选出目标基因，并将头颅性状与基因表达量相对应。在以鸡为材料的实验验证过程中，发现在调高目标基因的表达量后鸡胚胎中喙的长度果然变长了，从而揭示出达尔文研究背后的分子生物学故事。

以上几个成功案例也许可以反映目前大量的生物学热点研究中都有宏观生物学和微观生物学的影子。正如现代进化生物学的泰斗 Dobzhansky 所言：不从进化角度看问题，一切生物学问题都无法理解！

本文发表于《高校生命科学基础课程报告论坛文集（2007）》（高等教育出版社，2008年），收入本书时略有修改。

生命的节拍

许多生物的生命活动都表现出周期性的节律变化。例如，人体体温在24小时内常常呈现周期性波动；含羞草的叶片每到夜晚就会闭合，第二天又会重新打开，如此循环反复；有些虫媒传粉的植物总是在清晨或傍晚开花，而花香和花蜜的合成和释放又总是在一天中的特定时间达到峰值……

生物节律与生物钟

像含羞草的叶片这样在大约24小时内完成一个周期的生物节律现象，可称之为近日节律（或昼夜节律）。除此之外，还有周期显著长于或短于24小时的节律现象，如婴儿每分钟上百次的心跳，候鸟一年两次的迁徙，多年生植物的"一岁一枯荣"现象等。近日节律是最普遍也是最重要的一种生物节律。实际上，人们通常所说的生物钟在没有特别指明的情况下都是指近日节律。

尽管人们很早就注意到生物节律的存在，但直到近代才开始认识这些现象背后的机制。以含羞草叶片的昼夜运动为例，起初人们猜测这种运动是由外界光线的周期性变化的刺激引起的。直到1729年，法国天文学家德迈朗发现，即使将含羞草放在持续黑暗的环境中，其叶片依然保持有节奏的开合，似乎植物自身有一架内在的时钟使之能够按照固有节奏与外界的昼夜变化同步。这个简单的实验第一次证明了含羞草叶片的昼夜运动不是由外界光线变化所驱动的，而是生物内源的"时钟"所控制的节律运动——这种不依赖外界因素的"内源自主性"正是生物钟现象的一个基本特点。换句话说，即使在实验条件下屏蔽掉各种环境因素的周期性变化，生物钟现象也会继续保持，自主运行。其后，人们又发现生物钟的另一个特点——可调性。在外界环境信号（例如光或温度）的刺激下，有机体可以通过近日节律的重置，使内在的生物钟节律与外界环境变化同步。例如，乘飞机洲际旅行的人经常有这样的经历：跨越若干时区到达目的地之后，常常在白天疲惫不堪而晚上又无法入睡（生物钟的自主性），但经过几天的适应就可以将错乱的时差调整过来，重新适应外界的时间（生物钟的可调性）。生物钟的第三个基本特点是温度补偿性，就像生活中所用的时钟一样，生物钟可以在一定温度范围内维持正常运行。这一点很重要，保证了生物钟在变温动物或植物中也能正常工作，不会因为昼夜和季节引起的温度变化而造成周期紊乱。

生物钟是生物界中普遍存在的现象。一方面，从单细胞的原核生物到多细胞的高等动植物，大多数物种都存在内源性生物钟驱动的近日节律现象。例如，蓝藻是地球上最简单的一类生物，其许多生理过程包括固氮活性、细胞分裂以及氨基酸吸收等都具有以近24小时为周期的节律性。另一方面，生物钟贯穿生物体从孕育到出生到死亡的全部过程。以植物为例，生物钟涉及植物生长和繁殖的方方面面，包括种子萌芽、胚轴伸长、叶片运动、气孔开放、光合作用、避荫反应、开花时间、花蜜分泌、块茎形成、冬季休眠，等等。可以说，生物钟在生物体的整个生活周期的各个方面都发挥着极为重要的调控作用。

生物钟如何运转

持续黑暗条件下的含羞草如何感知时间？人们对此困惑已久。生物钟的运转机制是时间生物学关注的核心问题之一，迄今为止还不能给出清晰和完整的答案。然而，近年来所取得的一些突破性研究进展可以为我们大致描绘出生物钟运行的框架草图。

科学家根据生物钟的结构和功能特点，建立了一个简单的概念模型，它将生物钟系统分解为三个部分：振荡器、输入通路和输出通路。振荡器负责产生以近24小时为周期的昼夜时律信号，决定了生物钟的内源自主性；输入通路负责将外界环境的信号传递给振荡器，使得生物体可以将内在生物钟与外界时间同步化；输出通路能够将振荡器产生的节律信息放大，传导给专门的基因，从而实现对各种生命活动的调控。

许多单细胞生物中存在精确的生物钟，说明生物钟节律能够以细胞为单位正常运转。分子生物学研究表明，振荡器是生物钟系统的核心部分，由一组特殊的基因及其编码的蛋白质组成。每个基因都可以通过自身的蛋白产物激活或者抑制其他基因的表达，形成复杂的正反馈和负反馈相耦联的代谢网络，通过转录和翻译水平的相互调节，此消彼长，循环往复，从而实现基因产物的节律性表达，产生稳定的生物钟信号。

那么，在多细胞生物中是否存在特化的组织或器官来统一协调细胞水平的生物钟运转呢？在哺乳动物中，下丘脑的视交叉上核是生物钟的关键部位，其通过视网膜感受到的外界光线变化，传达信号到下丘脑，后者进一步作用于其他器官（如松果体），通过调节各种激素（如褪黑激素）的分泌来调控各种生命活动的生物钟节律。然而，生物钟在植物中情况则大不相同。一方面，植物没有发达的中枢神经系统，目前没有发现专门负责产生生物钟信号的组织或器官；另一方面，大多数植物细胞都可以通过一些特殊的蛋白（如光敏色素和隐花色素）来直接感知外界的光线变化。有人推测，植物的生物钟节律可能是由遍布全身的细胞水平的定时系统来共同控制的，但不同器官和组织之间如何协作目前仍然缺乏了解。

值得注意的是，表面上相似的生物节律可能是由不同的生理过程来实现的。例如，前面提到的豆科植物含羞草的叶片运动主要是由于叶枕的膨压变化引起的，而在另一种十字花科植物拟南芥中，子叶的近日节律运动与叶片不同部位的生长速率差异有关。

为什么会有生物钟

黑格尔有一句名言"存在即合理"。从进化生物学的角度来看，这句话可以理解为凡是普遍存在的生物现象一定有其进化上的合理性。生物钟现象以近24小时为周期，这个数字马上会让人想起地球自转引起的昼夜节律。对所有的生物来说，地球自转造成的昼夜变化意味着光线和温度等环境因素的变化，而且这种变化具有非常稳定的周期性。以近24小时为周期的生物节律现象的普遍存在，意味着生物钟不是可有可无的偶然现象，而是生命在上亿年的进化过程中在稳定的昼夜变化的自然选择作用下的产物。

为什么自主运行的生物钟对有机体的生存是有利的呢？简单地说，通过产生内在的信号节律，生物钟能够协调有机体各个部分的生理、代谢、发育、繁殖等各项生命活动，使之与外界环境的变化相协调，互相配合、避免冲突。这就好比在车水马龙的十字路口安装了自动化的红绿灯，通过周期性发送信号来保证道路的畅通。更为重要的是，生物钟的自主性使得生物能够利用昼夜节律的稳定性来"预期"下一时刻将会发生的环境变化，"未雨绸缪"地提前做好准备，而不是被动地对已经发生的环境变化作出响应，靠"亡羊补牢"来减少损失。俗话说"早起的鸟儿有虫吃"，能够准确预知天亮的时间以便在破晓之前醒来，其好处是不言而喻的。

为什么生物钟又具有可调性呢？这里的关键在于，生物体内源的生物钟必须与外界环境节律同步才能发挥生物钟的优势，否则就会弄巧成拙，产生破坏性后果。虽然昼夜节律具有稳定的周期，但是地球上除赤道外其他地区的昼长和夜长之比都在随季节而不停地变化。因此，生物必须不断地根据外界的信号对内源的生物钟进行校正，才能使得生物钟与外界昼夜变化长期保持同步。想象一下，如果生物钟不可调整，时差错乱将给长途旅行者造成怎样无穷无尽的烦恼。

上　编

如何研究生物钟

德迈朗对持续黑暗中的含羞草的观察可能是最简单的生物钟实验。在此后的三百多年来，科学家设计了许多方法和技术来研究生物钟。同其他所有生物学实验一样，研究工作的第一步是要精确地记录和描述生物节律的现象。对于叶片运动这样可以直接观察的宏观现象，记录起来相对比较容易，例如，著名的植物生理学家费弗尔曾经设计了一个简单的机械装置来记录豆科植物叶片运动的近日节律。他将植物的叶片与一条细线相连，细线的另一端穿过滑板，系在一支笔上。这样，当叶片运动时，细线就带动笔在纸带上划出一条纵向的指示线，从而自动记录下叶片在一定时间周期内的运动节律。进化论的创立者达尔文也曾经设计了一种装置来测量不同植物叶片的运动，并将研究结果总结在与儿子合著的《植物活动的力量》（1880年）一书中。然而，对于那些不能直接观察的微观水平的节律，例如某个基因的表达水平，应该如何记录呢？这里的难点在于：常规的方法都会中断研究对象的生命活动，只能测度某一个时间点的基因表达水平，而研究生物节律必须进行长达数天的连续记录。因此必须采用非破坏性的测量方法，避免干扰生物的正常生命活动。为此，科学家利用转基因技术，将从萤火虫[1]中得到的一种荧光素酶基因（LUC）与需要监测的目标基因连接起来，转入拟南芥中一起表达。这种荧光素酶在催化反应时能够释放出一个560纳米的光子，后者可以被仪器所识别；更妙的是，虽然荧光素酶蛋白本身不会被快速降解，但其催化活性只能维持很短时间，在几个反应之后就不再释放荧光。这样，转基因拟南芥所发出的荧光与我们感兴趣的目标基因的表达是同步的，利用高度灵敏的光学仪器可以自动记录下该基因的表达节律。目前，这种转基因技术已经被推广用于果蝇、小鼠等动物的生物钟研究。

在记录生物钟现象的基础上，科学家进一步研究是哪些基因及其产物在控制着生物钟的正常运行。利用正向遗传学方法（通过筛选表型突变体来

---

[1] 萤火虫，这个伴随我们童年成长的浪漫小昆虫，其身上的荧光素酶基因已经被广泛应用于生命科学研究，包括用于研究目标基因的表达节律。

确定基因功能）研究生物钟始于20世纪70年代。科学家相继在果蝇（1971年）、莱茵衣藻（1972年）、粗糙链孢霉（1973年）中发现了具有异常生物钟周期的突变体。然而直到1985年，果蝇中的这个基因才被成功地克隆。这也是人类发现的第一个生物钟基因，取英文"周期（period）"的前三个字母将其命名为per基因。1997年，哺乳动物的第一个生物钟基因在小鼠中被发现。近年来，生物钟研究快速发展，目前已分离出了一系列生物钟相关基因。

随着分子生物学技术的飞速发展，基因组学、转录组学、蛋白质组学等高通量大规模的技术被越来越多地用于生物钟现象的研究。此外，航天技术的发展使得人类可以完全屏蔽地球上的昼夜节律的影响，在太空环境中进行生物钟研究。目前，生物钟的运行机制已经成为生命科学领域一个非常活跃的研究热点，预计将催生出一个又一个新的发现。

调整好你自己的生物钟

人类是具有复杂行为和思想情绪的高级动物，除了前文提到的体温波动等人类共有的一些基本的生物钟节律，我们每个人都还有自己独特的生物钟现象。比如，有的人在早晨思维活跃（百灵鸟型），而有的人在晚上工作效率最高（猫头鹰型）。了解自身内在的生物钟特点，将有助于通过合理规划来提高工作效率，改善生活质量。同时，由于生物钟的可调性，长途旅行者可以口服褪黑激素来协助克服昼夜颠倒带来的时差错乱。

本文发表于《生命世界》2008年第10期，为该期"封面故事"，署名耿宇鹏、钟扬，收入本书时略有修改。

倘若生命失去节律

倘若生命失去节律，我们也许再也无法观赏到那五彩缤纷却又错落有致的景色：不再有春风桃李，也没有秋雨梧桐；不再有日月相约的花钟，也没有了如期而至的蝴蝶；不再有秋徙春归的候鸟，也没有了春华秋实的丰收……

倘若生命失去节律，我们也许再也无法聆听到那时而热闹欢腾时而清澈悠扬的旋律：不再有夏日蛙声，也没有秋蝉鸣泣；不再有悦耳动听的鸟啼，也没有了浅吟低唱的夏虫；不再有新莺出谷的清越，也没有了乳燕归巢的婉转……

倘若生命失去节律，我们自身——人类也许再也无法生存下去。每天，太阳照样升起，而我们已经不知日出而作、日落而息，人类生理和心理节律与大自然节律的和谐将不复存在；若不能认识、适应和调整自身的生物钟，我们丧失的又何止是飞行的乐趣？失去节律的人生如同行驶中皮带断裂的汽车，在紧张的快速运转中将遭遇灭顶之灾。

独特的生命节律是不同生物物种在漫长的进化过程中代代相传的宝贵财富。一次次无情的灾难，一点点微小的变异，自然选择和适应进化的力量将节律的印记深深镌刻在每一个生命的器官、组织、细胞乃至分子之中。那遵循最短路线生长或在传粉昆虫来访时适时散香生蜜的植物无一不是因这些节律而蓬勃至今，也谱写出一篇又一篇的进化乐章。

神秘的生命节律还为我们开启了生命世界通向数理世界的又一扇窗口。植物攀缘生长的曲线、叶茎中的黄金分割律、海螺和藤本植物的螺旋手性、花瓣和萼片的圈数数列、植物向性的模式、生物遗传编码的组成与结构……这一切蕴藏着无数数学的、物理学的、化学的和信息学的奥秘，多少年来吸引着一代又一代学者孜孜不倦地探索。

还是让我们用心去感受生命的节律吧。我们都曾为失聪的舞者演绎"千手观音"的精湛技艺所感动，也惊叹于他们"听见"音乐节拍的超常能力。其实，通过手语导引和地板震动，加之刻苦训练，残疾人士同样可以用自己的方式来感受音乐的旋律。是啊，只要生命节律永驻，不随听觉而逝，人生风采一样绽放如花。

本文发表于《生命世界》2008年第10期，为该期卷首语，署名钟扬、赵佳媛，收入本书时略有修改。

从战争中学习战争

2009年3月,一种新的甲型H1N1流感自北美暴发并很快在全球传播。目前,感染人数已超过17 000名,世界卫生组织(WHO)宣布的警备级别也迅速上升到5级,标志着人类与流感持续至少一个世纪的"军备竞赛"又进入了一个新的阶段。

就在6年前,严重急性呼吸综合征(SARS)在我国暴发。在医疗卫生预防和医学科学研究体系均准备不足的情况下,这场突如其来的、不知病因的流行病使我国医务工作者、科学工作者、政府官员乃至普通民众都经历了一次前所未有的考验,最终在举国抗击SARS的努力下还是打赢了这场"遭遇战"。随后,相关卫生预防体系有了极大的改善,科研投入大幅度增加。我国科学家陆续在国内外学术刊物上发表了一系列有关SARS冠状病毒分析的论文,标志着我国的病毒学及流行病学研究真正进入了后基因组时代。

此次甲型H1N1流感在北美暴发的消息一传到国内,对SARS记忆犹新的科学家们立即组织力量开展相关研究。5月8日,美国H1N1流感病毒联合研究组在 *N Engl J Med* 发表报告,对此次疫情及致病病毒进行了较为全面的分析,发现来自美国加州的新甲型H1N1流感病毒系由北美猪流感、禽流感、欧亚猪流感和人流感片段混合重排的产物。WHO直接领导的国际合作组也于5月11日在 *Science* 杂志上在线发表了对此次流感流行潜力的分析报告。基于已有数据的病毒遗传多样性分析预测,这次H1N1流感的传染性和严重性均不及1918年大流感,但有可能甚于1957年的H2N2流感。几乎就在同一时间，我国科学家对国际公开的分子序列及部分流行病学调查数据进行了生物信息学和进化分析，内容涉及甲型H1N1流感病毒基因组的进化、病毒的来源及去向、北美病毒株的分子特征、病毒的血凝素（HA）基因的突变网络结构、HA蛋白结构模建与构象表位、HA基因多变区与相对稳定区片段的确定及在此基础上建立的测序分型方法等。本期由这些工作汇集而成的专题"甲型H1N1流感病毒研究"反映了我国科学工作者在该领域所取得的一些研究进展。尤为可喜的是，跨学科和跨地区的合作优势已初步显现。

诚然，文章中的一些结论与国外同类研究不尽相同，部分观点不一致甚至相左，有待更多的数据支持。例如，研究发现此次流感病毒序列突变可分为墨西哥类型、过渡类型和纽约类型，HA蛋白的潜在空间表位与以往流感坐落位置相似但在静电势性质上明显不同等。然而，这些都反映了我国科学家在该领域独立研究的能力，也为我国控制和治疗流感以及研发流感相关药物提供了依据。尤其需要指出的是，针对HA基因多变区片段利用DNA测序方法进行亚型分型的工作，有可能为今后确诊A型流感病毒提供"金标准"，也为临床诊断、流行病学监测和分子进化分析的有机结合提供了基础技术平台。如能及时推广、坚持实施，再结合数据分析的生物信息学技术的发展，相信能将人类防治流感的水平迅速提升到新的高度。可以预料，随着研究者们掌握更充分的数据资料和国际合作的进一步深入，将有越来越多的甲型H1N1流感病毒研究成果在国内外发表，这将有力地支撑此次对新型流感病毒的"阵地战"，也将进一步促进后基因组时代病毒学、分子流行病学和感染免疫学等基础研究的发展，为人类抗击流感病毒（或其他病毒）传染病的"持久战"提供知识、技术和方法储备。

从SARS到甲型H1N1流感，频繁的流行病暴发对研究者们无疑是巨大的挑战，全球化社会对控制流行病的要求越来越高，而留给我们的反应时间似乎越来越短。唯有"从战争中学习战争"，我国与流行病预防治疗及其基础相关的众多科学技术和临床领域的研究水平才能不断提高。

本文发表于《科学通报》2009年第12期，署名赵国屏、钟扬，收入本书时略有修改。

美丽实用的分子进化

达尔文在19世纪中叶提出的生物进化论思想,使"进化(evolution)"成为100多年来生物学中使用频率最高的词语之一。生物进化论告诉我们,所有生物都是从远古的同一祖先,经遗传和变异演化而来的。在达尔文之后,重建地球上所有生物的进化历史成为了众多生物学家毕生的梦想。

生物DNA序列,构建全新的进化树

在达尔文发表《物种起源》六年之后,德国学者海克尔依据进化论思想画出了一颗"生命之树(Tree of Life)",开创了用"树"来形象描绘生物进化历史的先河。在这棵树上,不同的生物都有一个特定的位置,所有生物则具有一个共同的"根"。长期以来,人们常常用形态和生理特征辅以化石证据来构建生物进化树,然而,由于形态和生理特征的进化式样极其复杂,加上化石资料不够完整,因而所构建的进化树往往存在着不少争议,难以反映复杂生物进化历史的全貌。

20世纪分子生物学的快速发展极大地改变了进化生物学的格局。就在达尔文进化论诞生100周年之际,日本学者木村资生及英美学者金以及朱克斯等相继提出了分子进化的中性学说。该学说认为,分子(基因)的进化过程与达尔文所描述的宏观进化过程不同,"中性"的遗传变异(称为"遗传漂变")可能比自然选择发挥更大的作用。随后，人们相继建立了基于群体遗传学的DNA与蛋白质序列进化模型及分析方法，既能定量地描述和预测不同分子随时间变异的模式，也可以区分遗传和环境因素对基因变异的影响。更为重要的是，由于所有生命的蓝图都是用DNA（某些病毒则用RNA）来书写的，因而人们可以通过比较DNA序列研究不同的生物之间的进化关系，构建全新的生物进化树。这一新兴的学科领域被称为"分子系统学"，它为解决系统与进化生物学中的疑难问题提供了新的方法。

此后的40年，每当分子生物学出现一个新的概念或发展一项新的技术，进化生物学家都会千方百计地将之用于生物进化的研究。与传统的进化生物学相比，分子进化生物学方法具有一些明显优点：

第一，所有生物的DNA均由腺嘌呤（A）、胸腺嘧啶（T）、胞嘧啶（C）和鸟嘌呤（G）这4种碱基组成，因而可以通过分析碱基序列来阐明不同门类的生物之间的进化关系。目前，用碱基序列数据构建的"分子进化树"表明，地球上所有的生命体来自大约40亿年前的一个共同祖先。换句话说，如同达尔文所推测的，所有生物物种在进化历史上都是相互关联的。

第二，DNA的进化演变或多或少是有规律的。目前，科学家们已经建立了许多描述分子序列间DNA或氨基酸置换的统计学模型。相比之下，形态和生理特征的进化模型就要复杂得多了，难以精确地描述。

第三，一个基因组通常是指一种生物所有基因及序列的总和。对分子系统学研究而言，基因组中所包含的有用信息比形态特征要多得多，这将有助于提高统计推断的精确性。

第四，在估算生物进化的时间和速率方面，分子数据具有其他特征不可比拟的优势。例如，人们可以采用分子序列分析方法，推测来自同一祖先的生物彼此分道扬镳的时间，还可以检测不同生物的进化速率是否存在着显著差异。

美丽的分子进化树不仅生动形象地展现出生物物种之间或基因之间的亲缘关系，也吸引了不少生物数学家和统计学家去开发一个又一个模型与软件。这些计算工具先将分子序列进行对位排列，再通过软件计算（如遗传距离计算）得到距离矩阵构建可能的进化树，然后用统计方法挑选出一棵最具进化意义的树。

分子进化证据，提供新的解决之道

如今，越来越多的分子生物学家开始从进化途径来解决生命科学的前沿问题。其中，利用同源基因来预测未知的基因功能，已经成为基因组分析中的一项常规工作。这一工作通过构建进化树来确定不同生物的种间同源基因和种内同源基因以及基因家族成员。由于种间同源基因往往具有相似的生物学功能，因而可用于未知的基因功能预测。例如，原始血红蛋白基因经过基因复制和长期的进化过程，形成了血红蛋白 α 链和血红蛋白 β 链两个基因（家族），分别行使不同的功能。如果我们从一个物种中测定了新的血红蛋白基因序列，就可以通过生物信息学及分子进化分析，确定它属于哪个基因家族，从而推测它的功能，为今后进一步进行实验验证奠定基础。

目前，分子进化分析技术仍在不断发展之中，它的应用范围也在逐步拓展。由于分子生物学越来越趋向于大规模实验，分子进化模型预测将为研究人员节省大量的时间和经费。

分子生物学时代的进化研究还有一个显著特点，就是进化理论与实际应用紧密结合。众所周知的SARS冠状病毒分子进化研究是一个极好的案例。

2002年11月，我国广东佛山出现首例SARS，随后半年内此病迅速在国内及全世界范围内蔓延。科研人员在该病致病原的确定方面进行了大量工作，先后排除了肺炭疽、肺鼠疫、甲乙型流感、禽流感呼吸道合胞病毒、副流感病毒、腺病毒、鼻病毒、肺炎支原体等人类常见病毒，但仍然无法确定其可能的感染原。面对这一全新的病种，传统医学通过症状和常规筛选推测感染原的方法失效。2003年3月，世界卫生组织（WHO）组织了全球SARS联合研究合作网络，在4月16日宣布SARS致病原为新型冠状病毒——SARS冠状病毒，其依据是根据该病毒聚合酶基因的部分编码序列完成的进化分析。

我国科研人员进一步利用分子进化方法重建了SARS的流行病学过程。他们分析了来自中国SARS流行早、中、晚三个不同阶段病例的60余份样本，获得了每份样本的全基因组序列。在详细病史记录的基础上，研究了这些基因组序列中的变异和进化规律，构建了SARS病毒的基因型进化树。在结果树状图上标记出SARS流行的三个不同阶段后发现，所有的变异都可以归结为5个基因型的改变。

根据进化理论，当一种新型的病毒接触人体时，其生存环境的改变会导致不适应新环境的病毒被淘汰，而且这个数量是巨大的。但是，就是那些少之又少的病毒通过了"瓶颈效应"，在人体内繁殖并不断地累积突变，最终适应了新环境。那么，SARS冠状病毒是否适应了人体环境呢？科学家们在分析后发现，病毒在流行早期的确发生了适应性进化（称为"达尔文选择"），而晚期则倾向于不适应环境的淘汰选择。

尽管分子进化分析在SARS研究中发挥了重要作用，但仍有许多未解之谜，追踪SARS病毒的源头及其天然宿主的工作目前还在继续进行之中。

进入后基因组时代，生物进化理论正面临前所未有的机遇与挑战。机遇是大规模测序计划（如人类基因组计划）及各种组学（基因组学、蛋白质组学等）手段不断完善，人们可以通过这些工作所获得的大量分子数据来验证已有的进化理论，从而发展和深化传统进化思想和分子进化分析方法。对现有进化理论的挑战也来自于解释海量分子数据及探寻复杂生物学领域的巨大需求。例如，人们发现某些病毒的进化历史可能不是一种树状结构（如人畜共患病病毒位于不同的"树枝"上），而是一种网状结构，如同飞舞在进化树枝间的蝴蝶，因而用进化树来重建病毒进化历史时存在许多缺陷。如何构建更为全面的进化网络则是摆在我们面前的一项崭新的课题。机遇也罢，挑战也罢，它们都为生物进化论在分子生物学时代的进步展现了光明的前景。

本文发表于《自然与科技》2009年1、2月刊，收入本书时略有修改。

人类与病毒共舞

1918年大流感悲剧并不会重演，但下一次流感大暴发的危险从未消失。

现在《大流感——最致命瘟疫的史诗》之所以被人们提起，是因为书中的一些历史与现实产生了共鸣。早在三年前，作者约翰·M.巴里就明确地告诉人们：在不远的将来，流感肯定会有一次大流行，唯一不能确定的，最后会广泛传播的到底是哪一种流感病毒。

《大流感》并不是危言耸听的预言之书，它聚焦于1918年横扫世界的流感大暴发，细致入微地描写了那场致命瘟疫发生、发展及其肆虐全球的过程，被美国科学院评为"2005年度最佳科学/医学类图书"。1918年大流感是人类史上最致命的瘟疫，全世界约有1/5的人感染，数千万人死亡——超过了两次世界大战的死亡人数总和，也远远高于历年来命丧艾滋病的人数总和，更远超中世纪黑死病所造成的死亡人数。

一部医学的历史就是一部人类与流感共舞的历史，也是关于疾病的"认知史"，还是人与自然、人与社会关系的"共生与适应"的历史。1918年大流感带来的不仅是医学上的进步，而且引起了当时美国整个医疗体系和医疗教育乃至政治体系的变化。美国的现代医疗体系和科学研究体系，也正是从那时开始确立的。

在所有可能的瘟疫当中，潜在威胁最大的便是听上去似乎很不起眼的流感。日常时期每年死于流感的美国人达3.6万人，而1918年大流感在24周之内"杀"死的人比艾滋病在24年里面"杀"死的人都要多得多。我在这本书翻译后还有一个愿望，就是希望今后修改汉语"流感"一词，因为大家太容易将它理解成"流行性感冒"的简称而轻视其危害。

历史给我们提供了一些重要的借鉴。《大流感》中写到那场流感偏向青壮年，而这次甲型流感（H1N1）也是偏向青壮年。书中甚至已提到猪对今后流感传播的特殊作用，这就让人不得不佩服作者独到的眼光。更重要的是，政府要消除民众的恐惧与不安，积极采取信息公开等应急措施，这些描写都对现实有重要意义。

从"猪流感"到"H1N1"，更名只用了十几天时间，但大流感是在发生80年后才做基因组分析，表明到底是什么病毒的。1918大流感时期死去的很多人是因为家人和护士怕被传染而且不知道如何治疗，被隔离后活活饿死的。现在我们不会再这样，既有医学隔离，也有积极治疗。我们会在一次又一次的病毒袭击中，不断获得有效的经验。

由于H1N1的一个远亲还存在于世，这场流感引发瘟疫的可能性相对较小，倒是绝迹已久的H2N2可能更需要警惕。"只闻钟声滴答，而我们却不知道时间。"一位科学家道出了面对流感的忧患与无奈。流感的暴发是没办法预测的，流感病毒也和其他生物一样，是这个世界的一部分，是一个进化的产物。

本文发表于《中国企业家》2009年第11期，收入本书时略有修改。

分子进化

在达尔文提出进化论后的150多年间，生命科学经历了一次又一次的变革，其中一个重要原因就是物理学、化学、数学、工程学等很多学科的方法逐渐应用于生物学研究，使我们研究生命科学的视野不断扩大，研究的问题也越来越深入，我们开始有能力来处理比达尔文时代更复杂的生命科学问题。特别值得一提的是，1953年，DNA双螺旋结构的发现为我们打开了一扇崭新的窗口，开启了分子生物学的新时代。分子生物学已成为现代生物学的主旋律，我们可以从分子水平来重新审视原来公认的生物学理论。

分子生物学时代进化论面临的机会与挑战

在分子生物学时代，进化论面临什么样的机会与挑战呢？

首先，达尔文的进化论可以解释分子水平的生物进化现象吗？达尔文没见过双螺旋结构，他提出的进化论试图来解决整个生命进化的过程，但对分子水平的进化现象，达尔文的理论还适用吗？

其次，即使在分子水平进化论不能用，我们对它也不苛求，只要它的基本理论是对的，还是可以发展和完善的。分子生物学本身的发展速度日新月异，很多分子生物学的方法的确带来了新的结果，它可以促进现代进化生物学的发展吗？达尔文没有看到今天的分子证据，那么我们这一代的科学家能不能沿着达尔文的足迹前进呢?

达尔文进化理论的核心有两点:

第一是"物种可变"。达尔文通过加拉帕戈斯地雀(又称"达尔文雀")发现了变异的存在。若以达尔文的视角来看分子证据,"变异"就是一个关键词。

第二是"共同祖先"。这是达尔文一直在思考的问题,他在1837年的手稿上画了一个树状结构图,暗示各种生物都有一个共同的祖先。在《物种起源》发表后的1866年,德国学者海克尔(E. Haeckel)依据进化论思想画出了一棵"生命之树"(图1),第一次把所有生物放在了一棵"树"上,

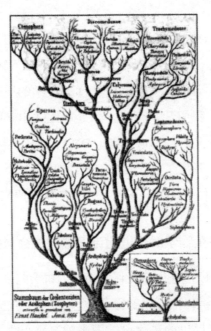

图1　海克尔的生命之树

开创了用"树"来形象描绘生物进化历史的先河。然而,由于形态和生理特征的进化式样极其复杂,加上化石资料不够完整,"生命之树"存在着不少争议,难以反映复杂生物进化历史的全貌。

在分子水平上,生物究竟怎么变异?分子水平上是否能将所有生物的分子,像达尔文一样用一棵树画出来呢?

上 编

分子水平的变异

先来看变异。在分子生物学时代到来的时候,许多人都在思考这个问题,到底从哪个角度用分子数据去检验达尔文的进化论？拔得头筹的是一位日本科学家木村资生（Motoo Kimura，1924—1994），他提出了中性进化学说。他本来是学遗传学的，在美国做博士后期间学习了生物统计学。生物统计学是一种有用的工具，可以对其他科学家所研究的成果加以综合和处理。同时，可以用统计学方法把一些不甚完善的数据放在一个框架中，计算其平均值，分析和寻找数据的规律性。

回到日本后，木村资生开始收集实验数据，当时只有很少的样本——8种蛋白质。他计算了不同生物（如人、狗、鸡、马等）在这8种蛋白质上的分子差异（变异）。他认为应该用物理学方法来思考问题，比如光考虑蛋白质之间的距离是不对的，这些距离差异有时不能说明问题。如果分子是蛋白质间的距离，分母是不同物种起源和分歧的时间，两者相除就是速率。根据速率值差异分析，应该可以判断自然选择正确与否。按照达尔文进化理论，经过自然选择而在各自范围内具优势的物种，其形态差异应该很大，那么是否可以推论其形态在分子水平上的进化速率差异也一定很大呢？

木村资生粗略地计算了一下，结果令他大吃一惊，不是想象的那么大。也就是某一个对人类起着关键作用的蛋白质，其进化速率可能与一只狗在相同蛋白质上的进化速率相当。这意味着分子水平的进化并不像达尔文说的"优胜劣汰"，而是可能有一个"中性"的分子钟[1]在里面起作用。

按照木村的分子钟计算，一个长度为100个氨基酸的蛋白质，平均每 28×10^6 年出现一次变异（不同物种的进化速率系数 K_{aa} 可能存在差别，但进化速率的"量纲"近似恒定）。

1968年木村教授首次在英国的《自然》(Nature)周刊上发表了他的理论，文章题为"分子水平的进化速率"，认为分子进化的速率大致相等，不快也

---

[1] 一种关于分子进化的假说，认为两个物种的同源基因之间的差异程度与它们的共同祖先的分歧时间有一定的数量关系。基于这个假说，可以估算生物谱系发育的年代表。

不慢。当时,另外一个科学小组也在做类似的工作,他们第一次直截了当地提出了"非达尔文进化"(non-Darwinian Evolution)。但是他们的文章被 *Nature* 退稿了,直到1969年才在美国的《科学》(*Science*)周刊上发表。经历了风风雨雨,达尔文进化论又面临着一场新的风雨。很多人在谈论这一科学成果时,没有用"非达尔文进化",而是用"反达尔文进化"来吸引眼球。

现在让我们回顾一下中性进化理论。木村假定大多数置换是中性,或者说只有小部分变化可能是由达尔文进化引起的。中性进化论认为,原来我们认为有害的突变比较多,有利的突变很少,在长期进化过程中,有利的突变经过自然选择终于占了上风,把群体中有害的突变淘汰了。但真实的情况却是:有害的突变并不是那么多,有利的也不是那么多,很多情况是不好也不坏,即中性突变。中性突变是指不改变生物体适合度的突变,即产生的新等位基因与群体中已有的等位基因的适合度相同的突变。也就是说中性突变(等位基因)可能与我们说的生存斗争中的占优势的野生型(基因)具有同样的适合度,没有好坏,只有适应。

中性进化理论提出后,科学家意识到在分子水平大多数进化(或者说变异)不完全是由选择造成的,而是与自然选择等价的随机遗传漂变[1]造成的,而且这种随机漂变的概率还很大(图2)。换言之,一个基因突变在群体中是否能固定下来,可能取决于群体的大小以及随时间变化的漂变"命运"。

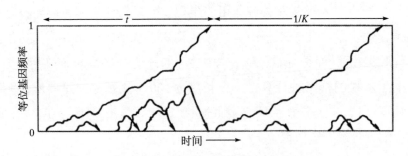

图2 中性进化理论下的基因随机漂变

---

1 由于某种随机因素,某一等位基因的频率在群体(尤其是小群体)中出现世代传递的波动现象。例如,在一个种群中,某种基因的频率为1%,如果该种群只有50个个体,那么就只有1个个体具有该基因。在这种情况下,可能因该个体的偶然死亡或没有交配,致使该基因在种群中消失。

当然，将达尔文进化论"优胜劣汰"的自然选择理论简单地套用到分子水平是不太合适的。事实上，达尔文原来的观点也不是绝对的。他也提出了在自然选择情况下，存在着各种可能的其他变异，只是他没有用"中性"这个词。

我们来看看DNA水平的真实变异（图3）有哪些？

（A）置换

Thr	Tyr	Leu	Leu
ACC	TAT	TTG	CTG
	↓		
ACC	TCT	TTG	CTG
Thr	Ser	Leu	Leu

（C）插入

Thr	Tyr	Leu	Leu	
ACC	TAT	TTG	CTG	
	↓			
ACC	TAC	TTT	GCT	G -
Thr	Tyr	Phe	Ala	

（B）丢掉

Thr	Tyr	Leu	Leu
ACC	TAT	TTG	CTG
		↓	
ACC	TAT	TGC	TG -
Thr	Tyr	Cys	

（D）反转

Thr	Tyr	Leu	Leu
ACC	TAT	TTG	CTG
ACC	TTT	ATG	CTG
Thr	Phe	Met	Leu

图3 DNA序列变异的方式

第一种可能性是"置换"。如果编码一个蛋白质的核苷酸是TAT。但在进化过程中会发生随机的变异，这种随机变异发生的概率有多大呢？木村当时只有很少的8种蛋白质的数据，但他估计出一个量纲——1 000个碱基对大约100万年随机变异一次，其变异概率也就是10^{-9}，TAT很有可能变成TCT。科学家把核苷酸改变但并未造成所编码的氨基酸发生变化的情况称为"同义置换"，而将氨基酸功能发生变化的情况称为"非同义置换"，两种可能性都存在。

有的生物其进化策略很有意思，它采取"丢掉"一部分核苷酸的策略。比如说病毒，病毒在重组过程中常常丢掉一部分，这也成为我们研究病毒进化的切入点。看一个病毒是不是新产生的，要看病毒序列与它原来的序列之间是不是存在缺失。还有"插入"，基因工程中常常人为地插入一个片断，这实际上是人工干预的进化。还有可能在重组过程中"反转"，等等，各种可能性都有。分子水平研究的"变异"与达尔文所看到的"变异"不一样，但是我们还是由衷地佩服达尔文，是他教会我们从"变异"入手。

分子生物学在近几十年取得了巨大进展。我们不仅认识到了变异，而且数学家和统计学家也开始加入这方面的工作。这里有一个最简单的进化模型，它告诉我们，我们可能很难知道一个核苷酸T如何变成A，什么时候又变成G，什么时候又变成C，但我们可以用概率去描述它（图4）。例如一个核苷酸变成另一个核苷酸的可能性都是α的话，这个核苷酸保持不变的概率就是1-3α，我们还可以估计出整个序列在一定时间内的变异率以及标准误差是多少。如果有大量化石和数据的话，就可以估计分子进化过程中的时间和速率了。

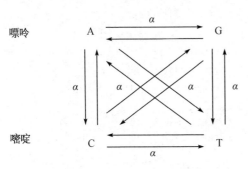

一个单位时间后：
A演变为3种其他任何一种核苷酸的概率为γ=3α
A保持不变的概率为A=1-3α

图4 一种核苷酸变成另一种核苷酸的概率

直至今日，我们发现即使达尔文的自然选择不在分子进化中占主导地位，但仍是一个很重要的方面，我们已不满足于讨论自然选择的可能性，而是要用统计学方法、测序技术和计算模型去发现到底在什么地方存在着自然选择。比如，有人通过比较黑猩猩全基因组序列和人类全基因组序列来研究人和黑猩猩之间是否存在自然选择，哪些地方存在自然选择。研究的数学工具就是统计学方法，统计核苷酸"非同义置换"和"同义置换"发生的概率，如果两者的统计值之比大于1，表明存在达尔文选择，即正选择；如果小于1，就成为"净化选择"或者负选择。结果，在计算过程中发现，达尔文选择是一个小概率事件，但的确存在。人类和黑猩猩之间在涉及大脑的部分基因存在选择，但这已不是简单的量和质之间的比较了。[1]

---

[1] 大自然对物种突变的选择可以分为正负两种类型。当一个群体中出现能够提高个体生存力及育性的突变时，具有该基因的个体将比其他个体留下更多的子代，而突变基因最终在整个群体中扩散，这种选择称为正选择或达尔文选择；群体中出现有害基因时，携带该基因的个体会因为生存力及育性降低而从群体中淘汰，这种选择称为负选择或净化选择。

分子进化树

生命起源于共同的祖先,如何从分子层面来看"生命之树",它的"根"究竟在哪里?在分子水平,可以"看"到这样一幅画面:经过一段时间 t,一个祖先基因在两个后代个体中分别成为 X 和 Y,又经过一个很短的时间 $t+1$,X 发生了一点变化成为 X′,Y 发生了一点变化成为 Y′(图5)。如

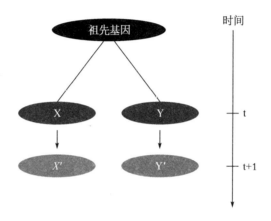

图5 祖先基因的变异过程

果有一段很重要的序列是 AAGACTT,经过 100 万年的演变,大部分还保留着,可是有两个位点发生了变化,一个位点由原先的 T 因某种原因变成了 C,一个位点由原先的 A 变成了 G。如果是中性进化的话,结果很可能是一个同义置换,但如果这个地方存在正选择的话,则很可能是非同义置换。

分子进化的过程并不都是对称的,其间有的 DNA 拷贝可能会丢失,比如说生物大灭绝导致的缺失,所以很可能在某一子代只有奇数个拷贝,而这些拷贝对祖先既有继承也有变化。到今天,我们找到了 5 个拷贝,存在于 5 种生物之中。可是作为一个分子进化生物学家,他要像达尔文画进化树一样,用分子测序的方法把这段基因测出来,然后用统计学方法把每个物种的这段基因建立起来,这就是自然历史的重建(图6)。

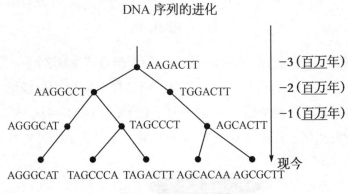

图6 DNA序列的进化

21世纪初,经过全世界科学家的共同努力,人们用核糖体RNA序列重建了一棵分子"生命之树"(图7),它非常清楚地告诉我们,现在的生物分

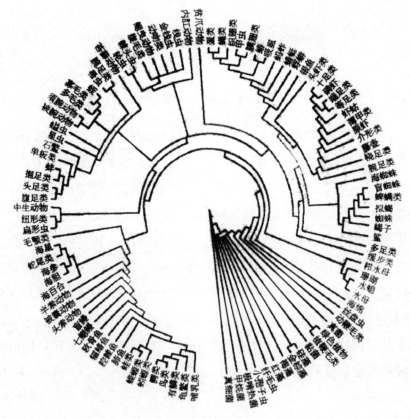

图7 用核糖体RNA序列重建的分子"生命之树"

成三大类（三界）——细菌、古菌和真核生物的解释是科学的。出人意料地，分子生物学数据显示古菌比细菌离人类的关系更近。当然，科学家还会针对其中的某些细节，倾其毕生精力来完善进化树。

SARS病毒的分子进化研究

分子进化已经从单纯的科学理论、数据搜集、古生物学证据分析，向实验验证的方向转变。今天更要强调，任何一门科学的理论和方法，不仅需要自身完善和发展，还要有用武之地。

2003—2004年，分子生物学家和医学家合作，共同对SARS病毒的传播展开研究。从2002年11月16日发现第一例感染病例，到2003年2月28日SARS在中国消亡，这么短的时间内，SARS传遍了世界各地。流行病学研究告诉我们，它经历了一个从零星传播到暴发，最后减弱的过程，但是仅有流行病学的证据是远远不够的。科学家试图通过分子测序来寻找病毒在这100多天内所发生的序列变异，并把这些变异的地方标注出来，进而找到其传播路径。分子生物学家用了64条全基因组序列的数据来认识SARS病毒，结果与流行病学研究的结果相似：SARS病毒的传播确实呈现了早、中、晚三个阶段的规律（图8）。

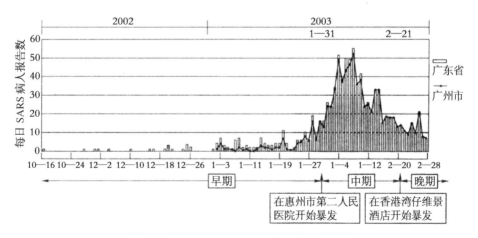

图8　SARS病毒在中国早期传播的三个阶段

如果想用达尔文的自然选择来解释SARS病毒进化过程的话，仅有理论论证还不够，还要用统计学的方法来验证，即看"非同义置换"和"同义置换"之比（K_A/K_S）在统计上是不是大于1。结果表明，SARS在初期的时候确实有一个正选择，有利的环境成为它进化的动力，因而进化速率很快；到了中间它变成了中性，进化速率减慢；而后期它经历了负选择。研究显示，SARS病毒在很短时间内就可以完成从正选择到负选择的过程。

古DNA的分子进化研究

进化论是一门研究从古到今的跨度非常大的学科。科学家尝试用分子生物学的方法检测古代生物的DNA序列来了解它们的变异规律。古老化石携带的DNA信息又能告诉我们什么呢？

图9　尼安德特人复原

1990年，美国加州大学一位科学家从一块有着1 000万年历史的植物化石上发现了一些可能残留着DNA的物质，并试着提取了叶绿体的DNA，获得了成功，证实这是与现今白玉兰同属的植物。到2007年，现代分子生物学可以把20万年前灭绝的"尼安德特人"（图9）的DNA全部组装起来。

尼安德特人（*Homo Neanderthalensis*）是一个大约12万到3万年前冰河时期居住在欧洲及西亚的人种，性格温驯。其遗迹最早于1856年在德国的尼安德谷（Neander Valley）被发现。根据最新发现，尼安德特人和现代人不同种，它是早期智人。

渡渡鸟的分子测序与进化分析

有一位科学家，他从小生活在一个古生物学家的家庭，一直对已灭绝动物感兴趣，想从动物形态学、动物心理学、动物生态学来研究渡渡鸟

（图10）是怎么灭绝的。他花了数年时间说服博物馆的工作人员给了他一点渡渡鸟羽毛的标本，提取了其中的DNA，最后成功地将渡渡鸟与其他现存的或灭绝的鸟类进行比较，构建了一棵进化树。这棵进化树不是简单地说明渡渡鸟到底与哪些物种有较近的亲缘关系，而是为了保护生物学的工作。

渡渡鸟的灭绝原因有很多，其中之一是人为因素。这种体积庞大的鸟没有翅膀、不会飞翔，所以当殖民者来到毛里求斯岛的时候，它就成了餐桌上的美味，很快就灭绝了。但是渡渡鸟会不会在进化上本身就存在某种弱势呢？基于分子生物学的进化分析发现，与渡渡鸟亲缘关系很近的一些鸟已全部灭绝，还有一些与之在分子水平上比较接近的鸟也已濒危了。所以，在用有限的人力物力来保护生物时，是不是

图10　渡渡鸟复原

可以先运用分子生物学方法得到一些启示，来确定保护的优先性呢？这位科学家用一个已经灭绝的古DNA材料给了我们一个答案。

新的挑战：基因树与物种树不一致

很多情况下，用分子生物学构建起来的进化树与达尔文构建的进化树是一致的，达尔文想构建的是物种之间的关系，而我们构建的基因树讲述的

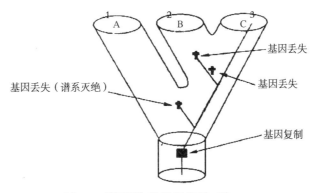

图11　基因树与物种树的不一致

是基因进化的历史。有时两者也可能不一致（图11），比如在物种水平上，B物种和C物种有共同的祖先，而A物种与它们的关系远一些，可是如果仅仅用分子生物学研究它们亲缘关系时会发生什么问题呢？

首先，在原始祖先那里确实只有一个基因，它通过复制产生两个拷贝传给后代。糟糕的是，在漫长的进化过程中发生了灭绝事件，其中一个拷贝丢失了，其他的继续在复制，之后进化过程中又发生了丢失。最后，分子生物学家研究告诉我们，A和B关系很近，而C则较远一些，这是一个错误的结论。因此，仅仅用一个分子生物学的证据，或者少数分子生物学的证据，很可能得出与物种水平分析不同的结论。

进化研究给生物学带来了无穷的魅力。达尔文以后，众多科学家前赴后继地接过他的火炬，继续沿着进化生物学的道路前进。被称为20世纪进化生物学大师的杜布赞斯基（Theodosius Dobzhansky）就是其中的佼佼者之一。他27岁时离开苏联来到美国，成为摩尔根的一名助手。1934年，他成为博士生导师，接受了他的第一位研究生——来自中国的谈家桢，也为中国造就了一位进化生物学的泰斗。这里，我们就用杜布赞斯基的一句话来作为总结："不按进化论思考，生物学的一切将无法理解"（Nothing in biology makes sense except in the light of evolution）。

本文为2009年2月20日钟扬参加"科普大讲坛"第一讲活动作的报告，后收入《科普大讲坛——从进化论到能源未来》（上海科技馆编，上海科学技术出版社，2013年7月），收入本书时略有修改。

番木瓜的故事

作为一名进化生物学家,我曾经多次到热带雨林进行考察。繁盛茂密的热带雨林不仅给人们展示了别样风情,而且提供了丰富的生物多样性资源。让我们从一个热带传奇开始吧!

热带雨林的长尾猴特别喜欢在番木瓜树附近活动,它们在寻找果实——番木瓜(图1)。

图1 番木瓜

番木瓜的英文是papaya,它给我们美味的享受,也给艺术家带来灵感。1994年,以其为名的越南电影(中文译做"香木青瓜")获奥斯卡外语片提名奖(图2)。

番木瓜的生产一直遭遇发展瓶颈,其致命杀手被称为环斑型花叶病毒。中国台湾地区的番木瓜农场也未能幸免于难,均由该病毒引发(图3)。

图2　同名电影获奥斯卡外语片提名奖

1992年的台湾番木瓜种植地（上）　　　　受病毒侵害的种植地
1994年的台湾番木瓜种植地（下）

图3　番木瓜的生产受到病毒困扰

以往科学家对付这种病毒以及培育抗病毒植物的法宝主要是通过杂交技术。从不同的植物上选择不同的亲体，如果有一种可以抗病毒，就可以通过杂交获得新的品种，并希望这个新的品种能够获得抗病毒的能力。

近来，科学取得了巨大的进步，我们又有了崭新的工具，就是转基因技术，或者说是DNA重组技术，通过一系列的生物技术在DNA水平进行基因重组操作，更迅速地将我们所需要的基因转移到作物的基因组中。转基因技术真是恰逢其时，美国康奈尔大学率先在实验室培育出了转基因的番木瓜（图4）。

转基因　　　　　　　　非转基因

图4　美国康奈尔大学率先在实验室培育出了转基因的番木瓜

但是光有实验室结果是不够的，康奈尔大学和夏威夷大学合作，进行了田间的对比试验，在野外将转基因番木瓜和自然传播的番木瓜品种进行田间对比，取得了巨大的成功。图5中我们可以看出，转基因（左）和非转基因（右）的番木瓜品种具有完全不同的抗病毒能力，直接影响了它们的产量。

图5　转基因与非转基因番木瓜的田间对比实验

田间对比实验的成功鼓舞了科学家们。1996年开始，转基因番木瓜终于进入大田的实验，到1997年结出了丰硕成果。历经多次艰苦实验和长期辩论之后，美国食品药品管理局（FDA）和美国农业部分别于1997年和1998年批准转基因番木瓜上市，这两个上市的品种最后取名为"日上"和"彩虹"，不仅表达了人们的喜悦之情，而且更寄望转基因番木瓜如雨后彩虹不断发展，蒸蒸日上（图6）。

图6　转基因番木瓜获准上市

　　然而我毕竟不仅仅是一名番木瓜的爱好者，作为一名科学家，我必须向转基因专家们表达我们深深的忧虑。我们可以准确地对作物进行基因操作吗？被转入的基因难道就不会影响基因组中其他的基因吗？转基因是否改变了生物原有的成分？如果是，对我们有意义吗？如果没有意义，甚至有害，我们能够像检测三聚氰胺一样将其检测出来吗？转基因植物即使现在可以抗病毒，但是转基因植物旁边生长的某种杂草如果也获得了这样转基因的抗病毒能力，我们能让这样一种杂草变成一种超级杂草吗？会不会由吃这种杂草的昆虫和其他生物通过食物链传播开来呢？

　　这是我们的忧虑，我们的忧虑远远不止这些……

　　本文为2009年4月13日钟扬主持"科普大讲坛"第二讲活动作的报告，后收入《科学大讲坛——从进化论到能源未来》（上海科技馆编，上海科学技术出版社，2013年7月），收入本书时略有修改。

学会"与流感共舞"

我们首先来回顾一下1918年的"西班牙大流感",从中得出一些跨越国界的医学和社会学上的教训与规律,然后再谈流感与人类的关系,最终确立今天我们应对流感所持的科学态度。

1918年"西班牙流感"是如何席卷全球的

1918年,第一次世界大战进入尾声,但人类所要面对的却是一场更为严峻的战争,对手叫作自然,它派出的主力军是流感。病毒序列的比对发现,今天的甲型H1N1病毒与1918年大流感的病毒仍然相似,只是存在着重排。

1918年的流感叫作"西班牙流感"。关于它的起源地曾有很多猜测,来自中国?来自驻法的英军营地?来自西班牙?根据考证都不是!在战争时期,每个参战国都认为公布存在流感这个事实很可能打击自己的士气,助长敌人的威风。所以,它们都不愿意承认流感存在的事实。唯独中立国西班牙不断报道相关新闻,所以舆论把矛头都指向了西班牙,认为是"西班牙流感"。其实,1918年大流感是从美国堪萨斯州的小镇赫斯克尔传播出来的。几十年后,科学家找到了埋葬在阿拉斯加、保存良好的1918年大流感死难者尸体,通过对尸体的基因组测序并据此进行基因变异速率估算,结果与文献记载吻合。

1918年，所有参战国都为最后的胜利调兵遣将，却最终点燃了大流感的导火索。随着美军登陆欧洲，短短几个月，流感从美国传染到法国、西班牙等国家；在不同港口停泊的船只也成为传染源，流感随之传到印度、缅甸、中国，等等，最后成为席卷全球的最致命瘟疫。

为何有5 000万到上亿人死于1918年大流感

1927年，首次回顾大流感时，正式的死亡统计是2 100万人。1940年，有科学家估计可能是5 000万人。还有人认为死亡人数可能会上亿，因为当时很多国家不报，还有一些国家无法确诊是否就是大流感。2002年再度不完全统计已经达到5 000万人。流感暴发时，几乎所有大城市里的棺材售罄，小孩去世时则用购物纸盒之类的东西装起来埋葬。在大流感暴发24周内，夺去的生命比艾滋病24年夺去的生命都多；大流感开始暴发的52周内的死亡人数，超过第一次世界大战的全部阵亡人数；而一年内的死亡人数几乎与黑死病绵延一个世纪造成的死亡人数相等。

为何扩散速度这么快呢？首先，RNA病毒变异非常快。其次，当时医疗资源十分匮乏，医护设备紧缺，且还被部队征用，如果在民间使用大量的医疗设备，对战争不利。第三，尚无疫苗和医学上的快速应对措施。作为我们今天应对流感行动指南的病毒学是1926年才确立的，而1918年尚无这门学科。

最后一点值得特别注意，政府相关部门出于种种考虑——或出于国家的利益，或出于本城市的利益——对大流感的真实情况都瞒而不报。

从医学上来讲，当时也不知道大流感从何而来：病原体是什么？是怎么传播的？尤其是无法确认它是不是病毒。我们知道抗生素是对付细菌的有效武器，但它对病毒往往不起作用。当时德国非常著名的细菌学家菲佛认为流感是杆菌，纽约市卫生局实验室主任威廉发现了它是病毒，但一直不敢确认。即使最后为流感研究献出生命的科学家刘易斯认为是一种离病毒很近的滤过性微生物，但考虑到权威们都认为它是细菌，也就不再坚持自己的观点。

在大多数流感肆虐的军营，医生们对病毒和致病菌的处理比平民医院有经验得多，但所采取的隔离措施也很简陋，仅仅是拿一块布将病床围起来。

1918大流感给人类带来了哪些"财富"

1918年的大流感给了人类一些意外的收获：一是发展了病毒学；二是1928年发现的"盘尼西林"即与流感相关杆菌研究有关，而1943年，艾弗里发现了DNA是染色体及基因的主要成分，更是直接催生了分子生物学；三是路易斯安那州一位参议员长期推动建立美国国立卫生研究院（NIH），1928年一场小流感来袭，美国国会终于批准了这个方案。如今，NIH已成为世界上最大的生物医学科研组织机构，几十年来的研究成果大大提高了人类公共卫生健康的水平。值得一提的是，美国开始重建自己的医学科学。一批有识之士针对当时哈佛大学医学院和耶鲁大学医学院不能培养出合格医学生的现状，借鉴国外的先进经验创建了体现美国现代医学教育体制的约翰·霍布金斯大学（Johns Hopkins University）和洛克菲勒研究所（现为洛克菲勒大学）。这对我国医学教育的发展或许有很好的启示。

疫苗研发的速度很难赶上病毒进化速度

我们经常要回答公众的一些问题，比方说为什么流感这么厉害？我们能不能控制它？到底我们相信的科学有何作用？实际上，通过对流感序列的分析，我们会发现：科学在进化，人类在进化，流感病毒也在进化。科学家研究了流感进化树，发现流感病毒株呈阶梯式前进，意味着药物再有效也赶不上病毒进化的速度。因此，疫苗研制能不能追赶上病毒进化的步伐是对人类的极大挑战。总的来说，我们这些年来还是有很大的进步，特别令人振奋的是，全世界已开始合作对流感进行监控。对流感监控已不仅仅是CDC、生物学家或者医学家的事，它已成为一个社会问题，成为各界都关注并为之努力的公共事业。例如，统计进化生物地理学之类的工具已经为掌握病毒传播的历史、病毒所在地以及病毒在迁徙过程中的变化作出了贡献。

随着信息的积累和数据分析工具的改进，计算生物学家和结构生物学家发现，病毒还是有一些缺陷的，我们可以以此作为新型药物的靶标。

对流感的关注已超越科学和医学界

对流感的关注已经超越了医学和科学界。美国有两批大学生搞过流感数据分析竞赛，一组用了谷歌，还有一组用了雅虎，通过这两个搜索引擎将全世界流感在网上的数据下载后进行分析，从最终的结果来看，分析是相当准确的。2009年2月，谷歌公司工作人员发表了研究文章，他们对流感信息进行了分析，发现从谷歌网上获得的数据与专业机构CDC最后发布的数据非常吻合。这一方面说明我们已真正进入了信息时代，互联网上有了足够的数据；另一方面，从统计学来看，如果数据拟合得好，就有希望做预测了。

5月8日，国际医学权威刊物《新英格兰医学杂志》上发表了一篇文章，作者来自美国许多州，他们对H1N1进行了分析，认为这次甲型流感病毒的8个片断是来自美国的猪流感和来自欧洲的流感病毒间的混合重排。重排就像"洗牌"，病毒通过这样的洗牌发生了变异，有可能获得了新的进化动力。而这种新的进化动力，如果跟环境因素结合在一起，就会产生巨大的影响。

人类与流感是一场不停歇的"军备竞赛"

我们要有很强的心理准备，人类与流感的斗争是一场永不停歇的"军备竞赛"。既然是"军备竞赛"，就不能指望科学家来消灭所有的病毒。相反，目前我们没有这样的能力。我们当然希望可以健康地生活，但不要寄希望于预先服一种药物就能百病不侵，其实药物本身也有副作用。如果有人告诉我，他已研制出比现在的抗病毒药效强100倍的药物，各位可能欢欣鼓舞，但我会忐忑不安。为什么？也许在不久的将来，抗100倍药效的新流感就会出现。这就是"军备竞赛"的真实含义。

"与流感共舞"的时代已经到来，流感也许会成为我们生活的一种常态。科学家还不能给大家带来更多的好消息，但我们要从中不断地学习。科学家不会停止脚步,他们一定会努力地工作,公众应当理解病毒也在进化,我们在与病毒赛跑,应当做好"与流感共舞"的心理准备。

本文发表于2009年5月16日《文汇报》第7版"文汇讲堂"总第22期,后被收入《智慧的声音》("文汇讲堂"栏目组编,上海人民出版社,2011年4月),收入本书时略有修改。

西藏，已不再遥远

　　1983年，19岁的藏族青年扎西次仁考上了华东师范大学生物系，一接到录取通知，他就在亲友们的祝福声中离开了雅鲁藏布江畔的家乡。在那个多雨的季节，他走走停停，整整用了23天才来到黄浦江畔，总算赶上了学校新生报到日。在四年的大学时光中，他时刻怀念那遥远的故乡，却一次次止住了回乡的脚步。22年后，已经是西藏大学讲师的扎西次仁考上了复旦大学生命科学学院的博士生，这一次他是当天从拉萨飞到上海的。当他以"一个藏族青年的求学之路"为题再次代表入学新生发言时，他的经历还是在众多师生的心中激起了阵阵涟漪。

　　在过去八年中，我和扎西次仁等人十余次在西藏开展野外考察与采集活动，领略了青藏高原独特的自然地貌和壮美景观，也目睹了近年来西藏建设和生态保护方面所取得的成就。在扎西次仁确定以"西藏巨柏的保护遗传学研究"作为博士论文并获得国家自然科学基金项目资助后，我们连续三年在藏东南地区沿雅鲁藏布江两岸调查该植物的分布与生境，直至将现存的三万余棵西藏巨柏登记在册，将数千颗种子贮存在我国最大的植物种子库——位于昆明的西南野生生物种质资源库，将十余个野生居群的遗传变异用分子标记来比较分析，为保护好这一西藏特有的濒危物种提供科学依据。

　　西藏巨柏生存现状调查与分析只是西藏生物多样性科学与生态学研究中的一个例子而已。越来越多的西藏特有生物和珍稀濒危物种及其生态环境引起了人们的高度重视。今天，由中美学者合著的《走进西藏》一书，以精美的图片和简洁的文字向我们呈现了一帧西藏自然历史的画卷。作为国际生物多样性的一个"热点"地区，西藏蕴藏着丰富而独特的生物资源。仅以高等植物为例，在已知约6 000个物种中，特有种竟达20%之多，表明了青藏高原隆起对物种形成与分化的强烈影响。一草一木总关情，这些鲜活的生物多样性宝藏会给所有热爱和向往西藏的人留下难以忘怀的记忆。

《走进西藏》真实记录了"世界屋脊"上生态保护事业的成就，再次唤起了我们的环保意识。目前，西藏自治区的自然保护区面积已超过40万平方公里，占自治区面积的近三分之一，居全国之首，也创造了世界自然保护的奇迹。然而，毋庸讳言，随着国际交流和社会经济活动的快速发展，人们对自然资源（尤其是药用动植物资源）的巨大需求，严重影响了高原生态承载力与生态系统的稳定性，再加上研究力量的相对薄弱，西藏自然资源与生态系统的保护工作正面临着新的挑战。在几代人艰难跋涉的自然保护之路上，还不断需要新的科学思想与多学科综合的方法、政策法规与资金投入的落实以及广大民众的积极参与。

是的，西藏已不再遥远。感谢《走进西藏》的作者们，他们用镜头又一次缩短了我们与高原的距离，也及时传递了重要的信息——保护高原自然环境依然任重而道远。

本文发表于2009年8月17日《文汇报》第十二版"走进2009上海书展"专版，为《走进西藏：生物多样性与保护事业》一书做推介，收入本书时略有修改。

进化论与进化生物学的发展

自达尔文1859年发表《物种起源》(*The Origin of Species*)一书以来，"进化（evolution）"已逐渐成为生物学文献中出现频率最高的词语之一，进化生物学(evolutionary biology)则成为当今生命科学中一个重要的前沿领域。

纵观150年来，随着科学界对生物进化现象的认识不断深化，人们对达尔文进化论的理解也随之不断深入，进化论自身也走过了曲折的发展之路。除了像其他任何一种科学理论一样需要补充和修正外，进化论还经受了来自科学领域之外的一次又一次挑战。今天，分子水平的生物进化研究正在蓬勃兴起，人们对进化论的兴趣有增无减，同时也提出了更高的要求，即以进化论为核心的进化生物学研究不仅应能够解释各种复杂生命现象，重建生物的自然历史，而且还应具有一定的预测性和应用潜力。因而，藉纪念达尔文（C. Darwin）诞辰200周年和《物种起源》出版150周年之际，回顾进化论与进化生物学的发展历程，将有助于我们全面了解该领域的科学理论与知识，并用于指导21世纪生命科学的研究。

进化论的科学本质

进化论从本质上改变了人们对地球生命现象的理解。进化论围绕生物多样性的起源与发展，引导人们探索各种生物之间的亲缘关系（或称"进化谱系"）。例如，作为地球生物的一员，人类究竟何时又是如何在地球上出现的？不同人种或不同人群之间关系如何？人类与其他生物（如细菌）有何种进化上的关联？如此等等，进化论为我们提供了科学的解释。

在进化论中，具有有益性状的生物存在差异的繁殖优势被称为自然选择（natural selection），因为是自然来"选择"提高生物生存与繁殖能力的性状。如果生物的突变性状降低其生存与繁殖能力的话，自然选择就会减少这些性状在生物群体中的扩散。人工选择也是一个类似的过程，但在这种情况下是人而不是自然环境使生物交配以选择理想的性状。最常见的莫过于通过人工选择来获得人们所需的家畜品系和园艺植物品种等。

迄今为止，支持进化论的证据层出不穷，从中华龙鸟化石的发现到酵母实验进化的分析，不胜枚举。近年来比较突出的例子有加拿大北部"大淡水鱼"化石的发现。科学家们根据进化理论和化石分析预测出浅水鱼类向陆地过渡阶段的大致时间，随后他们将目光投向加拿大北部努维特地区的埃尔斯米尔岛，那里有大约37 500万年前的沉积岩。通过四年的努力，科学家们终于从岩层中发掘出命名为"Tiktaalik"（因纽特人的语言中意为"大淡水鱼"）的生物化石，它既具有许多鱼类特征，又具有早期四足动物的典型特征，而它的鳍包含骨骼，可形成类似于有肢动物的肢体，用来移动和支撑躯体。"大淡水鱼"的发现证实了科学家们基于进化论的预测。反过来，对于进化论预测的证实也提高了达尔文理论的可信度。的确，每一种科学理论本质上都要具备对尚未观察到的自然事件或现象作出预测的能力。

另一个经典的例子是科学家们对特立尼达岛阿立波河中的虹鳉鱼进行的观察与实验。按照进化理论，不同时间尺度上的自然选择可能产生全然不同的进化效应。在仅仅几个时代的周期内，生物个体就有可能产生小规模的变异，可称之为微进化（microevolution）。科学家们发现，生活在阿立波河中的虹鳉鱼无论是其幼体还是成体均遭受较大鱼类的捕食，生活在河流上游小溪中的虹鳉鱼只有其幼体会被较小鱼类捕食，因而长期的进化过程导致该河流中的虹鳉鱼个体较小（更易于躲避捕食者），而溪流中的虹鳉鱼则个体较大（不易被较小的鱼类捕食）。科学家们将河流中的虹鳉鱼置于原来没有虹鳉鱼种群的溪流中，发现它们仅仅在20代后就进化出了溪流中虹鳉鱼的特性。

毋庸讳言，在科学上，我们不可能绝对肯定地证明某种解释是完美无缺的，或者是终结性的。然而，迄今为止，许多科学解释已经被人们反复检验，不断增添的新观察结果或新的实验分析很难对其作出重大改变。换言之，科学界已广泛接受这些解释，它们是以观察自然世界获得的证据为基础的。进化理论就是其中一个代表。从这一点出发，我们可以明确地将从科学上补充和完善进化论与从宗教上反对进化论的证据区分开来。

综合进化论

达尔文之后的进化论经过魏斯曼（A. Weismann）等人的"过滤"，去除了拉马克（J. B. Lemarck）"获得性状遗传"等学说的影响，形成了"新达尔文主义"。然而，更为重要的修正来自于20世纪的遗传学革命以及多学科的综合作用。

自孟德尔（G. Mendel）遗传规律被重新发现之后，摩尔根（T. H. Morgan）等人对遗传突变进行了深入的研究，以"粒子遗传"替代了"融合遗传"的概念。费希（R. A. Fisher）、霍尔丹（J. B. S. Haldane）和赖特（S. G. Wright）等人则相继发表了《自然选择的遗传理论》、《孟德尔群体中的进化》以及《进化的动力》等经典著作，创立了群体遗传学（population genetics），将粒子遗传理论与生物统计分析紧密结合，认为生物群体中包含有大量的遗传变异，而进化的方向和速率则由自然选择来决定。目前常用的适合度（fitness）概念就是在群体遗传学以生物繁殖相对优势来定义适应的前提下建立的，它修正了达尔文进化论中"物竞天择，适者生存"的概念，代之以生物个体或基因型对后代或后代基因库的相对贡献，即统计学意义上的生物适应。20世纪30年代后，杜布赞斯基（T. Dobzhansky）进一步将群体遗传学理论与实验生物学方法用于进化生物学研究，发表了《遗传学与物种起源》等著作，最终成为该领域的集大成者。他从理论和实验上统一了达尔文自然选择学说和孟德尔遗传学，创立了综合进化论学派。其后，动物学家迈尔（E. Mayr）、古生物学家辛普森（G. G. Simpson）以及植物学家斯特宾斯（G. L. Stebbins）等人相继从不同生物类群研究出发，阐述了生物进化的机制，发展和完善了现代综合进化论的理论框架，使其成为了当代进化理论的主流。

值得一提的是，迈尔本人堪称20世纪综合进化论学派的一棵"常青树"。他在长达80年的跨世纪研究生涯中，除发表了有关动物进化方面的重要论著外，还特别厘清了达尔文进化论的逻辑架构，将其划为五个既可分割又有联系的部分，即物种可变理论、共同祖先理论、渐变理论、物种增殖理论以及自然选择理论。其中，自然选择理论最为重要。总之，迈尔杰出的生物学哲学思想为我们全面掌握综合进化理论提供了有效的途径。

分子进化与分子系统学

20世纪分子生物学的快速发展极大地改变了进化生物学的格局。就在达尔文进化论诞生一百周年之际，木村资生等提出了分子进化（molecular evolution）学说。该学说认为分子（基因）的进化过程与达尔文所描述的宏观进化过程不同，中性的遗传漂变可能比自然选择发挥更大的作用。随后，人们相继建立了基于群体遗传学的DNA与蛋白质序列进化模型及分析方法，既能定量描述和预测不同分子随时间变异的模式，也可以区分遗传和环境因子对基因水平变异的影响。更为重要的是，由于所有生命的蓝图都是用DNA（某些病毒中则用RNA）来书写的，因而人们可以通过比较DNA序列来研究它们的进化关系。这一学科领域被称为分子系统学（molecular systematics）或分子系统发育学（molecular phylogenetics），它为解决系统与进化生物学中的疑难问题提供了新的方法论工具，对生物分类学的发展也产生了至关重要的影响。

最近五十年来，每当分子生物学出现一个新的概念或发展一项新的技术，进化生物学家都会千方百计地将之用于发展进化研究。同样，分子生物学家也开始从进化的途径来深入理解分子生物学以及发育生物学和免疫学等生命科学前沿领域。例如，分子生物学家现在常常构建进化树（evolutionary tree）以寻找不同生物的种间（直系）同源基因和种内（并系）同源基因。总的来说，分子生物学与进化生物学的有机结合具有下述优点。

（1）所有生物的DNA均由腺嘌呤（A）、胸腺嘧啶（T）、胞嘧啶（C）和鸟嘌呤（G）这4种碱基组成，因而可以通过分子序列分析来阐明大尺度、跨门类的生物进化关系。目前，分子进化分析已表明，地球上所有的生命体来自大约四十亿年前的一个共同祖先。换言之，如同达尔文进化论所推测的，所有有机体在进化历史上都是相互关联的。

（2）DNA的进化演变或多或少是有规律的。人们已经建立与发展了许多描述分子序列间DNA或氨基酸置换的数学模型。相比之下，形态性状的进化就要复杂得多了，难以精确描述。

（3）一个基因组是一种生物所有基因编码序列及非编码序列的总和。对分子系统学研究而言，基因组所包含的有用信息比形态性状要多得多，这将有助于提高进化统计推断的精确性。

（4）在生物进化时间估计和速率比较方面，分子数据具有其他性状不可比拟的优势。目前，采用分子序列分析方法可以推测生物类群（物种）间的分歧（起源）时间并检测不同谱系间的进化速率是否存在显著差异。

目前，分子进化研究通常是利用各种分子数据（主要是DNA序列和氨基酸序列），针对以下三个方面的问题进行统计分析。

（1）进化式样

应用系统发育分析方法，可以重建物种或基因间的进化（系统发育）关系，结果则用进化树或进化网络的形式来表示。在解读系统发育树所蕴含的物种间关系时需要特别关注单系类群（monophyly），因为这种类群中的所有物种（或基因）具有一个共同的祖先。

（2）进化速率

分子进化研究与分子钟（molecular clock）概念是分不开的。分子钟假说认为，氨基酸或核苷酸置换速率在进化过程中近似地保持恒定。如果这一假设成立，则我们可以用恒定的速率来估计生物类群间的分歧时间。然而，分子钟概念长期存有争议，焦点是其准确性和变异机制。近年来，人们在实际工作中常常将其作为非严格分子钟假定条件下估计分歧时间的统计工具。

（3）进化机制

分子进化研究的首要问题是基因突变。由核苷酸置换、插入和（或）缺失、重组等引发的突变基因或DNA序列，可以通过群体水平的遗传漂变和（或）自然选择进行扩散并最终在物种中得以固定。用分子序列来确定生物特定的进化机制通常有两条途径。一方面，群体遗传学家对一个基因座上的不同等位基因进行测序并构建同一个物种内不同等位基因的进化树，这种多态等位基因的进化树可以判别不同进化机制的相对重要性，也可以为研究两个群体间的基因交流程度提供重要信息。另一方面，分子进化生物学家对所感兴趣的基因检测非同义置换（引起氨基酸改变的置换）和同义置换（不引起氨基酸改变的置换）的比率。若该比率大于1，则认为非同义置换是由达尔文正向选择（positive selection）造成的；若比率等于1，则认为该基因处于中性进化；若比率小于1，表明存在负选择（淘汰选择）。大量研究表明，分子水平的达尔文选择对阐明遗传与环境因素在生物进化过程中的作用极为重要，而统计检测更是分子进化研究中不可或缺的工具。

后基因组时代的进化生物学

进入21世纪后，生命科学研究也进入了后基因组时代，诸如蛋白质组学、功能基因组学、药物基因组学以及系统生物学之类的交叉科学迅速成为生命科学的前沿领域，而后基因组时代的进化生物学研究也呈现出若干新的态势。

进化基因组学无疑是基因组学中最为活跃的方向之一。由于测序技术的迅速发展与普及，人们已测定了大量生物物种（尤其是微生物）的全基因组序列。利用这些序列信息，辅以统计分析工具及计算机软件，可以重建生物类群间的系统发育关系或者进化历史，这将有助于人们解决生物进化研究中长期存疑的问题，深化我们对生物多样性起源与发展规律的认识。

近年来，进化基因组分析方法的应用已极大地提高了分子进化研究的水准。例如，很长时间以来，围绕我们人类自己与进化上的近亲——黑猩猩之间的差异存在不少争论。人类基因组计划和其他基因组计划的实施为我们解决相关问题提供了新的证据。对全基因组信息的进化分析表明，人类与黑猩猩在全基因组序列上的差别并不大，但人类大脑发育相关基因的确存在着适应性进化。

对一些较小基因组（如病毒）而言，进化分析方法更有其得天独厚的优势。以严重急性呼吸道综合征（severe acute respiratory syndrome，SARS）致病原——SARS冠状病毒为例，科学家们在测定一批病毒样本的全基因组序列基础上，应用分子进化分析及生物信息学方法，鉴定出这是一种新的冠状病毒，它与已有的三组冠状病毒都不相似，在进化上是较为独立的一枝。进一步的分子进化分析还发现，SARS流行强度与其冠状病毒序列特征存在明显的关联。对达尔文正向选择的检测表明，SARS冠状病毒在SARS流行早期发生了适应性进化。该病毒在SARS流行的中期呈中性趋势，而在流行晚期则处于淘汰选择。

在生物医学研究中，进化生物学正在发挥着愈来愈重要的作用。以癌症研究为例，传统方法中很少涉及进化分析思想。然而，在新兴学科方向——进化医学（evolutionary medicine）中，可以通过分子水平的进化错配（evolutionary mismatch）分析和达尔文正向选择作用来探索癌症形成和演变的机制，而免疫进化模型和模拟计算也将为癌症治疗提供新的思路。可以预料，进化医学将为后基因组时代分子医学的基础与应用研究开启一扇新的窗口。

本文发表于《科学》2009年9月61卷5期"纪念达尔文专题"栏目，收入本书时略有修改。

达尔文进化论的科学本质与贡献

自达尔文1859年发表《物种起源》一书以来,"进化"已逐渐成为生物学文献中出现频率最高的词语之一。随着科学界对生物进化现象的认识逐渐深入,人们对达尔文进化论的理解也在不断深化和提高。从魏斯曼等人提出"新达尔文主义",到摩尔根等人以"粒子遗传"替代"融合遗传"的概念以及费希尔、霍尔丹和赖特等人创立群体遗传学,达尔文进化论中"物竞天择,适者生存"的概念一直在修正和完善之中。更为重要的是,杜布赞斯基将群体遗传学理论与实验生物学方法用于进化研究,创立了综合进化论学派;迈尔、辛普森及斯特宾斯等人相继从不同生物类群研究出发,发展和完善了现代综合进化论的理论框架,使其成为当代进化理论的主流学说;木村资生等人提出的"中性进化"学说则以分子变异速率为突破点,强调中性突变和遗传漂变是分子水平生物进化的主要机制,这些都为现代进化生物学的蓬勃发展开辟了道路。

达尔文进化论在过去150年间走过了曲折的发展之路。如同任何一门科学理论一样,它必须经受不同时期科学家们的质疑和检验。与许多科学理论所不同的是,进化论从创立伊始就接受了来自科学领域之外的一次又一次挑战。今天,人们对进化论的兴趣依然有增无减,同时对进化生物学也提出了更高的要求,即进化研究不仅应当能够解释各种复杂的生命现象,重建不同生物的自然历史,而且还应具有一定预测能力和应用潜力。藉纪念达尔文诞辰200周年和《物种起源》出版150周年之际，回顾达尔文进化论的科学本质与贡献，将有助于我们全面了解该领域的科学理论与知识，并用于指导21世纪的生命科学研究。

进化论的科学本质

总的来说，达尔文进化论从本质上改变了我们对地球生命现象的理解。它围绕生物多样性的起源与发展，引导人们探索各种生物之间的亲缘关系（或称进化谱系）。例如，从孩提时代起，人们就对各种各样复杂的生命现象与过程充满好奇，有关我们人类自身——地球生物中的一员也是疑问连连：人类究竟何时又是如何在地球上出现的？不同人种或不同人群之间关系如何？人类与其他生物（如细菌）在进化上有何关联？如此等等。进化论为我们提供了科学的解释。

然而，学术界对达尔文进化论的理论体系和科学本质的讨论一直争议颇多，持续至今。值得一提的是，有"20世纪达尔文"之誉的恩斯特·迈尔在长达80年的跨世纪研究生涯中，除发表了有关动物进化方面的重要论著外，还特别厘清了达尔文进化论的逻辑架构，将其划分为五个既可分割又有联系的组成部分，即物种可变理论、共同祖先理论、渐变理论、物种增殖理论以及自然选择理论。其中，自然选择理论占据了最为重要的地位，堪称进化论的灵魂。迈尔杰出的生物学哲学思想为我们全面掌握进化论的理论体系提供了一种有效的途径。

在达尔文进化论中，具有有益性状的生物存在差异的繁殖优势被称为自然选择，因为是自然来"选择"提高生物生存与繁殖能力的性状。如果生物的突变性状降低其生存与繁殖能力的话，自然选择就会减少这些性状在生物群体中的扩散。事实上，人工选择也是一种类似的过程，但在这种情况下是人而不是自然环境使生物交配以选择理想的性状。最常见的例子莫过于农牧业和园艺栽培，它们都是通过人工选择方式来获得人们所需的家畜品系和园艺植物品种的。

达尔文进化论为进化生物学研究提供了比较分析途径，它基于以下原理：现存生物在结构复杂性及与之相关的功能完善程度上的差异是进化过程中不同进化单位的不同变异速率和程度所致；倘若各进化分支的进化速率相同（恒定），各类群之间相关性状的差异程度就是该性状分异时间（进化时间）的函数，因而可以作为衡量类群之间亲缘关系的尺度；从连续地层中发现属于同一类群的化石类型在相关器官或性状上若呈渐变序列，则可据此推断这些化石所代表类群的进化历史。

迄今为止，支持进化论的证据层出不穷。从中华龙鸟化石的发现到酵母实验进化的分析，寻找可填补进化缺失环节的关键证据以及确定不同生物类群进化速率的工作不胜枚举。近年来，比较突出的例子有加拿大北部"大淡水鱼"化石的发现。科学家们首先根据上述进化比较原理和已有化石分析预测出浅水鱼类向陆地过渡阶段的大致时间，随后将目光投向加拿大北部努维特地区的埃尔斯米尔岛，那里有大约三千七百万年前的沉积岩。通过四年的艰苦努力，科学家们终于从岛上岩层中发掘出命名为"Tiktaalik"（因纽特人的语言中意为"大淡水鱼"）的生物化石，它既具有许多鱼类特征（如鳃、鳞片和鳍等），又具有早期四足动物的典型特征，它有简单的肺，鳍包含骨骼，可形成类似于有肢动物的肢体，用来移动和支撑躯体，适应离水生活。"大淡水鱼"的发现具有很高的科学研究价值，它证实了科学家们基于进化论的预测。反过来，对于进化论预测的证实也提高了达尔文理论的可信度。这一重大发现再次说明，每一门科学理论本质上都要具备对尚未观察到的自然事件或现象作出预测的能力。

有关自然选择的另一个非常经典的例子是科学家们对特立尼达岛阿立波河中的虹鳉鱼所进行的观察与实验。按照进化理论，不同时间尺度上的自然选择可能产生全然不同的进化效应。在仅仅几个世代的时间周期内，生物个体就有可能产生小规模的变异，这一过程可称之为微进化。科学家们发现，生活在阿立波河中的虹鳉鱼无论是其幼体还是成体均遭受较大鱼类的捕食，但生活在河流上游小溪中的虹鳉鱼只有其幼体才会被较小鱼类捕食，因而长期的进化过程导致河流中的虹鳉鱼个体较小（更易于躲避捕食者），而溪流中的虹鳉鱼则个体较大（不易被较小的鱼类捕食）。在实验进化生物学研究中，科学家们将河流中的虹鳟鱼置于原来没有虹鳟鱼种群的溪流，发现它们仅仅在20代后就进化出了溪流中虹鳟鱼的特性。

毋庸讳言，在科学上，人们不能绝对肯定地证明某种理论或某种解释是完美无缺的，或者说是终结性的。然而，迄今为止，许多科学解释已经被人们反复检验，不断增添的新观察结果或新的实验分析很难对它们作出重大的变更。换言之，科学界已广泛接受这些解释，它们是以观察自然世界获得的证据为基础的。进化理论就是其中一个突出的代表。从这一点出发，我们可以明确地将从科学上补充和完善进化论与从宗教上反对进化论的所谓证据区分开来。

进化论的科学贡献

达尔文进化论对人类社会的主要贡献包括两个方面：思想贡献和科学贡献。首先，达尔文是一个伟大的思想家，他将宗教从自然科学领域中驱逐出去。在达尔文之前，"神创论"占据了绝对的统治地位，大多数人相信世界是由上帝有目的地设计和创造的，因而协调有序、完善美妙、永恒不变。达尔文进化论则展示出另一幅景象：既没有预定目的也没有预先设计，变幻无穷而又充满竞争。正因为如此，进化论作为神创论的对立面一出现就给科学界乃至整个人类社会带来了一场暴风骤雨式的思想革命，它瓦解了传统的思想观念，使人们认识到自然界是变化和发展的产物。从这个意义上讲，进化论导致了人类自然观的巨大转变，不愧为人类文明史上的一件大事。

然而，即使在生物学界内部，进化论也不能为一些现代生物学家所接受，他们认为进化论并不完全符合现代科学理论的标准（如生物进化过程不能在实验室重复）。随着进化研究的深入，生物学中的进化概念已逐渐扩大到广义演化概念，从孤立的个体和物种进化研究发展到对复杂生物系统中相互作用与联系的协同进化研究。进化思想已渗透到生物学及众多的相关学科领域，如同迈尔所言："进化论是生物学中最大的统一理论。"杜布赞斯基则认为"如果不按照进化思想思考问题，生物学的一切将无法理解"，真是一语中的！

值得指出的是，进化论除了对生命科学研究具有重要意义外，还为整个自然科学的发展提供了新的认识论。传统认识模式本质上是还原论，即复杂性是简单性的叠加，而复杂过程可视为简单过程的特例。基于此，科学研究强调"分解"和"还原"，即将复杂过程逐级分解为相对简单的、小的组成单元或因果链上的片断，然后将这些单元或片段叠加起来。这种模式导致了自然科学各分支学科之间的"隔离"。更为严重的是，研究复杂系统和复杂过程的生物学逐渐沦为物理学和化学的附属学科。

达尔文进化论揭示了生命进化历史从简单到复杂的规律，而物理学和化学在1969年普里高京提出"耗散结构"理论（普里高京因这一理论荣获1977年诺贝尔化学奖）后才真正接受了进化（演化）概念，认为自然界的基本过程是不可逆的、随机的，非生命系统也有类似生物进化的、从混沌到有序的演化行为。如今，演化已成为整个自然科学的核心概念和普遍规律，正如普里高京所言："无论向哪里看去，我们发现的都是演化。"可以毫不夸张地说，进化论使物理学与生物学实现了基本理论原则的统一。

本文发表于《生命世界》2009年11期，为该期"封面故事"，收入本书时略有修改。

进化论的分子时代

当DNA的双螺旋结构被发现后,人们的视野从宏观世界深入到微观世界,进化论也随之迈入分子时代。此时,达尔文提出的物种变异、万物同源、自然选择三个观点还站得住脚吗?

达尔文去世前两个星期,他在家中的园子里发现一只小型蛤被一只水生甲虫牢牢夹住。为了纪念这只小型蛤,他专门写了一篇短文,这也成为他的最后一篇著作。这只水生甲虫是一位年轻的鞋匠兼业余博物学家沃尔特·克里克送给他的。鞋匠有个孙子叫弗朗西斯·克里克。1953年,弗朗西斯·克里克和一位年轻的美国小伙子詹姆斯·沃森等,发现了DNA双螺旋结构,为达尔文的生物进化思想提供了强有力的证据。

这份证据不是来自化石,不是来自现代生物标本,也不是来自生物器官的解剖学知识,而是来自一本"生命天书"。沃森和克里克等发现,每种生物的细胞里都携带一份生命密码,它的语言对所有生命来说都是一样的:简单的由4个碱基组成的DNA密码。

达尔文曾经写下这样的真知灼见:"古往今来,生活在地球上的全部有机生命体,都是由某种单一的原始形态传衍而来。"坦白说,这只是他的猜测。事实证明,这份猜测是正确的。当破译了DNA分子组成的生命密码后,我们便可以从分子水平验证达尔文的生物进化思想,不断修正和完善它。

从基因水平验证物种变异

加拉帕戈斯群岛上的地雀给了达尔文太多的启示。经过细致观察，达尔文发现，这些地雀的喙形状和大小各异。这样的例证显而易见，所以，当达尔文提出物种变异时，几乎得到所有科学家的认同。不过，当我们破译了生命密码后，物种变异的宏观现象在微观世界里还能解释得通吗？

哈佛大学的霍诺夫和哈佛医学院的塔宾等人给出了肯定答案。他们从加拉帕戈斯群岛上的地雀入手，找到了一些掌控喙形状和大小的关键基因。在大嘴地雀的胚胎发育过程中，一种特定的基因BMP4开始在胚胎的下巴部位表达。这个基因合成的蛋白质能够促使大嘴地雀的喙宽厚有力，帮助大嘴地雀破开大块头的种子和坚果。而另一种特定的基因CaM在仙人掌地雀的胚胎中表达，合成钙调蛋白，促使仙人掌地雀的喙细长小巧，帮助仙人掌地雀在仙人掌果实里寻找种子。

自然界中还有许多物种间存在变异的事实，科学家们也在寻觅分子水平的实证。在佛罗里达州沿岸的沙滩上，一种沙滩小鼠的毛色比生活在陆地上的小鼠的略显苍白。这是沙滩小鼠伪装自己的一种策略，因为猫头鹰、苍鹭等猎手很难将沙滩小鼠和沙滩区分开来。通过研究，科学家们发现，沙滩小鼠的毛色之所以略显苍白，主要是因为编码皮毛色素的基因中的一个碱基发生了改变。正是这个不起眼的微小变化降低了这个基因的表达量，合成的皮毛色素相应减少，所以，沙滩小鼠才有了略显苍白的皮毛。

其实，作为世界万物中的一种生命形式，人类也存在可见的变异。比如，蓝眼睛和黑眼睛。2008年，丹麦哥本哈根大学的伊尔伯格和他的同事宣称，他们破解了蓝眼睛和黑眼睛的奥秘。拥有蓝眼睛的人的15号染色体上的一个碱基发生了改变，减少了一种名为OCA2基因的表达。巧的是，这个基因刚好合成眼睛的黑色素，帮助眼睛颜色变暗。可想而知，当这个基因表达量减少后，眼睛的颜色自然变浅，于是便成就了一双迷人的蓝眼睛。

不过，眼睛颜色的基因调控方式有点与众不同。无论是地雀的喙还是小鼠的皮毛，这些特征的变化是由不同的基因掌控的，或者是同一个基因内部发生改变引起的。而蓝眼睛的出现不是色素基因本身发生改变，而是色素基因附近的DNA片段上的一个碱基发生改变引起的。看来，为了实现物种多种类型的变异，需要借助多种基因调控手段，不仅使用不同的基因，还可改变基因的表达方式——何时开启，何时关闭，以及表达时间是长还是短。

就拿长颈鹿的长脖子来说。长颈鹿的体内并没有特殊的基因促使它长出长脖子。编码长颈鹿长脖子的基因与编码小鼠的短脖子基因是一样的。不同的是，在长颈鹿体内，这个基因表达的时间稍长，合成的用于构建脖子的蛋白质略多，这样长颈鹿才拥有一个长脖子。

至此，物种变异在宏观现象和微观视角里得到了统一。从宏观现象来看，物种间的变异是性状的多样，比如，地雀的喙、小鼠的皮毛、眼睛的颜色。而从微观视角来看，物种间的变异是基因的多样，基因种类可以不同，基因本身可以不同，就连基因的表达方式也可以不同。正是这些多样的调控方式，成就了物种间的变异，让我们一目了然。

以"生命之网"完善"生命之树"

当你肯定物种间的变异时，也是在肯定物种间的相似。因为如果两者没有共同点，也就没有可比性了。达尔文的聪明之处在于他不仅看到了物种可变的事实，还看到物种之间的联系。当达尔文看到加拉帕戈斯群岛上的众多地雀时，他不仅看到了这些地雀的喙的形状和大小各异，还看到了这些地雀的相似性，大胆提出这些地雀源于同一个祖先的推测。通过分析这些物种的遗传密码，科学家们确定了这个推测，加拉帕戈斯群岛上的地雀是从一个祖先进化而来的，这个祖先是一种毛色灰暗的曹雀。

不过，在达尔文那个年代，人们还不知遗传密码是何物。但是，这并没有妨碍达尔文构筑他心中的那棵"生命之树"。1837年，在伦敦家中的书房里，达尔文翻开红色皮面笔记本，写道，"我认为"。然后，他勾勒出一棵树的形状，并写下这样的注释："同一纲的物种之间的密切关系有时可以表现为一棵巨大的树。我相信这个比喻在很大层面上说明了事实。"

"生命之树"的猜想堪称达尔文进化论的核心。达尔文通过"生命之树"

的概念把自然界中的万物统一了起来，传达了万物同源的思想。加拿大达尔豪斯大学的杜利特尔说："没有生命之树，进化论永远也不可能诞生。"我国著名进化学家张昀也肯定了"生命之树"的功绩："现代生物进化概念的核心是万物同源及分化、发展的思想。"

自达尔文以来，"生命之树"便成为人们了解自然界中生物统一的原则。树的根部是所有现存物种的共同祖先。从根部延伸出树干，树干再分枝分杈，形成一棵大树。每一根树枝代表一个物种，分支点代表物种分离的地方。有的树枝终结，表明物种灭绝；有的树枝继续生长，代表现存的物种。这棵树就像是一个纪录片，你可以探访过往生命，追溯生命起源。

但是，达尔文只为我们勾勒出"生命之树"的美好蓝图，它的枝繁叶茂还有待后继者的添枝加叶。DNA双螺旋结构的发现为这些后继者开辟了一方新天地。随着物种的DNA序列、RNA序列和蛋白质序列的相继破解，科学家们相信，这些数据能够为达尔文的"生命之树"提供确切证据。根据物种的DNA序列、RNA序列和蛋白质序列之间的差异，可以判断物种之间关系的远近。

事情开始时进展顺利。科学家们最先掌握了大量物种的核糖体的RNA序列。1977年，通过比较多种植物、动物和微生物的核糖体的RNA序列，美国微生物学家伍兹勾画出这棵大树的基本轮廓，为其确定了3个分枝，细菌、古细菌和真核生物。看到这样的成果，科学家们更加坚信"生命之树"的伟大蓝图。

20世纪90年代早期，科学家们开始测定细菌、古细菌的DNA序列。意料之外的事情发生了，通过DNA序列构筑的"生命之树"与RNA版本的"生命之树"无法重合。比如在RNA版本上，物种A更接近物种B，而不是物种C，但是，在DNA版本上，结论刚好相反。到底哪个正确，科学家们开始寻找产生分歧的原因。

他们在审视细菌和古细菌的DNA序列时，发现了一个难以置信的事实，细菌和古细菌经常和其他物种交换遗传物质。这种基因"水平"转移的现象曾被科学家们否定。按照常理，物种繁衍是"垂直"的，一个物种只会把遗传物质传给它的后代。达尔文也正是根据这个思路，认为物种之间不会发生联系，所以才有了树枝只会向上发展，不会水平扩展的"生命之树"的猜想。

但是，细菌和古细菌之间的遗传物质交换客观存在。事实胜于雄辩，当这种基因水平转移的事实存在时，树枝便会勾连在一起，这时的蓝图不再是一棵简单的树，更像是一张交织在一起的网。一些科学家提出，生命之树上的细菌和古细菌应该是一张网。有的科学家直接否定了"生命之树"的存在，认为这只是人类为自然界中的生物分类的一种方法。这个观点并没有得到所有科学家的响应，不少科学家希望借用更多的数据还原"生命之树"的蓝图。

2006年，德国海德堡欧洲分子生物学实验室的伯克研究小组检测了来自细菌、古细菌和真核生物的191个物种。他们发现，这些物种共享了31种基因，而这些基因没有显露出水平转移的迹象。他们还就此绘制了一棵接近完美的生命之树。但是，这棵树却遭遇了许多科学家的攻击。因为31种基因只是一个典型细菌基因总量的1%，仅是一只动物基因总量的0.1%。这棵树充其量只能算是生命之网中的局部分枝，如果这张网真的存在的话。

后来的实验结果越发显示出生命之网的存在，因为科学家们在真核生物界找到了基因水平转移的证据。加拿大大不列颠哥伦比亚大学的植物学家瑞森伯格发现，有14%的现存植物是两个不同的物种混合的产物。很多例子也表明，动物会"水平"地从细菌、病毒甚至从其他动物身上获取基因。牛的基因组中包含小部分蛇的DNA片段，大约是在5 000万年前从蛇那里借来的。而果蝇应该是一个细菌-昆虫杂交体，因为在果蝇的基因中发现了沃尔巴克氏体的全部基因。

至此，"生命之树"的概念已经不能解释物种间千丝万缕的联系，取而代之的是一张错综复杂的"生命之网"。不过，"生命之网"并没有完全否定"生命之树"，因为网中的局部结构还保留有树的痕迹。在这个意义上，达尔文功不可没，他的"生命之树"蓝图让我们看到了万物同源的事实，也指引着我们发现物种间更丰富的联系。

用"中性"选择补充自然选择

当我们肯定了物种变异、万物同源的观点后，我们看到了物种进化的事实。这不禁又引申出一个新的问题，什么力量在支配着物种的进化。达尔文看到了海鬣蜥为啃食海底的海藻而长出的盐腺和趾蹼，认为这是生物适应环境而产生的变异。加上马尔萨斯"生存竞争"的启发，达尔文提出，自然选择是物种进化的机制。

他在自传中写道："由于长期不断观察动物和植物的习性，我已经有充分的准备来评价无处不在的生存斗争的意义。我顿时领悟到，在这些条件下，有利的变异将保存下来，不利的变异将消失，其结果将是新物种的形成。"为了更明确地表达自然选择的思想，达尔文在《物种起源》中借用了英国哲学家斯宾塞提出的"适者生存"的说法。这不仅使斯宾塞的名声大振，也使得自然选择学说备受关注。

但是，受到关注并不表示全盘接受。一些科学家认为自然选择促成的物种间的变异不足以导致新物种的形成，而把物种进化的机制归结为某种内在动力。1903年，荷兰植物学家弗里斯根据美洲夜报春花的几种变异体，提出物种进化的突变学说，认为突变导致新物种的形成。美国遗传学家摩尔根发展了弗里斯的学说，认为突变的作用大于自然选择的作用，突变是在创造变异体，而自然选择是在保留有利的变异体。20世纪30年代，一批群体遗传学家根据数学模型，再次肯定了自然选择的作用。

就这样，自然选择学说不断受到质疑，又不断得到肯定，在批判中走过了近半个世纪。直到DNA作为生命密码的事实被揭开，许多基因的DNA序列以及蛋白质的氨基酸序列被掌握后，科学家们开始从分子水平理解物种的变异，用分子手段研究物种进化的机制。首先，科学家们肯定了DNA在复制时发生突变的事实，而正是这种突变引起了物种间的差异。那么，它对物种进化有没有推动力呢？

我们知道，DNA的突变无外乎4种方式：碱基的替换、缺失、附加和倒位。其中最常见的当属碱基的替换。因为3个碱基决定一个氨基酸，所以碱基的替换意味着氨基酸的替换。氨基酸可以替换成不影响蛋白质功能的、性质相似的氨基酸，这称为保守替换；还可以替换成影响蛋白质功能的、性质明显不同的氨基酸，这称为非保守替换。

1968年，日本群体遗传学家木村资生研究了不同物种中若干蛋白质的氨基酸序列，发现这些氨基酸序列中的替换大都是保守替换，也就意味着分子水平的突变大都是中性的，无所谓有利或是不利。1969年，美国分子生物学家金以及朱克斯也得到了与木村资生相同的结论。这样的结论无疑再次挑战了自然选择学说。

按照自然选择学说，有利的变异保留下来，不利的变异淘汰出局。但是，如果这些变异从分子水平来说都是中立的话，那么，自然选择无法作用于它们，哪些变异可以流传，哪些变异趋于消失，全靠机遇。科学家给这种随机变动性起了一个专业名称——遗传漂变，并据此提出了分子进化的中性学说，认为中性突变的遗传漂变才是分子进化的主要原因。

随着越来越多的分子证据证实中性突变的存在，达尔文的自然选择学说一度陷入了万丈深渊。中性学说的追随者甚至把自己的学说标榜成"非达尔文学说"，出现两个理论对峙的局面。究竟孰是孰非，经过近30年的发展，科学家们慢慢领悟到，中性学说并没有打倒自然选择学说。

科学家们发现，一部分中性突变能够在环境改变后变为有利或不利，受到自然选择的作用。比如，编码蛋白酶的基因发生突变，尽管是中性突变，没有影响蛋白酶的功能，却赋予蛋白酶不同的失活温度。一种在33℃以上失活，一种在44℃以上失活。如果环境温度是30℃的话，对这两种蛋白酶没有选择作用。但是，如果环境温度是35℃的话，在44℃以上失活的蛋白酶便具有选择优势。

另外，尽管证据显示大多数分子突变是中性的，但还有少部分突变是有利的或是不利的。就拿人的血红蛋白来说，这种蛋白质的大多数氨基酸的替换都是保守替换，不会影响血红蛋白的功能。这些中性突变不受自然选择的作用，可以在群体中随机漂变。但是，如果其中一个氨基酸由谷氨酸替换成缬氨酸，影响了血红蛋白的功能，这种有害突变将在自然选择的压力下被淘汰出局。

由此看来，在特定情况下，分子进化的中性学说可以用自然选择学说来解释。只不过，自然选择学说以宏观环境作为依据，而分子进化的中性学说则将微观世界的分子突变当作标准。不过，自然选择学说只看到有利或不利的变异，却忽视了大量存在的中性突变。从这个角度来看，分子进化的中性学说更为全面，它是对自然选择学说的有力补充。

在迈入分子时代后，达尔文的生物进化思想经受了批判和质疑，得到了修正和发展。20世纪的进化论大师杜布赞斯基曾经高度评价进化论："除非按照进化论来解释，否则生物学毫无意义。"其实，进化论的贡献远远超出了生物学界，它已经成为人们理解其他科学领域，乃至整个世界的最高准则。因为没有物种的进化，也就没有丰富多样的生命，更没有五彩缤纷的世界。

本文发表于《科学画报》2009年第12期，收入本书时略有修改。

世界之巅上的断思

"耐心"标准

改革开放带来中国经济的快速发展，但是，作为一个有五千年文明的民族，我们的耐心在快速地丧失。好像国际有一个"耐心"标准，是一个人到餐馆去吃饭，他从坐下来到服务员上菜所能忍耐的时间。我们现在已经缩短成八分钟了。而法国人一个小时不上菜也不着急，大家坐下来喝喝红酒，说说话有什么不好呢？我们总喜欢催问服务员："怎么还不上菜？"表面上看好像是人家服务质量不好，其实是你的忍耐力没了。我们老笑印度人做事拖拉，其实人家的哲学是，在一个历史长河中这几个小时算得了什么呢？在西藏，我们有一次路过一个边境小镇，去最大的饭店，叫成都饭店，才两张桌子。那菜上来也是一个小时以上。你抱怨什么呢？在上海，如果一个司机说前面修路要绕路，绕出四公里我们就火冒三丈。在西藏跑野外，路断了，我让司机绕一下，680公里。这就叫西藏。

进化与革命

新西兰裔的一个著名教授（现居澳大利亚）拒绝了我学生的留学申请，嫌我们底子太薄。他说，你们研究生去野外训练的时间太多。我说："生物学家怎么能不去野外呢?"他说:"我高中的时候南极、北极都去过了。"后来我才知道他父亲是新西兰古生物学会主席。于是我想起来我父亲,他出生于湖南偏远山区的一个农民家庭,读了一个不要钱的师范院校。到我这一代,不用像他那样为离开农村而放弃一切了,但我还是把在国外受教育作为理想之一。到我儿子这代,也许就视美国为平常物了。我记得杨振宁先生说过,他父亲是芝加哥大学毕业的,他不会只是为了到芝加哥大学而放弃一切,他有更高的追求。这个就是 family tree(谱系树),一代一代地积累,可以叫evolution(进化)吧。但我们很多人只相信暴发,教育上 revolution(革命)。自己大学都没读过,儿子一定要上剑桥,标准定得过高,心态就不容易平和。对待财富也是。这些可能都需要反思。

进化上的适应度

小时候我天真地以为,高血压可怕,那低血压该多好啊!后来我母亲患了低血压,走路晕、摔跤、吃不好饭,除了名字叫低血压,其他症状跟高血压差不多。所以我从小就知道了,任何事情不能走极端。比如你说氧气少了,所以咱们在西藏有高原反应。可是藏族人到上海来也是成天晕晕乎乎的,氧气太多了也不行。人活在这个世界上,要考虑进化上的适应度问题。但我们经常追求极致。本来只是想挣点钱,让生活好一点,久而久之就要做大房地产商,要打造中国第一。其实,到了那个时候,钱已经不能让你愉快了。科研也是一样,做些研究不是蛮好吗?非要追求极致,一心想得大奖,往往就把事情做过头了。

父 与 子

很多中青年人自己不爱学习,但给儿女的目标定得很高。今天要读北大,明天想读哈佛,恨不得儿子会五门外语。我有个湖南亲戚去年全家来看我,让我好好教育一下他十岁的儿子。他儿子凑过来要听我们聊天。"快去做作业!"他把那小孩往里屋一推,门都还没关严,就把我肩膀一搂,说:"哥,下次到长沙来我带你去歌厅。上海连个像样的歌厅都没有,长沙有几百家呢。"我说:"我现在知道你儿子学习为什么这样了。"他说:"小孩要好好教育,你看我都35岁了嘛。"我说:"你哥都45了呢,还要学藏语。"

100年前的哈佛医学院

今天我们对哈佛已经崇拜到:一个人只要从哈佛拿博士出来了,大家认为一切都可以交给他了。从我主译的《大流感》一书中可以看到,大约100年前,美国行医的人要在门口挂一块大字招牌,声称"本人曾在德国进修过",而"哈佛大学医学院毕业"是没有什么信誉的,以前的哈佛大学医学院交点钱就能上。那段时间,德国人每年夏天甚至要办补习班,专门接收美国医学院的学生来进修。就像我们今天好多人根本不提自己是中国著名的医学大学毕业,经过了多年的医生训练,张口闭口"我到美国进修过一年"。但是仅仅过了一百年,美国医学教育就一下子改朝换代了。

本文发表于《杂文月刊》2010年第5期,收入本书时略有修改。

生物信息学专业规划的理念与实践

一、生物信息学的诞生和发展

生物信息学是1991年左右才在文献中出现的（美国国立医学图书馆——NLM生物医学文献数据库——Medline中1993年才在正式文献中出现）。该学科雏形可以追溯到20世纪50年代，1956年在美国田纳西州盖特林堡召开了首次"生物学中的信息理论研讨会"。1987年，林华安博士正式把这一领域称为生物信息学（Bioinformatics）。到20世纪末期，随着分子生物学与计算机技术的迅猛发展和数据资源急剧膨胀，迫使我们必须摆脱手工劳动的束缚，转而寻求强有力的工具去组织它们。与此同时，蕴藏在这些生物学数据资源中的生物学规律已无法继续沿用传统手段以人脑来加以分析和归纳，因此人们同样需要寻求强有力的工具。

生物信息学正是通过它独特的桥梁作用和整合作用，以数学、信息学和计算机科学为主要手段，以计算机硬件、软件和通信网络为主要工具，对浩如烟海的原始数据和纷繁复杂的生命信息进行存储、管理、注释、加工、解读。生物信息学对21世纪的生命科学和医学科学的发展具有非凡的推动作用，是当今生命科学的重大前沿领域之一。

生物信息学的发展可以分为以下三个阶段：

第一阶段，前基因组时代。早在1962年，美国的Pauling和Zuckerkerkandl就将DNA序列的变异与其生物进化联系起来，从而开辟了分子进化的崭新研究领域。20世纪60年代，美国建立了蛋白质数据库。1964年Davies开创了蛋白质结构预测研究。1970年，Needlman和Wunsch发表了序列比对算法。美国洛斯阿拉莫斯国家实验室1979年就建立起Genbank数据库。欧洲分子生物学实验室（EMBL）1982年就已经提供核酸序列数据库的服务。日本也于1987年开始提供DNA数据库服务。

第二阶段，基因组时代。标志性工作包括基因发现和识别、网络数据库系统的建立和交互界面的开发等，建立和发展了表达序列标签数据库以及电子克隆。

第三阶段，后基因组时代。标志性工作是大规模基因组分析、蛋白质组分析以及各种数据的比较与整合。

二、生物信息学的学科性质

生物信息学到底应归属于生物学范畴，还是属于以计算机为中心的信息领域，是生物信息学科体系所无法回避的问题。字面上理解"生物信息学"是"生物"加"信息"的交叉学科。广义地说，生物信息学通过对基因组的研究，获取、加工、储存、分配、分析和解释相关生物学数据的应用范畴。这一定义包括了两层含义，一是对大量数据的收集、整理与服务，也就是如何管理好这些数据；另一层是从中发现新的规律，也就是用好这些数据。前者表明"生物信息学"是以研究生命科学为对象的"信息学"，研究重点在于分析工具的发展、数据库的建立与使用者接口的开发。后者则强调整合、分析数据库中的生物信息，寻找致病基因，预测基因的功能等。是利用生物信息学技术，解决生物学问题的学科，因此它是"生物学"的一种，或称之为"信息生物学"。实际上，我们可以认为这两个方面的内容属于生物信息学包含两个层次的内涵。具体地说，生物信息学发展出精致的算法或完善的分析方法使我们能更好地利用数据进一步开展疾病相关基因寻找，蛋白质结构和功能预测，分子进化等研究；生物信息学的最终研究目标是为了破译隐藏在 DNA 序列中的遗传语言规律；在此基础上归纳、整理与基因组遗传信息释放的数据，认识蕴藏于其中的有关生命代谢、发育、分化和进化的规律。例如，就疾病而言，生物信息学有助于系统地理解导致机体功能异常的生理机制，从而得出科学的治疗方案，为人类疾病的防治提供崭新的途径。就生物进化而言，生物信息学有助于系统地解释生物进化中从微观分子到表观个体水平的基本规律，从而使人类更清醒地给自己在自然界中定位，科学地认识和改造自己的未来。

三、发展生物信息学教育的意义

过去人们对生命的认识有很大的局限性，对生物学的认识仅仅停留在观察上。到了 19 世纪，达尔文发表《物种起源》之后，生物学开始总结出有重大哲学意义的普遍规律。此后，孟德尔发现了遗传学规律，沃森和克里克发现 DNA 双螺旋结构以及核酸是生命本质，为生物学发展奠定了坚实的基础，从而生物学正式摆脱了那种仅靠观察和比较的方法，发展成为一门实验科学。伴随着 21 世纪后基因组时代的到来，生物学的重点和潜在的突破点已经由 20 世纪的实验分析和数据积累，转移到数据分析及其指导下的实验验证上来。

人类科学研究史表明，科学数据的大量积累将会导致重大的科学规律的发现。例如，对数百颗天体运行数据分析导致了开普勒三大定律和万有引力定律发现；数十种元素和上万种化合物数据的积累导致了元素周期表发现；氢原子光谱学数据积累促成了量子理论的提出。生物学确实有非常重要的问题等着我们去解决，比如脑信息学、生命起源，这些重要问题就像当初 20 世纪初期遗传学家面临的问题，看着染色体却不知道要用怎样的技术去解开基因的谜。最重要的遗传学上的突破，是发现如此复杂的生命居然只是由四种不同的脱氧核糖核苷酸（A、T、C 和 G）按照特定的编码规则串联成的脱氧核糖核苷酸串（DNA），其中却蕴藏着生物体中所有的结构信息和控制信息，着实让人诧异。借鉴历史的经验，有理由相信，当今生物学数据的巨大积累也将导致重大生物学规律的发现。

放眼世界，一些发达国家在生物信息学上已先行一步，争相投入巨资，争取通过生物信息研究与开发获得知识产权。中国科学家正联合起来，共同推进中国生物信息学的研究和发展。充分利用网络上大量免费的数据资源，去发现新线索、新现象和新规律，关键在于迅速培养一批在数学、信息科学、计算机科学以及分子生物学方面均有造诣的跨学科人才。

四、生物信息学人才的培养

进入21世纪，全国高校都按照"基础扎实、知识面宽、能力强、素质高"的人才培养要求开展了新一轮专业课程体系改革。生物信息学既是一门学科也是生物领域研究的工具，其课程教学内容符合当前课程体系改革中人才培养"厚基础、宽口径"的要求。首先，生物信息学是一门交叉性、平台性学科，其研究的内容相当广泛，涉及计算机科学、数学、信息学、统计学、生物学、物理学等众多学科，能培养学生从整个自然科学的高度和不同角度分析问题的能力。其次，生物信息学发展日新月异，知识不断更新，学习生物信息学有利于学生了解自然科学研究进展和方向。最后，生物信息学也是一种研究工具，学习和掌握这门工具为今后从事生物、医药、农业、环境、食品领域的研究提供了新的研究手段和途径。

目前，生物信息学在我国尚处于起步阶段，对研究人员要求很高，需要生物大分子结构和功能方面的背景知识，需要扎实的应用数学或统计学知识，还需要精通计算机。但我们面临的实际情况是大部分从事生物学研究的人不熟悉计算机，而从事计算机科学的人员多数又缺乏对生物学的了解。尽管知道生物信息学及其重要性的计算机科研人员为数不少，但对于计算机学科如何与生物学进行有效的融合尚未给予足够的思考与研究。

新世纪国家与国家之间经济、科技实力的竞争从本质上讲是人才的竞争，人是一切科学研究的主体因素，离开研究者的创造性思维，科学也就无法进展。生物信息学对从事其研究者提出更高的要求，不但要具有广博的知识，更要具有在此基础上进行比较、创新、归纳以及上升为理论高度的科学研究能力；同时还要求研究者具有敢于突破前人的气魄，即高素质的复合型通才。他们能够将自然科学与社会科学沟通起来发展生物信息学教育、培养综合性高水平的生物信息学专门人才，这必将为我国的学科和产业的发展注入强大的动力。

五、课程体系

课程体系是学校教学工作的中心环节，同时也是人才培养的模板。在某种意义上，有什么样的课程体系就能培养什么样的人才。加强知识能力和素质的培养，提高学生创新意识和创新能力是21世纪人才培养的基本要求。

培养生物信息学专门人才是发展生物信息学的关键所在，现阶段交叉学科人才培养工作中要明确学习动机和实际需求的关系。专门人才的培养可以从本科、硕士、博士三个层次展开。本科层次课程体系设置中，在短短的本科学习的四年时间内既要达到坚持基础，强调技能，拓宽专业口径，但又不增加学生的学习负担，有利于学生个性发展，还要充分体现生物信息学交叉学科的特点，对整个课程体系的规划尤其重要。

值得注意的是，这些课程的教学内容必须按照生物信息学学科内容的要求设计，有增有略，有所综合，在保证知识体系必备性和衔接性的同时，避免填鸭式的教学方式。让学生有更多时间关注自身感兴趣的研究方向，培养学生向以兴趣为出发点的主动学习方向发展。专业课程体系设置在学科人才培养成功与否中起着决定性作用，国外一些高校课程设置值得借鉴，如美国加利福尼亚大学洛杉矶分校的生物信息学专业课程设置，包括专业先导课程（Premajor）和专业课程（Major）两个模块，每个模块规定了最低学分数和必修课程要求，同时又留给学生很大的选课自由度（图1）。

研究生的培养规划强调培养跨领域之人才。由于生物信息学多学科交叉特点，学生招生面应尽可能广泛，可以是数学、信息、物理、化学、计算机、生物等专业，这既有利于学科融合，又可以防止学科过于朝单一方向发展。在课程设计上，根据学生背景，兼顾高校和研究所现有研究领域。比如针对数学和计算机背景的学生，开设生物类课程的必修课；而针对生物背景的学生，则开设系统的数学和计算机课程的必修课。研究生层次的培养还需要与其他学科的研究领域和产业相结合，及时掌握研究方向，提高研究成果转化力。

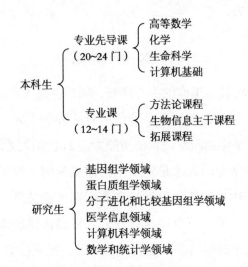

图1　美国加利福尼亚大学洛杉矶分校的生物信息学专业课程设置

六、教学内容和手段

在教学内容组织方面，除了要注意对各门课程间的内容协调外，还要注意到生物信息学在理论研究和应用研究处于不断的发展完善中，同时随着新的应用领域和新问题的发现，也不断地渗透到生物信息学领域，进一步增加了其多学科交叉融合的深度和广度。因此在教学工作中必须紧跟学科发展方向，随时进行知识更新，了解最新的前沿动态，掌握新方法，将最新的知识和方法教给学生。在高年级本科生和研究生交叉学科课程中增加案例分析，可以在很大程度上弥补交叉学科人才培养缺乏系统性的问题。在教学中鼓励学生进行探索式学习，培养其终身学习的能力和创新意识，正是生物信息学教学的特点。

1. 生物信息学给学生提供探索式学习的基础

用生物信息学培养探索式学习，取代枯燥的教学模式。人类基因组爆发式数据的产生，产生了生物信息学，聚焦在对大量涌现的DNA和蛋白质数据的获取、储存、分析、建模中。科学的基本原则就埋藏在探索式的学习中，探索是一个理解科学基本原理非常有效的进程。探索式学习是学生直接从教室面向社会，学生需要去解决他们在学校或者社会上遇到的问题。

生物信息学课程强调理解、推理和解决问题，而不是纯粹地记忆词汇解释和运算规则。教师应该鼓励学生有针对性地利用计算机或其他工具。生物信息学在科学活动中通过新技术的运用来帮助老师和学生实现这个目标。

2. 运用真实的数据去解决问题

学校只要有计算机和网络，就可以免费获得大量可靠的教学资源。用生物信息学工具，学生可以得到真实的数据，更重要的是可以用这些工具去独立解决问题。在这个过程中，学生实际上经历了从被动的学习到积极的学习，再到有效的学习的过程。现在我们只是需要引导学生强烈的好奇心和求知欲，就可以带领学生进入真实的科学世界进行探险，把他们从教科书中带出来，奔向科学的殿堂。

在课堂上整合生物信息学知识的目的是学生易于接触到真实世界。各种在线指南帮助学生和老师学习怎样利用这些工具去收集数据。一些尚未解决的问题，如人类进化历史、系统发育树和种系起源等问题。这些问题会出现在学生的头脑中，通过这些数据，引导学生开展进一步的探索。这些技术的合理运用，学生有能力进行相关的探索，增加他们学习的动机，并且与真实世界联系起来。在教室里就可以让学生都加入到过去只是老师才关心的领域。

3. 理解完整的生物学概念之间的关系

生物信息学是一个很好的整合工具，它可以起到穿针引线的作用，把你所学的课程都联系起来。生物信息学提供了一个思路把你所学的主题都连接起来：分子生物学、遗传学、遗传疾病、进化、细胞生物学、植物学、动物学、微生物学等。而这些课程一直都是独立授课的，尽管老师也会强调这些知识的统一性，但学生还是很难将这些知识连接起来。生物信息在讲授分析和解释知识时，产生一种学习环境。先将整个中心法则融会贯通，DNA和复制、RNA和转录、蛋白质和翻译，然后再运用它去解决遗传疾病、生理和进化等问题,从而激发学生将视线聚焦到这些内容上,产生强烈的求知欲。学生可以很好地理解学习内涵,极大地提高了把知识串接起来的能力。要提高学生在工作、学习和今后的科研工作中分析解决问题的能力,培养复合型、交叉型人才,提高学生综合素质进而提高就业竞争力。

本文发表于《教书育人·高教论坛》2010年11期,署名张晓艳、钟扬,收入本书时略有修改。因篇幅所限,参考文献不再一一列出。

从艰苦中提取欢乐

——访植物学家钟扬教授

钟扬教授高中未毕业,就考入了中国科技大学少年班;大学学的是无线电,却在毕业后搞起了植物学;在中科院的植物所干到了"副厅级",却跳到复旦大学,从普通教授开始"从头革命";在复旦干到"正处级",却"飞"到世界"最高学府"——西藏大学,当上了普通教授,对高原生物情有独钟;在领略藏波罗花独特风姿的同时,却又寻思着写一部关于一代"情僧"仓央嘉措的小说……

在致力于青年科普的科学松鼠会的科学嘉年华活动上,听了钟扬关于进化论的讲座,我们如痴如醉;半个月后,瞅着他来北京开会的空隙,我们"逮"着他聊了一个下午,同去采访的女同事当场成为他的"粉丝"。钟扬说:"科学研究是一项艰苦的事业,科学家的特质便是从中提取欢乐。"这一点上,他真的做得很棒。

那个年代的少年班

《学习博览》:钟老师,您在1979年怎么考上科大少年班的?

钟扬:我1978年初中毕业,打算1979年提前参加高考,就把高中的东西看了一遍。不料国家明令禁止提前考,已决定提前参加高考者必须提前高中毕业。因此,只要是考上了,就算随便一个师范学校或中专都要去,没有复读的机会。考科大少年班有一条好,没考上可以回去再读。所以我冒险一搏，就搏上了。现在有人说少年班怎么好或怎么不好，但作为亲历者当时根本不可能考虑这些问题。

《学习博览》：您在少年班，有什么感触？

钟扬：有人讲过：少年班可能是全中国最能让你知道自己还不够聪明的地方——人要去错了地方就以为自己特聪明。少年班每个人似乎都有一招，很邪门，现在的教育很难筛选出来。

从无线电到植物学

《学习博览》：您大学学的是无线电，毕业后却被分配到中科院武汉植物研究所工作，那大学学的专业是不是浪费了？

钟扬：你首先要回答我，什么叫专业？

前一阵子丹麦驻上海总领馆一个中国雇员带丹麦人去西藏考察。完了她告诉我说非常喜欢西藏植物，我说："那明年珠峰考察你继续来就是了。"她说："不行，我要去美国读博士了。"我说："去学什么呢？"她说："英美文学。"以我对她的了解，我觉得她文学肯定比我差，学英美文学好像前途不大。但她说了一句很有代表性的话："我大学的专业是英美文学。"我就说："你现在不是才23岁吗？怎么会有专业呢？那个所谓的专业，不就是高考前你爸和你妈两个，拿着铅笔在报纸上戳来戳去戳出来的吗？说不定他们戳定的是计算机，后来被学校录取成英美文学呢，这不是意外嘛！你说能理解爸妈，他们不是害你，那你熬四年也就够了。可是，在美国读个博士，那是人生很大的一笔投资，怎么一辈子的专业就变成英美文学了呢？"她听了以后很震惊。因为每个人都跟她讲要有专业思想。专业思想是对的，问题是很多年轻人现在其实没有专业，怎么树立专业思想呢？后来她半真半假地问我："跟着您学可以吗？"我说"当然可以啊"，不过后面加了一句话，"如果你肯拿出你到美国留学一样的十年时间的话"（没告诉她我自己用了15年）。当然，她到底没有跟我学什么生物学。她不爱文学，但爱美国学位，但愿这是她人生成功的起点，而不是悲剧的起点。

《学习博览》：看来美国博士学位还是比钟老师更具吸引力。

钟扬： 不过，人与人还是有差别的。有一次在美国和一个女博士去采样，也没发现她有什么特长，只是她在山间石头上比我们蹦得快。我终于忍不住问她："你大学学什么的？"她说本科主修芭蕾舞，研究生才读植物学的。

另外，我为什么跟其他植物学家不一样呢？也许是我从生物信息学做到进化分析，有一些统计学和信息学的知识背景，别人缺这一课，我大学正好学了。而且，我在植物园待了十五年。现在的孩子一听十五年就摇头。其实，很多研究生听几年音乐、复习几年外语、学几年计算机，再搞几年金融，合起来十五年，也许一事无成。

"押宝式"的教育和科研

《学习博览》： 您在中科院和大学工作多年，对咱们的教育和科研现状有何评价？

钟扬： 我的第一个美国导师讲过一句话：美国教育不一定先进，但比较成熟。成熟的一个标志是，美国没有一所一流大学在所有的方向都保持一流，也没有一所二流的大学在所有的方向都是二流。反观中国，许多单位的学科布局和研究水平都差强人意。

这与教育部导向有关，说我们钱不够，怕浪费，就重点投到一流大学吧。结果，一流大学原来优势的学科要保住，新兴的学科要发展。中国近两千所大学，人家还活不活了？个别"985"大学的人为什么比较狂呢？他们认为自己所有的地方都是一流的——上面都是这么说的，他们也这么要求自己，久而久之就当真了。我到西藏大学去，说藏大的藏学已经很不错了，我们再努力，把高原植物学搞上去吧，其他大部分学科没法超过复旦，无所谓。

我们与加拿大女王大学合作时，发现一个很厉害的教授是挪威毕业的，在做湖泊沉积物分析。这个在复旦谁搞啊？他毕业的地方在挪威都排不到第一，但人家就是干这一件事情很出色。

现在国家把宝押在少数学校上，但愿没押错。

《学习博览》： 我们习惯于"集中力量办大事"。

钟扬： 中科院提倡"凝练科学目标"：大家做的事太杂了，稍微做一两件事就好。本意虽好，但有时确定科研目标又成押宝了。比如一个研究所有八个研究员，应该做八件事，但要只做一件事，往往就是做所长的事。如果错了呢？那就都错了。

我做的是交叉学科，不宜隔开。比方说，我搞了一套比较好的算法，正准备用于植物数据计算，但一个微生物所的老师找来说用微生物算，我觉得这样算更好。但我属于植物所，要敢去微生物所，那就是不务正业。在学科界限上，大学毕竟要模糊一些，这是我决定去大学的原因之一。

集体思考却导致个人行动

《学习博览》：在复旦待了几年，怎么又到西藏大学去了？

钟扬：做植物研究，你必须到资源多的地方，西藏就是这样的一个地方。上海的生物多样性全国排倒数第一，北京也在倒数几位，可是许多一流的植物学人才待在北京、上海。

我去的是西藏大学，而复旦对口支援的学校是西藏民族学院，所以我一直没列进正式渠道。教育部分配西藏大学理科由南京大学对口支援。哪个人该支援西藏哪个地方一般是组织派的。

《学习博览》：我们一直强调服从组织决定。

钟扬：为什么我们能生产这么多博士，却很难产生杰出的科研成果？可能是我们的科学研究强调独立思考不够,过于强调集体思考。国家先有目标，再是校长怎么想，科研处长怎么想，最后才轮到你怎么想。

想法不一致很苦恼：从小你跟爸妈不一致很苦恼，读书以后跟老师不一致很苦恼，上班以后跟领导不一致很苦恼。久而久之一个能够混到今天的人，在某种程度上与上下左右都已经一致了。但是，由于人的本性不是纯集体的，就会导致一个问题：个人行动。一起合作做一个课题，但每个人都在打自己的小算盘，把力量分散了。

美国是相反的。你跟他说："我们来做这个课题吧。"他会问："这个课题跟我个人有什么关系？"个人思考，但最后是集体行动。最后能合在一起的人，显然是经过独立思考，认为只有跟你合在一起才能搞好的人。

《学习博览》：孰优孰劣，显而易见。

钟扬：你一看开会就知道了。我们一群人开会，你不来开会就不对，因为你违背了集体意志。但是坐在那地方想什么，那就是你自己的事了。很多外国人会说：对不起，我对这个事情没兴趣，不来开会。来开的，一般是要把这个事搞成的。

《学习博览》：您觉得自己能做到个人思考么？

钟扬：援藏就是一次个人思考。组织上也没派我，现在顺便将我纳入到援藏轨道。但到了西藏大学我就要参加集体行动了，比如要和他们一起争取植物学博士点。

现在我也不怕看学生走眼。一问"你热爱植物学吧？"他说"是"。我说"跟我去西藏吧"，他说"那就算了"。他爱的是复旦大学，而不是植物学，更不是钟老师。我跟北大生科院院长饶毅说过："很多学生来找你，少半是奔着饶毅，大半是奔着北大。"

生物学家援藏的生物学依据

《学习博览》：在藏大，奔您来的学生又是怎样的情况？

钟扬：我招过藏族学生。在我的研究生里面，一个西藏来的跟一个北京来的比，在知识方面显然差一点，但聪明程度还要强一些。为什么？现在的北京孩子中可能前三个优秀份子出国了，前几十个去北大清华了，最后轮到我手上的说不定排一百多名了。但在西藏跟我的人一般是前三名。我对藏族学生不考察他们英语，也不看知识。我认为只要你给他们一段时间，肯定能有好的表现。我带出了藏族的第一个植物学博士，也带出了哈萨克族的第一个植物学博士。他们所起的作用是其他人达不到的。

在一个群体里面，适合做一件事情的优秀份子比例大致是固定的。就算每个小孩都去学小提琴，但最后能成为小提琴家的人还是几个人，它有个百分率。

《学习博览》：这个"百分率"很有启发意义。

钟扬：我们现在的一些做法破坏了这个百分率。大多数人上大学，结果大学生水平下降，只好告诉大家：现在大学不是当年大学了。当年大学是培养国家栋梁，现在大学只是让每个人都受一下高等教育，导致学识的总体下降和学位的升高，是典型的"通货膨胀"。

更糟糕的是，越来越多的孩子读大学，但其实是越来越多的城市孩子读，农村孩子所占比例反而下降了。我们选择人才的范围实际上是相对收缩了。

《蜗牛，快跑！》

《学习博览》：咱们聊聊科普的事情吧，毕竟咱们还是因为科学松鼠会结的缘呢。

钟扬：第一，科普毕竟"姓科"，你的科学素养要高。如果你很久不在科学一线工作，思想就脱离主流了。现在，一流的科学家一般不写科普，许多写的人科学思想又比较老旧。

第二，科普的载体属于文学范畴，你的文学素养也要高。现在好多科学家写科普太直白了。我大学的时候写诗，感觉助学金不够了，就写首诗去发表，一首两块钱。写得好一点有十块钱，可以请三个朋友喝酒。我知道做人要直，但写诗要"曲"。天上管写作的叫"文曲星"，不叫"文直星"。

《学习博览》：钟老师有没有写过科普作品呢？

钟扬：我正在写一个关于"全球变化"的寓言小说，暂名《蜗牛，快跑！》。我没有写"全球变暖"，什么北极熊热得受不了了，你一猜就猜出来要写啥了。我写的是"蜗牛历险记"。蜗牛怕什么，全球变咸。你看小孩折腾土蜗牛（鼻涕虫）就给它撒把盐。对蜗牛来说，全球变暖一点问题都没有，打仗也跟它没关。它一直过着和平美好的生活。突然全球变咸了，它身体一点点地缩小。当然，我的小说有一个极光明的尾巴，因为蜗牛特聪明，海蜗牛居然进化成了第一个光合作用动物。我拐了一道弯来写全球变化。其实，写科普也得要创新，这跟搞科研一模一样。

"断点续传"

《学习博览》：钟老师忙乎的事可真不少。

钟扬：我喜欢利用"废"的时间。我跟朋友约会早到了五分钟，就一定要干一件事，这样就不会冒火，甚至会觉得朋友怎么不再晚来一点呢？我能在飞机上写文章，在主席台上也写，在出租车上睡觉。

我一般先把事情切碎。我写个一百字的东西，如果写到五十字必须去干另一件事，我就停到这里，干完回来马上从五十一个字接着写。这和生物信息学一样。基因组太长，不是没法整个儿一下子测出来吗？就用鸟枪法（shotgun）打断，一段段测，再用信息学方法拼接。这个原理在上传大文件时也会用，叫"断点续传"。

《学习博览》：这个"断点续传"的能力可不一般呢。

钟扬：我看电视剧不太挑，无论流行剧还是韩剧，看过都能讲出来。同事问："你那么忙，啥时候看的？"我说，我走到任何地方把电视机一开就看，每次都随机地看一个片段，看多了再拼接。当然，你脑子容量要大，看了40多段要记得。另外，每一段要有明显特征才能正确拼接。有朋友开玩笑说，钟扬最清楚编剧是否高明。因为如果台词重复，甚至跟别的电视剧重复，我就会产生拼接错误——我头脑中同时存了好几部片子。生物信息学中也尽量避免"重复序列"。

45岁后还有梦想

《学习博览》：这么"断点续传"地利用时间，还打算干点别的什么？

钟扬：我给自己定了一个目标：到离开西藏的时候，一定要会说藏语。45岁开始，50岁肯定学会。我只说了这个目标，就引起藏族同志的好感。他们说，那么多援藏过来的，有人都待20多年了都不学。我倒觉得这纯属个人兴趣，不好勉强他人。

《学习博览》：除了学藏语，还有什么梦想？

钟扬： 我曾经是文学青年，现在还在学写小说。我原来想写六世达赖仓央嘉措，收集了很多资料，但考虑比较敏感，也许写五世达赖比较好一点。我没什么写作任务，全凭兴趣，也不指望拿这个卖钱。

我还有个梦想是办一所私立大学。美国最好和最差的大学都是私立大学，全凭你办。我国现在的民办大学很多还处于补习班的水平，因为教育部控制了学位授予权。我跟很多人说过，只要有一天全面开放私立大学，我一定辞职去办。

本文发表于《学习博览》2010年第1期，是该刊对钟扬教授的一篇专访，署名李勇刚、黄渡海，收入本书时略有修改。

生命之树常青

2010年是联合国国际生物多样性年。"生物多样性（biological diversity）"一词据考证是美国生物学家达斯曼（R. F. Dasmann）于1968年在其通俗读物《一个不同类型的国度》（*A Different Kind of Country*）中最早使用的。1985年，罗森（W. G. Rosen）在美国国家研究理事会（The National Research Council，NRC）举办的生物多样性论坛上建议采用缩写形式"biodiversity"。其后，在著名学者威尔逊（E. O. Wilson）和雷文（P. Raven）等人的大力推动下，生物多样性研究与濒危物种保护等热点话题一起逐渐家喻户晓。目前，生物多样性科学作为一门独立学科，也已在生命科学和环境保护等领域占据了十分重要的地位。

尽管人们已普遍意识到生物多样性的重要性，但其科学概念的复杂性与理论本身的多样性还是让普通大众摸不着头脑。在笔者看来，生物多样性理论主要针对两大类科学问题：一是生物多样性是如何起源的；二是生物多样性是如何丧失的。当然，介乎起源与丧失之间的还有生物多样性的维持机制等问题。而从公众角度看，人们对第二类问题的关注明显甚于第一类问题。更简单地说，人们关心的是地球生物多样性的未来。在这一点上，绝大多数大众媒体一直持悲观的态度（例如，估计全世界每天有多少个物种消失，等等），而生物多样性理论是否又能给出一个稍微乐观一点的预测呢？

先说一个坏消息。这些年来，我们并没有建立起准确估计生物多样性丧失速率的模型，甚至远远没有弄清建立这一模型的要素和条件。换言之，发表在各类学报上的生物多样性理论预测结果，也许并不比各种电视节目上的报道更为科学。尽管我们可以笼统地讲，全球变化、生境片段化以及人类活动（例如，让人们生活更美好的城市化）等等可能都是导致生物多样性丧失的原因，但要具体落实到威胁某一类生物的生存，甚至导致其灭绝的关键因子时，事情就没那么简单了。以两栖类为例，这是公认的当今地球上物种数量下降最快的一个门类，也一直作为人类活动对动物栖息地造成破坏并最终导致物种濒危的典型案例。但近年来的定点观察和深入研究表明，厄尔尼诺现象所引起的海平面与水温变化以及水面紫外线强度的降低，会导致一些微生物（如水霉）暴发，而这种急剧增长的微生物常常侵染滨海蟾蜍的卵，致使其胚胎发育不全，这才是当地两栖类濒危的主因。没有厘清影响生物多样性的遗传和环境因子间纷繁复杂的关系，就不可能建立合理的定量分析模型，当然也就无法获得有说服力的预测结果。

再说一个好消息。随着分子生物学技术的发展和普及，应用分子标记来发现新的物种并进行生物多样性编目的工作方兴未艾。尽管分类学专业队伍在不断减少，但鉴定生物物种的速度似乎并未减弱，这得归功于刚刚发展起来的生物DNA条形码（DNA barcoding）技术。这种类似于超市货品条形码的技术主要是利用不同物种具有不同特定的DNA序列这一性质，通过将一段或几段特定的DNA序列与已知物种的DNA序列数据库进行比对，从而实现对物种的快速、精确乃至自动化鉴定。例如，科学家利用DNA条形码技术对哥斯达黎加热带雨林中蝴蝶的"隐存种"进行了鉴定，获得了令人振奋的结果，既改变了科学家以往对这一地区蝴蝶类群濒危状况的认识，也为其他地区的同类研究提供了一个可资借鉴的范例。

生物多样性保护实践迫切需要更为牢固的理论基础。例如，由于人力、物力和财力的限制，一项有意义的生物多样性保护活动必须考虑优先原则，即究竟哪一个物种（种群）比别的物种（种群）具有优先保护的"特权"？这方面有不少研究，最常用的并且具有指导意义的概念当属进化显著单位（evolutionarily significant unit，ESU）。一个ESU可以指一个物种、亚种、地理宗或种群。在实际工作中，它由若干标准（或显著差异）来确定，如单位间的地理隔离，用中性标记所反映的种群间遗传分化，以及因选择压力不同所导致的表型性状适应，等等。在种群水平确定ESU，需要计算遗传分化及多样性指数；而在亚种和物种水平，则要借助进化分析（构建进化树）以及谱系地理学等方法。

说到"进化树"，让我们先回到了第一类问题——生物多样性是如何起源的？生物多样性起源的理论基础就是进化论。20世纪著名的进化生物学家迈尔（E. Mayr）曾总结过达尔文进化论的逻辑框架，它由五个理论组成。其中，"物种增殖理论"以及"自然选择理论"无疑解释了生物多样性起源与形成的机制。在达尔文出版《物种起源》（*The Origin of Species*）一书的六年之后，德国生物学家海克尔（E. Haeckel）绘出了世界上第一棵"生命之树（Tree of Life）"，开创了用进化树方法描述地球生物多样性之先河。近20年来，分子系统学（molecular systematics）快速发展，基于大规模DNA测序继而构建分子"生命之树"的工作取得了较大进展，有力地支持了达尔文进化论中的"共同祖先理论"。

毋庸讳言，尽管随着生物多样性科学的发展，越来越多的理论、方法和模型还会不断出现。但理论研究与保护实践的严重脱节现象依然令人忧虑。先看以保护生物多样性为己任的联合国。它所聘请的专家委员会一直在制订各种生物多样性行动计划，而实际行动上又一直无法完成这些计划。2008年，笔者作为中国代表团成员参加了在德国波恩举行的第九次联合国《生物多样性公约》缔约国大会（COP9），与会专家所讨论的各种理论与技术问题给笔者留下了深刻印象。不过，印象最深刻的是，所有专家基本上对任何一个国家实现他们所承诺的2010年拟达到的生物多样性保护目标都持否定态度。由此可见，在这个还不够成熟的领域，科学家的作用仍然十分有限，理论与实际相结合的研究方向就显得尤为重要。

回到我们的学校、研究中心和实验室。可喜的是今天的学生、未来的学者对学习生物多样性科学及进化生物学的热情有增无减，但他们对大自然的了解实在太少，对现实存在的生物多样性问题常常把握不准，更遑论保护实践了。即使是构建进化树这样的计算工作，虽然可以采用大量的分子序列和复杂的教学模型，但真实世界中还是存在许多"缺失环节（Missing link）"。如果一个研究者对于这些领域的了解并不比一个普通的自然爱好者多，怎么能凭借现有理论或方法去打开复杂的生物多样性问题之门？就连目前人们热衷追求的理论创新也将是无源之水、无本之木。

"一切理论都是灰色的，唯生命之树常青"——二百年前的歌德借《浮士德》中魔鬼之口所表达的思想，是否还能给今天的生物多样性研究者以启迪？

本文发表于《科学》2010年第4期"百草园"栏目，收入本书时略有修改。

"影响因子"与科研创新无关

汪品先生在建议讨论"创新障碍"的公开信（见《文汇报》1月9日头版）中提出了一个重要的问题：我们用来促进科研发展的举措中是不是存在一些误导性，如在科技成果的评价和奖励中拿文章数量和影响因子作标准。记得几年前，有关SCI及其影响因子的利弊在国内就曾引发过争论，其结果对抑制当时已兴起的"唯SCI论"产生了一定作用，科学界由此形成了一些共识：SCI已成为国际自然科学领域最权威的文献检索工具之一；发表SCI论文有助于提高博士生科研水平和英文写作能力；SCI的影响因子是一份刊物所有论文被引用情况的相对均值，并不代表在该刊发表的某一篇科研论文本身的创新性和影响力，等等。

不料，最近一段时间追求SCI影响因子之风竟愈演愈烈，不仅在成果申报、项目申请、职务聘任等诸多场合提出的科研论文都被要求标注，还有人将不同刊物甚至不同年度的影响因子相加得出一个所谓的"总点数"来对单位或个人进行排序。此外，与SCI类似的SSCI以及由SCI衍生出的ESI等指数也陆续登场，成了影响科研创新活动的"推手"。由此看来，进一步探讨汪先生所提出的"影响因子问题"依然有很强的现实意义。

近来，笔者趁在日本国立统计数理研究所担任国际访问教授之机，与日本同行就SCI影响因子对科研创新的正负效应作了探讨。总的印象是，影响因子在日本科学界也有一定市场，但在项目申报和绩效管理等方面并没有我国那么盛行,因而对多数科学家(尤其是致力于基础理论创新的科学家)未造成明显的压力。而在国际公认的、我国科学家梦寐以求的创新研究巅峰——争取诺贝尔奖方面,日本已经积累了相当多的经验。他山之石,可以攻玉,对以下几个成功案例的分析,也许可以给我们新的启示。

汤川秀树是日本第一位诺贝尔奖获得者,也是一位完全由日本本土培养的理论物理学家。1932年,25岁的汤川秀树担任大阪大学讲师,在从事科研教学活动的同时,为日后申请博士学位做准备。1935年,他将研究成果写成"论基本粒子的作用"一文,发表于本国的专业学术期刊——《日本物理-数学会刊》,论文中首次提出了著名的介子学说,预言了介子的存在。这一新理论使汤川秀树于1938年获大阪大学物理学博士学位,并于1949年荣获诺贝尔物理学奖。

1946年,汤川秀树在京都大学基础物理研究所创办了一份本土刊物《理论物理学进展》,致力于向国外推介日本的理论物理学研究成果,帮助日本科学家克服因国际竞争和语言障碍等对发表创新思想不利的因素。1973年,这家研究所两名年轻的助理教授小林诚和益川敏英合作在该刊发表了一篇题为"弱相互作用可重整化理论中的cp破缺"的论文,提出了著名的小林-益川模型,以解释弱相互作用中的电荷宇称对称性破缺。2008年,小林诚和益川敏英因此获得诺贝尔物理学奖。顺便提一句,益川敏英因英语成绩太差曾险遭名古屋大学研究生院拒绝录取,其后他自己也放弃了一切出国学习和交流的机会。可想而知,他的成功离不开坂田昌一和汤川秀树等科学大师慧眼识才,《理论物理学进展》也功不可没。

更极端的例子莫过于2002年的诺贝尔化学奖得主——田中耕一。田中耕一自1983年本科毕业于日本东北大学电气工程学系后一直供职于日本岛津制作所(公司)。1987年他在日本京都举行的一个专业学术会议上提交了一篇论文,在高分子研究领域提出了性质界定和结构解析的想法。这篇会议论文未被SCI收录,因而也没有影响因子,但引起了国际学术界的重视,其想法被欧美科学家逐渐发展成一套高灵敏度和高精确度的生物大分子分析方法,对今天的蛋白质组学及相关领域的研究起到了决定性作用。田中耕一的创新精神和成功范例曾一度引起日本科学与教育界的热议和反思。

笔者并没有从严格统计学意义上分析影响因子与科研创新活动的相关性，不过从上述实例不难看出，影响因子等各种定量指标也许在很大程度上简化了科研评价的复杂性，试图将这些指标作为标准来提升我们的科研创新能力，无异于舍本求末。

本文发表于2011年2月6日《文汇报》第二版"创新障碍在哪里"栏目，收入本书时略有修改。

生命的高度

分布于美国加州的北美红杉株高可达150米以上，在现存植物界中令人叹为观止，且已接近数理学家们理论预测的高度极限。在生态环境适宜的原始森林中，每一棵北美红杉都算得上是一个真正的成功者。

从我到青藏高原第一天起，就一直在寻觅这样的成功者。我们对西藏巨柏和沙棘植物进行了多年的研究，特别是后者，引发了我对青藏高原植物株高的思考——沙棘近缘种的株高依海拔高度从10米到10余厘米呈梯度变异，取决于每一株植物的生境。那些生长在海拔5 000米以上、株高不足20厘米的沙棘更令人感怀。我们在向它们致敬的同时，不禁要问：它们是怎么来的？为什么要在如此恶劣的环境中生存？

随着我们对喜马拉雅山的雪莲——鼠麴雪兔子的研究不断深入，答案渐渐清晰起来。1938年，德国探险家在海拔6 300米左右的珠穆朗玛峰南坡采集到一棵几厘米高的鼠麴雪兔子，将其记载为世界上分布最高的高等植物。随着全球变化和冰川消融，植物的分布可能有新的高度。为此，我们一次又一次去珠峰考察，今年6月终于在海拔6 100米以上的北坡采集到了宝贵的样品。这些样品使我们在南坡找到突破现有世界记录的最高海拔分布植物的信心倍增，而进一步的分子生物学分析将为揭示其种群来源和动态及其与全球变化的关系提供科学的依据。

然而，鼠麴雪兔子并不是一个青藏高原特有种，它广泛分布于我国西南和西北各地以及邻近国家,其中不乏低海拔和环境优越的种群。仅从生物学特性看,青藏高原种群明显要差得多,但这些矮小的植株竟能耐受干旱、狂风、贫瘠的土壤以及45℃的昼夜温差。生物学上的合理解释是:它之所以能成为世界上分布最高的植物就是靠这一群群不起眼的小草承担着"先锋者"的任务,向新的高地一代又一代地缓慢推进。

在一个适宜生物生存与发展的良好环境中,不乏各种各样的成功者,它们造就了生命的辉煌。然而,生命的高度绝不只是一种形式。当一个物种要拓展其疆域而必须迎接恶劣环境挑战的时候,总是需要一些先锋者牺牲个体的优势,以换取整个群体乃至物种新的生存空间和发展机遇。换言之,先锋者为成功者奠定了基础,它们在生命的高度上应该是一致的。这就是生长于珠穆朗玛峰的高山雪莲给我的人生启示,它将激励我毕生在青藏高原研究之路上攀登。

本文发表于《复旦校刊》2012年7月6日第9版,是钟扬作为上海市教卫党委系统创先争优优秀共产党员,在复旦大学创先争优表彰大会上的发言摘登,收入本书时略有修改。

中国适应性生存与可持续发展

对一个国家，一个民族，一个个体，生存与发展都是永恒的主题。但我们每个人从不同的角度对这同一个话题都可以谈出完全不同的思想。我是研究自然科学的，我的思想（包括科学研究中的哲学）不可能脱离我的专业。

你如果接受过科学训练，无论你本身学习的专业是什么，后来又做什么，你一定会跟没有受过训练的人有一些差别，可能有两个差别与我们今天这个讲座有关。

第一，要用证据说话。不是简单从概念和理论上讲怎么生存好，怎么发展好，而是要罗列证据，最后还要把证据加以分析才能得出结论。这是科学的一个本质，就是不唯上，而要唯实。

第二，要透过现象看本质。我发现很多人并不喜欢透过现象看本质，他们更喜欢表面的东西，喜欢形式。如果你是科学家，没有别的办法，你必须钻研事物本质。

最近，我和学生们谈论最多的一个话题是"对转基因生物的看法"，我觉得它可以用来判定一个学生是否可以成为好的生物学家。一是因为这个问题并没有标准答案。在我心目中，我从不认为谁讲的是正确的，别人讲的就是错误的。我的评价标准是一个学生讲出来的话里面有多少证据，再就是他讲出来的话和一个没有学过生物学的或仅仅只是对转基因感兴趣的人讲出来的话到底有多大差别。如果差别不大，我认为他至少不应该是一个好的生物学家。也许他的立场是对的，但在科学上，对错并没有我们想象的那么分明，受过与没有受过科学训练则完全不一样。二是这个问题中包含了科学和技术两个方面的考虑，即容许大家（包括我自己）从科学上肯定转基因，但从技术上质疑转基因，而一般人会将科学和技术当成同一件事。因此，我今天的报告里面有很多的内容可能在你们看来是错的，不一定是对的，但它是一个生物科学家真实的思考。我会做一些分析和比较，将我自己领域的东西稍微向外推演一下，甚至可能推演到我自己的人生，推演到你们今后的发展问题。这个推演有可能错得很厉害，大家见仁见智吧。我不光是想告诉大家一个结论，还希望大家了解我为什么会这样思考。在目前的研究生教育中，我发现老师们太注重告诉你所谓正确的结果，这样的话，为了考试，你当然要去记忆正确的结果。但是，这种教育忽略了我们的前人是怎么得到这个结果的，他们为什么能走向正确的道路。如果研究过程没有告诉你的话，虽然你学了很多很多的知识，但到你自己做科研的时候，你就不会做了。另外，我们有时太注重结果了，很多时候并没有看到结果后面更本质的东西。

进 化 理 论

让我们先从达尔文讲起吧。所有进化生物学（包括我们今天作为重点的分子进化）都起源于进化论。2009年是一个重要的年份，它是达尔文诞辰200周年，以及达尔文50岁生日时出版的一本书——《物种起源》问世150周年。这本书给全世界带来了非常大的影响。马克思主义的创立与达尔文的进化论也有一定关系。达尔文的一生给我们许多重要的启示，也在很大程度上决定了我的人生道路。达尔文没有正规读生物学，他先是学医学，后来改学神学，但他对大自然的热爱却从来没有停止过。后来发生了一件重要的事情——参加贝格尔号旅行。当时英国军舰贝格尔号上需要一位大学生参与远航，顺便采集各地动植物的标本。达尔文参加了。我个人认为要改变一个人的世界观，最好是让他离开他死活都不想离开的地方，去的地方越远越好，时间越长越好。达尔文的四年旅行给他的人生带来极大的收益。他回来以后，凭借他采集的矿物等标本，当上了英国皇家科学院院士（皇家学会会员）。达尔文亲手绘制的加拉帕戈斯群岛的地雀十分细致。一个学习神学的人能绘制出如此精致的标本画，你不觉得他就该学动物学吗？我们好多同学获得了动物学学位也绘不出这样的图，学位只是证明你上过这个课，证明你有这个知识，不证明你就具有这个能力。达尔文在这个方面做得非常好。他还是一位分析大师，他做事情，第一是找证据，第二是分析。加拉帕戈斯群岛地雀不同的喙让达尔文领悟到，就像人类的工具一样，这些鸟喙是不断进化出来的。如果是上帝在加拉帕戈斯群岛上创造出各种地雀，这种解释未免太愚蠢了。较为合理的解释是，最初较少的物种生活在这一区域，随着地质活动，岛屿分开了，鸟类个体间产生了生殖隔离。最后在多重因素的影响下，鸟喙变得不一样了。达尔文很会组织分析材料，他将所有的生物用一个树状结构表现出来，我们今天称之为"进化树"。《物种起源》比起一本现代生物书，更像是一本哲学著作。因为现代生物书里的插图特别多，而《物种起源》中的插图就是这一幅。书中很多内容都是思辨的产物。

　　达尔文的人生也是这么抉择的。他29岁的时候，遇到一个让他很纠结的问题——是不是和自己的表姐结婚。和做研究一模一样，他拿了一张纸，中间画了一条线，左边写上结婚的好处，右边写上结婚的坏处。经过一番论证，最后得出的结论是"结婚！结婚！结婚！"说到达尔文成功的因素有很多。有人总结了达尔文成功的七项因素，其中有几条让人感触很深。一条是达尔文一辈子没找工作。我们现在的研究生动不动就说，"老师，现在找工作困难啊"，找到工作以后也不满意，总是要换。这种事达尔文一辈子没干过，他一直就在那儿安心地搞研究。稳定的家庭生活为他创造了极为优越的条件。另一条就是达尔文家里非常有钱。我曾经去英国达尔文故居看过，达尔文家里为了他在植物学方面的兴趣，为他建了两个温室，这比我们复旦大学今天的生物实验温室还要强。想想一百多年前，就有这样的温室，里面种了那么多的植物，光是豆科类植物就有大概30种之多。达尔文自己家里很有钱，结婚以后发现太太家更有钱。

　　有一位著名的科学家恩斯特·迈尔，被称作"20世纪的达尔文"。他非常推崇达尔文，自己的人生也和达尔文差不多。他离开大学后去非洲待了很多年,差一点送了命。后来他到哈佛大学工作,最后在哈佛大学动物博物馆馆长的位置上去世。今天,哈佛大学的动物博物馆就叫"恩斯特·迈尔博物馆"。迈尔102岁去世,90岁以后他专心致志地对达尔文进化论进行哲学思考和解读。他整理出达尔文进化理论的五个框架,其中第五个最重要,叫"自然选择理论"。

许多科学理论的发展都有这样一种情况:一开始大家都不信,过了一段时间大家都信了,再过一段时间,它就变得有点像宗教了。我在外国大学讲课的时候,曾问过外国大学生一个问题——你信进化论吗?大约有一半的同学举了手。对于那些不相信进化论的人,我也不敢等闲视之,曾经有一个外国大学生跟我说《物种起源》他读过六遍,而且读的是原文,我肯定比他差,他把字里行间都看了个遍,因为他不信,他一定要找出达尔文在哪个地方犯了错。在我看来,这更像是搞科研的方法。在国内,我从不做这样的测试。我如果问大家"你信进化论吗?"所有的手都举起来了。再问"看过达尔文《物种起源》吗?""没看过。"结果基本上都是这样。因为我们对待信仰都是这样一个态度:既然都信了,还看它干嘛?科学家们却不一样。

日本科学家木村资生在大学时期受的是遗传学教育,他也是达尔文的忠实信徒。他后来到美国去做博士后,接触了更多的统计学。于是,他开始用生物统计学去分析当时不多的蛋白质数据,同时将分子生物学数据和古生物学数据整合到一起,发现了一个非常奇怪的现象。每一个蛋白在不同生物的进化过程中速率竟然大致是恒定的。这个结论违背了达尔文进化论中的自然选择理论。人们一提到自然选择,大脑马上会反应出"适者生存"。比如人和狗的嗅觉基因差别很大。但木村分析的时候没有选嗅觉基因这样受到强烈选择的基因,他选的是一些生物共有的基因(俗称"看家基因")。分析结果并不能用达尔文的自然选择理论直接进行解释。但木村思想上不敢突破,他写了一篇文章,题为"分子水平的进化速率",没有直接说达尔文不对,但他展示了证据,说明在分子水平可能并不是我们所想象的那样强的自然选择。这就是所谓的中性进化。在原有的达尔文进化论框架下,无论是性状,还是基因,其中一部分是有害的,另一部分是有益的。一段时间后,有益的最终胜过了有害的，在群体中扩散了，留存下来。但在新的框架下，木村资生告诉我们有害的基因没有那么多，有益的基因也没那么多，起决定性作用的、能导致进化速率恒定的因素是不好也不坏的。中性进化的科学表述是"中性突变与野生型具有相同的适合度"。如果你学过进化生物学，就不会轻易说一个基因是好的或者是坏的。在我头脑中已经没有那么多好坏之分。最好的进化就是最适应环境的。

有些基因在核苷酸上发生了改变，但在蛋白质水平并未改变，生物学家将这种改变叫做"同义置换"，或者叫做"非功能性改变"。然而，在另一些情况下，其产物（蛋白质）会发生变化。生物学家把这两种变化区别开来，然后对"非同义置换"与"同义置换"的出现频率进行统计，将它们相除，结果大于1的情况就是自然选择起了决定作用，叫"正选择"，有利于这个变化在群体中固定；如果结果小于1，叫"淘汰选择"，又叫做"负选择"，无论现在这个基因在群体中的频率如何最终总是要被淘汰的；等于1就表示中性进化，或者说随机突变和遗传漂变起主要作用。这个公式所蕴含的哲学意味足以令我们回味，它将达尔文进化论与非达尔文的进化理论统一起来了。这个公式所揭示的就是生物生存与发展的策略，或者说生物生存与发展的哲学。我们假设在一千万年的进化过程中，给某种生物的一个重要基因十次改变的机会。有一种策略是不管外界环境怎么变，我都不变。那么，非同义置换为0，同义置换为10，两者之比为0，小于1。这一策略的好处是这种生物的功能一直没变，以不变应万变。但如果设想一下，这种生物原来只能生存在陆地上，后来海平面上升，陆地全部进入海洋，那它肯定要灭绝。对这么重大的环境变迁，它一点适应能力都没有，肯定要被淘汰，我把这种策略称之为等死型策略。另一种生物不采取这种策略，它把十次机会全部用作功能性改变，因此非同义置换为10，同义置换为0，两者之比为无穷大，远大于1。这种策略的好处是，十次机会中只要有一次抓住机会，哪怕环境由陆地变成了海洋，成功者就是它。但如果这一千万年内环境没有大的变化，它的十次变异就很可能有害了。我把这种策略称为找死型策略。一种是等死，一种是找死。实际上，生存到今天的生物常常采用的是中性策略。不可一味地折腾，也不可一味地等死。在分子水平，生物就是这么中性进化的。

组合复杂性

事物的复杂性是怎么来的？很多复杂性是掩盖在表象下的组合复杂性。外国人学中国汉字特别难，主要难在他们把每一个汉字看成一幅画，他如果学两千个汉字相当于头脑中存了两千幅画，这太困难了。中国人为什么不觉得呢？因为我们知道汉字由偏旁部首构成，记起来就简单了。外国科学家发现兵马俑每一个都不一样，但如果兵马俑是一个一个由工匠做成的，工作量就会大得惊人。后来，人们发现了兵马俑的奥秘。兵马俑是由几个部件构成的，有几种头，几种身子，手和腿的姿势也有好几种，就这么简单的几个部件，烧制结束后，工匠们把它们拼在一起，会产生多种多样的兵马俑了，复杂性和多样性就此产生。我们就讲一讲同学们找工作吧，同学们临毕业的时候总是喜欢来找老师咨询职业方面的问题。我会问，你有什么要求。我最怕听到的就是"我要求不高"，这其中很可能含有巨大的复杂性。他说：我每月工资只要5 000元。我一想：在上海月薪没有5 000元挺难生存的。他接着说：我的工作地方不要太偏远，最好离复旦5公里之内。我一想也有合理性。他又说：工作时间不要太长，只要8小时就行了。我一想：是啊，还有很多人不工作呢。当他说到第五句的时候，我就叫他打住了。这个工作太难找了。他所提出的每个条件似乎都很正常，组合到一起产生了很大的复杂性。生物靠这个办法来形成多样性和复杂性，而我们人生的困境很大程度也来自这一点。了解了这一点，有助于我们发现解决困境之道：消除一个因素的复杂性，可能会极大地简化全局复杂性。相反，如果不经意间加上了一些貌似合理的因素，往往会带来很大的代价。

可持续发展之路

我们的本科生、研究生如何走可持续发展的道路？最重要的一点就是学习。进入大学以后，同学们的学习积极性明显下降。之后，人生道路上谁能成功呢？和大学所学的专业不一定有关系，和工作的内容也不一定有关系，成功全靠自主学习。中国人很奇怪，对小孩子抓得特别紧，因为我们有一个理念叫"不能输在起跑线上"。为什么很多人不可持续发展呢？因为我们的重点在起跑线上，但终点撞不撞线没人在意。此外，整天教育孩子学习的家长自己往往不学习。我问过一些家长自己怎么不去学习呢？他说："我都三四十岁了，还学习干嘛？"我们的学习之路在一个节点上突然中断了。现在的知识更新太快了，我们一定要跟上时代发展的步伐。我非常信奉一个秘诀，就是"练习一万个小时"的定律。一本有关成功者的书中写道，无论是音乐，还是写作，只要你肯花一万个小时进行练习，你就一定能达到一定的高度。有许多硕士生毕业后来考我们的博士，我们有时都不敢相信他们读过研究生。他们的学位是真的，也真的读了很多年书，但上研究生的几年中到底花了多少时间在学习上？当你觉得自己的知识还不够，当你觉得自己到社会上找工作时不能引起他人重视的时候，我建议你先问问自己在学习方面花了多少小时。

开 放 问 题

我接下来谈谈四个开放问题。它们没有标准答案，更没有对错之分。主要是请大家琢磨一下，我们思考问题的过程。

压力 vs 动力

在自然选择的力量中，什么是导致生物进化的压力？什么是动力？生物学家如何思考环境对于生物的作用？环境的变化尺度往往很大，比方说全球气候变化，作用时间很长，范围很广，一个人终其一生也难以观察其全过程。生物学家的思维是通过观察生物对环境的适应，来理解环境的变化及其带来的压力与动力。在英国有一种野生羊，叫埃索羊，主要分布在一个岛上，其食物结构和生殖规律等各种情况都被科学家掌握。经过前后三十余年的研究，科学家们发现埃索羊的腿变短了，并以此作为全球变暖的一个证据。一般人认为，如果气候变暖了，草长得更茂盛了，那么羊应该长得更健壮，腿也应该长才是。实际并非如此,这个现象被生物学家们称为"自然选择的压力放松了"。以前腿短的羊吃不上草,就饿死了,被淘汰了。气候变暖以后,资源变充足了,过去健壮的、腿长的羊自然是有草吃的。可是当年那些腿短的、要被淘汰的羊也有草吃,都留下来了。这样,统计平均以后,发现羊的腿变短了,放松和加强自然选择的压力会导致生物进化不同的后果。

丧失vs获得

很多人以为生物在适应环境的过程中要更多地去获得,实际结果未必如此。在进化假设上,获得和失去基本上是等价的。很多生物是依靠失去而并非获得来进化的。蝙蝠拥有飞翔的能力,这在哺乳动物中是很少的。在很长一段时期的研究中,人们探索的是蝙蝠是如何获得这种能力的。后来,人们发现并非蝙蝠获得了飞翔能力,而是同一人类的其他动物失去了这一能力,其中一种现存的该类生物就是我们今天看到的马。科学家们给这类动物起了个名字"飞马兽类"。这类动物最早都是有飞行能力的。在进化过程中,只有蝙蝠保留了这一能力,而其他动物都失去了。丧失也有好处,进化上还是有一定优势的。这样的例子不胜枚举。在进化生物学家看来,丧失与获得都是为了适应。

数量vs质量

与进化生物学关系最紧密的学科之一是发育生物学。发育生物学有一个重要的研究方向,就是对寿命的研究。生物学家寻找寿命基因的办法就是看哪一个基因妨碍长寿。他们用的是敲除法,将一个基因敲除,然后观察生物的寿命。科学家们利用线虫进行实验。线虫的生命周期在5~7天左右。美国加州大学伯克利分校的科学家找到了一个貌似妨碍长寿的基因;把这个基因敲除之后,线虫居然活了30天以上。他们报告说:这里有个好消息——我们人类也有这个基因。这里还有一个坏消息,把这个基因敲除以后,线虫就不能生育了。这个妨碍长寿的基因居然和影响生命质量的基因是同一个基因。这就演变成了一个哲学问题了——是要数量,还是要质量?大家可能要说:能不能先过生育期,把生命的质量保证了,再敲掉这个基因来延长寿命。科学就是这么无情。经过研究,发现要敲除这个基因得早敲,晚敲就没用了,刚出生就敲除才有用。另一方面，当我们的寿命延长很多以后，生命的节律也许会变得很慢，生活质量发生很大变化。

个体 vs 群体

大部分成功的生物在进化过程中都理解了个体与群体之间的关系。一个好的基因必须在群体中扩散。如果一个优秀的基因不能在群体中扩散，实际上就是我们所说的淘汰。

结　语

我相信我想做的大多数生物学实验，大自然已经做过了。作为进化生物学家，我们所要做的就是去探索，去领悟，去分享大自然所创造的成果。

本文为2012年11月29日晚钟扬在复旦大学校团委"中国道路大家谈"主题活动作的报告，后收入《中国道路大家谈》（高天、滕育栋主编，复旦大学出版社，2013年4月），收入本书时略有修改。

获得，还是失去，这是个进化问题

作为万物之灵，人类总会给身边的动物定性。那些有利的，慢慢被豢养成了家畜家禽；那些有害的，人们或是避而远之并冠以害虫猛兽之名，或是需冒生命之险才能取其入药。但若要说人类总会对某些动物怀有难以言喻的特殊情感的话，蛇理所当然是其中之一。

人们对于蛇的情感，可谓经历了种种复杂和纠结的心路历程，从惧怕、崇拜蛇到利用、养殖蛇，再到保护蛇，这之间的渊源恐怕可以追溯到远古。从世界各地的神话故事或是民间传说中，我们都能见到蛇的身影。远到南美玛雅人的羽蛇神库库尔坎，埃及神话中眼镜蛇神艾约，或是近邻印度神话中的七头蛇神那迦，更不用说中国神话《山海经》中女娲、伏羲的人面蛇身，以及某些少数民族对蛇的图腾崇拜了，还有古希腊医神手中的长杖，上面盘踞的就是一条蛇。蛇在人类祖先的心中是智慧和力量的化身，不可侵犯而具有法力。这兴许反映的就是蛇冰冷阴暗的外形以及动辄致人于死地的危险性给人类祖先带来的恐惧和压迫感吧。

而让人们对蛇有所改观的则当属一条叫白素贞的白蛇和一条叫小青的青蛇，也正是她们，让蛇的形象变得善良生动起来。据传，这个耳熟能详的故事源于唐传奇《白蛇传》或南宋话本《西湖三塔记》，直到明代冯梦龙的《警世通言》中的《白娘子永镇雷峰塔》才将这人蛇畸恋的故事定型。其中两个女主角化作人形报恩，她们一反人们对蛇的一贯认知，以有情有义、敢爱敢恨的性情令所有读者印象深刻。

由此看来，蛇类所承载的文化涵义极为丰富。而最近有不少科学研究表明，这些植根在人类文化中的蛇类形象，其实是同包括人类在内的灵长类伴随着蛇类威胁而进化的经历有关。通俗点说，人们对蛇的恐惧可以说是世代相传、与生俱来的，而灵长类的诸如视力等感官的发展，似乎也要感谢蛇类给它们带来的威胁。但这是题外话，因为蛇自身的进化，才是一个更加错综复杂的故事呢。

在漫长的生命历程中，生物究竟是丧失还是获得某种性状，无论在科学上还是在哲学上都还真是个问题。"用进废退"算得上是将进化论阐释得最为直观的一种概括。这里，功能与器官的退化和性状的丧失完全是同义的。乍一看，"退化"似乎是一个略带消极的词，没什么讨论的意义，殊不知"退化"亦是"进化"的一种形式，而且有时候，器官退化和性状丧失甚至能帮助生物更好地适应新的环境。

让我们回到蛇，看看它们的进化吧。尽管蛇一直是人们最为惧怕和厌恶的动物之一，不过，生物学家们对它始终兴致盎然，而各路学者关于蛇类起源和进化的争议也从未停息过。其中，最具争议的两个焦点，莫过于蛇与蜥蜴这两个爬行动物近亲的关系以及蛇毒这一致命武器的来源和演变。

从现有的化石证据看，蛇的祖先更接近于蜥蜴类动物，而且是陆生的。之前，确有人提出过蛇类是海洋起源的假说，证据是蜥蜴类动物中有一种已灭绝的巨型沧龙曾在海洋生活过，人们认为正是这种巨型沧龙在大约一亿五千万年前爬上了陆地，最终进化出蛇类。但这种说法并不十分站得住脚。为了解开蛇类的起源之谜，美国的一个科学家小组收集了现存蛇类和蜥蜴类的大量样本，测定了它们的DNA序列。这些序列的比对分析结果表明，除沧龙外的所有蜥蜴类在蛇类出现之前均已在陆地生活，而蛇类的祖先就在它们中间。顺便提一句，巨型沧龙虽然早已灭绝，但现存最大的蜥蜴——生活在印度尼西亚小巽他群岛（科莫多岛及邻近岛屿）的科莫多巨蜥（或称"科莫多龙"）算得上是它的近亲。看过《动物世界》和《发现》（Discovery）节目的朋友们一定不会忘记科莫多巨蜥这种争抢腐肉、吞吐着分叉舌尖、相貌怪异的庞然大物吧！谁又会想到它跟蛇在进化上也有亲属关系呢？

既然蛇类是陆生蜥蜴类进化而来的，它们又是何时失去了四肢？科学家们在这一点上基本达成了一致——蛇是与蜥蜴在进化之路上分道扬镳后逐渐失"足"的。对人而言，四肢中如有一肢残疾也会给行动带来不便，然而四肢全无的蛇非但未受限制，反而从中得到了不少好处：足的丧失使得蛇能更好地在草丛中快速滑行；而其修长灵活的身体没有四肢的障碍后，就能轻易钻入许多其他动物无法到达的地方来拓展领土；蛇类虽然最终失去了四肢，但它们所具有的大型腹鳞、肋骨和连接这两者的肌肉使它们能够采用更加多样的运动方式来移动，如蜿蜒式、侧行式、直蠕式以及风琴式等。这些特殊的运动方式和更为宽泛的生存环境的选择，使得蛇类在同其他生物竞争时可以迅速避开天敌并选择具有更多生存资源的地方来定居。与它们的爬行动物近亲相比，蛇类可谓是另辟蹊径，最终"蛇行"出了一条专属自己的生存之道。

在陆地生活，就要面对各种天敌，还得有获取食物的本事。毒牙和毒腺堪称蛇类防御天敌以及捕杀猎物的"绝杀技"。众所周知，蛇类可以被分为毒蛇和无毒蛇。之前人们一直认为，毒腺也是蛇类在进化过程中逐渐获得的一种抵御天敌、捕猎或是威胁其他生物的有力武器。但出乎意料的是，近期对爬行类毒腺的科学研究发现，蛇毒其实也是进化道路上逐渐丧失的一个性状。换言之，并不是毒蛇类通过进化"获得"了毒腺，倒更像是无毒蛇类逐渐"失去"了毒腺（更确切地说，无毒蛇也有毒，只是其毒性和浓度对人类造成的伤害不大而已）。

五年前，一个由多国科学家组成的联合小组通过基因序列比对，结合化石资料分析结果，终于确定了蛇类和蜥蜴类等有鳞目爬行动物的进化关系及分歧时间。最后，他们找到了毒蛇和毒蜥蜴的共有祖先，并发现那时期的此类有毒物种，都很规律地集中在爬行动物的某一分支上，于是，研究者们建议在"生命之树"上为爬行类增设一个新的分支——有毒亚目。

对有毒亚目的研究表明，蛇的进化历史差不多是一亿多年，而蛇毒的历史则在两亿年以上。换言之，现存的毒蛇和毒蜥蜴都是从两亿多年前一种有毒动物进化而来的。这个共同祖先有几种毒液，都是偶然借用其身体其他部位所产生的蛋白质而形成的。在之后的进化中，由于生态环境的长期变迁，环境和气候条件等因素的变化各有不同，有些蛇扩充了它们的弹药库——发展出了十几种新型毒液，使它们可以更有效率地杀死猎物；同时，蛇类家族的另外一些成员所受到的进化选择压力却在逐渐放松，从而失去了它们的蛇毒，变成了所谓的无毒蛇。这些为数本就很多的无毒蛇获得了新的进化动力，很快发展成迄今为止世界上最大的蛇类类群，在全世界2 500种蛇中占了大半壁江山，有1 500余种。丧失了蛇毒的无毒蛇类当然还进化出两种新型的方式来杀死猎物，要么缠绕猎物致其窒息而亡，要么将猎物制伏后吞下。

无论是从进化历史还是现存分布来看，蛇类无疑都是适应环境最为成功的生物类群之一了。其在生物性状适应环境的方式上既有获得，也有失去，各有千秋，相辅相成。也正是这种性状失去与获得的奇妙组合，进化出了如今各式各样的蛇类。同人类和其他物种一起，构成了这个世界缤纷繁复的生物多样性。

在生命历史的长河中，进化和退化就像是舞步的进退一般，在大自然划定的舞池中舞蹈。一曲终了，有些物种退场了，又有些新鲜角色随着新曲目而登场亮相，还有些物种从头舞到尾，中间更换了面具。在这样不断变化却又稳步向前的进程中，那些刻意或者无意的每一小步，可能都是未来创造生命神奇的一大步。

本文发表于《艺术世界》2012年12月第270期，署名钟扬、赵佳媛，收入本书时略有修改。

播种未来（钟扬院长文字配音稿）

  我曾经有过许多梦想，那些梦想都在遥远的地方。

  我独自远航，为了那些梦想。我坚信，一个基因可以为一个国家带来希望。一颗种子可以造福万千苍生。

  初到这片土地，只为盘点世界屋脊的生物家底，寻找生物进化的轨迹。在漫长的科考道路上，我慢慢意识到，这片神奇的土地需要的不仅仅是一位生物学家，更需要一位教育工作者。将科学研究的种子播撒在藏族学生的心中，也许会对未来产生更为深远的影响。

  "生物多样性的排名，西藏在全国一直是前三，西藏还有很多没有被探索的地方。"高原植物学人才的培养，不仅仅在课堂，也在雪山脚下，荆棘丛中。这一路总是充满艰辛，而我的学生，未来的植物学家，必须要学会克服困难，迎接挑战。

  "采样是很辛苦的，为了采样我们每年至少走三万公里。"（学生语）

  采种子的路上经常会发生各种各样的状况。高原反应，差不多有十七种，在过去的十三年间，每一次我都有那么一两种。我们也不能因为高原反应，我们就怕了是吧！科学研究嘛，本身就是对人类的挑战。

  海拔越高的地方，植物的生长越艰难，但是越艰难的地方，植物的生命力越顽强。我希望我的学生，就如这生长在世界屋脊的植物一样，坚持梦想，无畏艰险。我相信，终有一天，梦想之花会在他们的脚下开放。

"这里有一片难得的香柏林，这种生活在海拔四千米以上的植物很有用处。我们今天多采一点。"生长在海拔四千米以上的香柏，在复旦大学药学院提取出抗癌成分，并得到国际权威认证。

梦想，无论多么遥远，总驻守在我们心底。创新的心永远无法平静。只要心在不断飞翔，路就不断向前延伸。

我这十三年在西藏干了三件事：为国家和上海的种子库收集了上千种的四千万颗种子，它们可以储存上百年；培养了一批藏族科研人才，我培养的第一个藏族植物学博士，已经成为了教授；为西藏大学申请到第一个生态学博士点，第一个国家自然科学基金项目。我希望打造一种高端人才培养的援藏新模式。

任何生命都有其结束的一天，但我毫不畏惧，因为我的学生会将科学探索之路延续；而我们采集的种子，也许会在几百年后的某一天生根、发芽，到那时，不知会完成多少人的梦想！

不是杰出者才做梦，而是善梦者才杰出。我是钟扬，一名工作在青藏高原的生物学家，一名来自上海的援藏教师。

本文据2013年9月"播种未来"视频整理，视频网址：http://v.youku.com/v_show/id_XNjA3NzMzNTQ0.html。

创新文化是一种"试错"文化

在我们从小学到研究生的课程教学中,普遍存在一种现象:老师们针对某个科学问题,通常重点讲述该问题的答案,而忽略解决这一问题的历史过程;学生们则因应对考试需要,只关心答案,而不关心前人是如何获得这一答案的。

这一教育学的模式产生了一系列的后果。最严重的莫过于大多数的人都认为,这个世界上存在两类科学家:一类特别聪明(再加上一点汗水和灵感),他们总是能在众人迷茫的时候找到通往正确答案的道路;而另一类科学家则比较愚笨,辛辛苦苦,到头来总是获得一些错误的结果、错误的观念和错误的理论。如果你在中学学习了氧气的发现,你就会认定,拉瓦锡就是前者,而其他提出"燃素学说"的人都算是后者吧!然而事实上,所有相对正确的理论都来自谬误,而即使今天看来是错误的研究中可能也包含正确的方法。

由于我们人为地将正确和错误对立起来,并且暗示学生只有不断学习才能发现正确的道路,这就使得在学习期间,许多成绩优良的学生总以为掌握了正确的科研方式和正确的科学理论。令他们迷惑的是,在参加真实的科研时,就感觉困难重重。尤其是他们把自己获得的一些结果拿给导师或其他学者评审后被认为是错误的结果时,往往沮丧至极。如果这样的过程经历几次或者持续时间长一点,这位年轻的科学家往往会对自己的能力产生怀疑,甚至不少人放弃了今后从事科学研究的念头。

如果我们仔细地检审科学史，就会发现科研创新远没有课堂教学中描述的那么清晰，正确和错误的研究时刻混杂在一起，科学家更没有聪明和愚笨之分。不仅一个领域的所有科研创新只是阶段性正确的，其实所有理论都是阶段性正确的，甚至可以认为是可证伪的假说而已。对一位成功的科学家而言，他通往正确的道路也是由错误铺就的，这是科学研究的本质，更是创新活动的基本范式。可以说，只要不畏艰难，勇于探索，承认错误，吸取教训，随着时间的推移，每一位科学工作者或多或少总能取得相对正确的结果。

不知为什么，我们的科研文化中有意无意地忽略了试错的过程，鼓励成功，宽容失败也只是贴在墙上的标语而已。我们渴望创新，但没有为失败留下足够的空间。也许形成这一文化的原因过于复杂，但我们不得不说，至少在科研领域，缺乏试错的文化不是健康的文化。

试错对科研工作来说为什么如此重要呢？第一，不能崇拜权威。因为有错的可能恰恰是专家、权威。第二，不能唯上。行政和管理能力与科研水平没有必然的联系。无论一个人知识多么渊博，思考问题多么缜密，反应多么灵敏，在复杂多样的大自然面前，都只能是小学生。对大多数科学研究而言，尤其是实验科学，必须反复地实验而不是寻求捷径。

因为我们对自然现象和科学规律实在知之甚少，如果我们轻易地把现实生活中的人际关系、管理规则和伦理道德等文化特性带到科学领域，必然会影响到我们的创新文化。尤其需要指出的是，科学研究并没有大家希望的那么高的收益。因此在科研管理中过多地强调科研产出比，重成果转化、轻技术研究等商业文化和急功近利的奖励考核方式都在事实上妨碍了试错文化。一味追求见效快和花费省显然和试错文化是不相容的。

以遗传学为例，基因和环境导致遗传变异是近百年来反复争论的话题。由于正常的学派之争和不正常的政治干预使得强调环境作用的米丘林学派在我国成为正确的科学理论，而基因学说被认为是错误的、应该被抛弃的科学理论，那一时期的学生无法开展基因水平的科学研究。随着李森科主义的破产和分子生物学的兴起，基因学说成为了主流理论，而环境论则被归为错误的理论。直到近年来我们发现了表观遗传的作用，人们才重新审视所谓错误的环境学说，并且试图将环境说和基因学说融合在一起。

另一个例子是，18世纪彼得森曾经提出了平权原则，即对每一个生物性状赋予平等的权重，但由于计算能力的限制，这一想法无法实施，无法用于生物分类学实践，因而被作为错误的想法而被人遗弃，而代之以林奈的以繁殖性状为核心的生物分类法。到了20世纪50年代，Sokal等人发现了两百多年前平权原则的可用之处，将其用于计算机分类，形成了数量分类学这一新的流派。这在一定程度上反映了科学研究的曲折以及错误与正确并存的现象。还有其他一些经过反复试验才能获得正确结果的例子。即使在应用性研究和工程技术发明方面，反复试验、不断从错误中筛选正确目标的过程也非常常见。

强调试错的文化可能要从小抓起。在科学上不可能只有平坦大道，试想一个在小学、中学乃至大学学习中只知正确答案的"优秀学生"，在今后崎岖的科学山路上怎么能披荆斩棘，直至攀登上顶峰呢？他们更不可能传承创新文化。普及试错文化其实并不难，只要多给科研人员一些信任，多创造一点机会，多忍耐一些时间，从群体水平而不是个体水平上寻求成功。

附录

为什么要试错？在科学研究中即使正确的方法也不能保证正确的结果，某些创新性极强的理论和结果在当时看来很容易被专家和同行们看成是错误的结果。但只要社会有宽容度，科学家本人有不怕错误的精神，是金子总会发光的。所有正确和错误的东西都要经过实践的检验，如果我们的文化是只允许正确而不允许错误的，往往会误伤一些包含正确的结果，从而妨碍科学道路的探索。

本文写于2014年12月30日，未正式发表。

怎样真正提升研究生的能力

编者按：研究生的能力是研究生培养过程中不容忽视的话题。什么是研究生的能力？当前研究生能力现状如何？如何在研究生教育中提升研究生的能力？是研究生、高校及社会共同关注的问题。带着这些问题，我们走访了复旦大学研究生院院长钟扬教授，倾听他对这些问题的看法。

记者：研究生培养的目的在于提高研究生的能力，您认为研究生的能力主要包括哪些？您是如何理解研究生能力的呢？

钟院长：在我看来，研究生的能力至少分为两种，第一种与他的个性、自身内在的因素密切相关，大部分是在上研究生之前就已经有的、无需培养的能力。第二种则是跟科学研究密切相关的，需要在研究生阶段进行培养的能力。

我们需要在研究生招生过程中，结合不同专业的特殊性，把这部分有特殊才能的人发掘出来

记者：您认为这两种能力是如何在研究生教育当中相辅相成的呢？

钟院长：研究生教育只是教育接力赛中的一个环节，不能把一切成功和失败都归结为研究生教育。我所说的第一种能力，其实是一种已经形成的，大多数情况下需要我们导师去引导和发掘，而不是从头进行培养和教育，因此需要在招生环节把有这种能力的人挑出来。

长期以来，特别是在扩招以后，上学机会增多，我们对这个问题有所忽略。学生自身也存在这个问题，最直观的体现在专业选择上，学生和家长往往过多地看重分数而忽视自身的能力。一般认为，在一个固定的群体中，拥有某种特殊才能的人数是有限的。举个例子来讲，如果在上海办一百所音乐学院，大量招生，你猜结果会是什么呢？结果会让上海民众唱卡拉OK的水平得到大幅度的提升，但上海的音乐家数量呢？还是那么多，绝不因为你多办音乐学院而产生更多的音乐家。我们需要在招生过程中，结合到不同专业的特殊性，把这部分有特殊才能的人发掘出来。我的教学科研经历中，有的同学认识植物的能力很强，如果让他去学植物学能更好地发挥他的才能，但如果被录取到分子生物学专业，可想而知这种能力很容易被埋没。而由于对分数的过多重视，以及招生数量的增多，导致我们在以往的招生中往往不注重这种本来已经具备的能力。

第二种能力我认为它至少包含了四种具体内容：发现问题的能力、课题选择的能力、团队组织和协作的能力以及表达和写作的能力。这四种能力应作为我们研究生教育阶段培养的重点，也可以在一定程度上体现出研究生教育质量。

发现问题的能力：培养发现问题的能力实际上是对我们当前研究生教育的挑战，作为研究生导师，我们培养学生发现问题的能力必须有别于原来的培养模式

记者：您能否跟我们详细谈一下第二种能力，咱们先聊聊发现问题的能力，有些人认为当前的研究生都是导师给课题，自己发现问题的能力很差。

钟院长：我们认为"问题"一经得到较为准确的发现，就已经解决了一半，所以发现问题的能力是十分重要的。而在我国的教育中，学生发现问题的能力非常弱。我国教育大多数的情况是老师发现问题，学生去回答，这便在学生的头脑中埋下了这样的种子：问题就应该是由老师发现的。更糟糕的是，我们还有所谓的标准答案，这个标准答案引导大家认为，解决问题的终极法宝（正确答案）一定在老师手上，不在这个老师手上，也一定在另一个老师手上；不在博导手上，也一定在院士手上。因此，我国研究生发现问题的能力并不理想,这在学生的论文中便有所体现。在论文写作的过程中有一个引言部分,引言的意思便是问题的提出,很多同学写不出引言就是因为他并不知道问题怎么提出,缺乏发现问题的能力。

我认为发现问题的能力包含两个方面,一方面是能从散乱的数据、复杂纷繁的自然与社会现象中,找出真正起决定作用的主导因子。我们所观察到的事情并非就是要研究的"问题",一件事情的问题可能有十几二十个。比方说研究食品安全问题,影响食品安全的因素可能有几十个。但是作为一个研究生,每篇文章中只能谈到一个,如果这个问题选得好,便有可能引起公众的关注;如果选择的问题是比较次要的方面,论文的水平往往会略低一筹。问题与问题之间也有着大小、主次之分。另一方面,要明确不能解决的问题就不是问题。在当前研究生身上有一个常见的情况,他找到了一个问题,但是甚至连试图解决它的想法都没有,他的论文就是罗列了这个问题,最后提出了毫不相干的另一堆东西。所以我们在研究一开始便要经常问自己这样几个问题,这个问题是不是主要问题?我能不能提出解决的办法?我的解决办法跟这个问题有没有直接的关系?否则不能称之为研究。

记者:要怎样培养与提升研究生发现问题的能力呢?

钟院长:我们跟导师强调,在对研究生学术指导时,要告诉研究生科学研究中要多一些本质的东西,往往一些本质的东西容易在无意中被掩盖。其次,要进行不断的实践,从前那种由老师提供标准答案,学生一问一答,为了考试而背知识的过程,在研究生阶段即使不是全部抛弃,也必须得以改变,这是根本性的变革。现在的研究生习惯于经常向老师询问"这个问题应该怎样看?"表面上是学习,实际上是寻求答案。我遇到这种情况就会不客气地告诉他"对不起,我不知道"。我之所以当教授是知道怎么样去研究这个问题,我要知道这个问题的全部,我早就自己去写论文了。但是学生肯定不这么认为,学生会想老师是不愿意告诉我,他肯定知道,要不他怎么当教授呢。如果老师告诉了他某种答案,他据此写了一篇论文,也能顺利毕业,但可能出去之后难有所成,因为他在能力上有所欠缺。发现问题的能力实际上是对我们当前研究生教育的挑战,作为研究生导师,我们培养学生发现问题的能力必须有别于原来的培养模式。我们要明白，若我们轻易交给学生一个问题，可能会忽略了对他发现问题能力的培养。由此，要强调他们独立进行选题。

课题选择的能力：因研究生培养时间有限，又有学位方面的要求，在研究生课题选择时，我们一般是要选新材料老方法，或新方法老材料，通过帮助研究生顺利完成论文，培养他们科研的乐趣，让他们尝到科研的甜头，而不是轻易去尝试没有把握的新材料、新方法，让他们尝到科研的苦头，打消科研的积极性

记者：您是怎么看待研究生课题选择的能力的呢？

钟院长：在课题选择的能力中，非常重要的一点是对课题性质的判断。这个和办一个公司或搞一个企业投资是一样的道理。从来没有一个企业家走到大街上，看见卖水果的就决定摆个水果摊，看见卖包子的就决定开个包子铺，他是有选择的。凭什么你走进力学系就做力学的论文，走进生物系就做植物的论文呢？你为什么觉得科学研究有这么大的随机性呢？没有的，它跟你们课题的平台有关，跟你付出的时间有关，它由时间、效益各方面的综合指标所决定。很多同学在做课题时经常存在这样一种现象，刚开始信心百倍，做到中间想打退堂鼓，结题的时候匆匆忙忙建造"烂尾楼"。很大一部分原因在于，我们的研究生在选题时具有很大的随意性，没有从时间、资源、个人能力等方面考虑清楚题目的可行性，加之研究过程中缺乏合理的规划和方案，便经常会出现这种问题。所以我们强调抓好开题报告和中期考核等环节就是这个道理。

记者：您认为这种情况怎样获得改善呢？

钟院长：我觉得现在有几个问题值得注意，一是在本科阶段，许多选择继续读研的大学生在思想、知识及能力储备上并没有做好准备，本身就缺乏课题选择的能力，如果本科阶段对这一能力有所培养，这个情况就好多了。第二点是关于老师的指导理念。因研究生培养时间有限，又有学位方面的要求，在研究生课题选择时，我们一般是要选新材料、老方法，或新方法、老材料，通过帮助研究生顺利完成论文，培养他们科研的乐趣，让他们尝到科研的甜头，而不是轻易去尝试没有把握的新材料、新方法，让他们尝到科研的苦头，打消科研的积极性。你怎么能在不对一个新兵进行训练的情况下，就让他上战场杀敌呢，他首先要能保全自己才敢上战场，他当了三年兵之后决定再也不上战场了，你觉得这个军队就培养好了吗？所以导师正确的指导理念非常重要。

我自己有两个例子，当我去做红树的时候我就去尝试新方法，因为红树这个材料我心里有数，而我去做拟南芥的时候，一定是老方法，因为那个材料对我来说是新的。很多导师追求创新，让学生选择两方面都新的课题，这样学生很难完成。我们借用别人的软件，在很大程度上是比较同样的软件在做别人的事情和自己的事情上的差别，为了创新你自己编个新软件，没什么意义。你在研究中使用了新软件与新数据，大家怎么知道你的结果是由于软件新呢还是数据新呢。这也许能解释为什么中国的科研水平上不去，一是追求时髦，二是追求跟国外不一致，不敢使用相同的材料跟别人硬拼。

团队组织和协作的能力：长期的科学研究表明，科研合作是科学研究中最困难的过程

记者：我国研究生团队协作的能力又是怎样的情况呢？

钟院长：这十年来，我从科学院转到大学里面，我发现团队协作的能力在退步。这可能和我们的独生子女政策，也和社会上过于强调个性有关，这个是此消彼长的问题。我曾经在我所上的生物信息课上做了一个实验，课程临近结束时对学生成绩评定的方式，既不是考试，也不是要求同学们写篇论文，而是要求他们组成一个团队，最多五个人最少三个人，在一起完成一个类似project一样的东西，并且我来组织一个由几名老师组成的评审组，听他们每个团队中的同学讲5~10分钟，这样我就能清楚地知道他在团队中的贡献，以及团队合在一起解决一个问题的能力。我做了这样一个训练，结果教育了我和学生。班上有的同学成绩特别不错，反复来向我申诉，强调他只能一个人干，担心别人会拖他后腿；有的同学过于自卑，认为自己是最笨的，因为所有的团队都把他踢出来了。所以很多同学在课程结束的时候很有感触地说，做了什么project已经不记得了，但钟老师这个事是震撼性的，说这让他们知道原来在宿舍一起打游戏的哥们,在一起做课题的时候是多么的无力。女生平时都是闺蜜,但做这个课题在踢你的时候一点都不含糊,生怕影响成绩。有的同学说她两次都被别人踢了出来,我问她该不该反省一下自己?我告诉他们,你们从小学考上重点中学,重点中学考上重点大学,最后再考上研究生,每一个步骤都似乎在向你暗示那是你个人奋斗的结果,跟其他人无关;恰恰相反,如果其他人表现弱一点,你成功的可能性更大。但当他们需要合作完成一个课题时,问题就出来了。他们非常习惯于想要去显示我真的比别人的孩子强,而不是去协作。现在的科研中是讲究合作的,但现在的研究生却总是认为自己比别人厉害,这种情况非常严峻。

此外,长期的科学研究表明,科研合作是科学研究中最困难的过程。一位院士曾经告诉过我这样一句话"科研合作甚至违背了科学家的本性"。我国科学家当前所寻求的合作,大部分都是在自己一个人无法完成的情况下进行的,当自己能够独立完成时,很少会与他人分享成果。我很注意培养我的学生这方面的能力,尤其是在实验室中培养团结合作的氛围,使之成为一种实验室文化。比如我经常组织实验室同翻译一本书,一人一章,并相互修改。我还会在我的实验室里尽可能招收外国学生和少数民族学生,通过对不同文化的吸纳,增加实验室的包容性,促进实验室的团结合作,让学生在相互帮助中彼此受益。最近,我们也在探索一种通过微信群进行合作的方式。我挑选了一个我自己也从来没研究过的题目,通过微信群与我十多个研究生以及校内外同行一起讨论,要求大家通过在微信上为期12个月的讨论,写出有一定水平的论文。这一方式的效果还有待观察,但我们研究生参与的积极性无疑得到了提升,我认为会对提升其团队协作能力有所帮助。

表达和写作的能力:大多数英文论文水平低的原因,其实很多情况下并不单单是英语的问题,而是你语言表达就有问题

记者:表达与写作能力我们都不陌生,您对这一能力有什么看法呢?

钟院长:写作问题本质上也是一个综合问题,我首先想要强调的一点是不要把中文写作和英文写作截然分开,它不是一个对文字与语言的简单应用。现在英语写作对理科生可以说是必备的,而大多数研究生的英文论文水平并不好，其实很多情况下并不单单是英语水平的问题，必须从中文写作抓起。

我们曾举办过一次写作竞赛，结果并不理想，当前研究生在写作上是有着很大障碍的。我想有以下几点原因：一是在我们小学、中学教育阶段对写作能力的培养上，总是抓数量，而忽略质量；二是现阶段研究生普遍在写作上投入的时间太少；三是我们的导师对写作逻辑性引导得太少；四是现在传播工具的片段化。我想重点强调一点，我们现在倾向于依赖手机网络进行阅读与写作，而在手机上写作往往都有字数限制，你总不能在手机上写出长篇，通常都是一百字左右，所以学生经常以一百个字为指标，这就叫片段化或碎片化。在阅读上也是同样的情况，而且表现得更为明显。在学生面试的时候我经常会问他们读过哪些重要专著，很多学生甚至连本专业基本的专著都不读，而花很多时间去大量阅读网络上那种简单的介绍性文章。另外，写作是一个艰苦的事，大部分伟大的作品都和寂寞、艰难以及不幸有关，而现在我们的学生缺乏一种必要的磨练或历练的过程，所以导致整体的写作水平低下。这一问题也是学生、学校和社会等多方面原因综合导致的。

这里我想特别谈一件事，就是科普写作。现在国家越来越注重向民众普及科研成果。而我们科普写作的水平是很差的，科普写作实际上是文学写作和科学思想的一个综合，难度更大。首先，它在字数上有所限制，比如要求在四百字内描述清楚某件科学事件。同时，这类文章既要满足科学性，又要让更多的人能看懂，而我们甚至有的时候自己写的东西自己都看不懂。今年我在上海市承担了一些科普写作的工作，进行这项工作时我的体会就非常深，在这个过程中我发现我们有些同学是很有这方面才能的，而他的这一才能在之前就被埋没了。而也有些之前科研能力很强的同学，写作科普文章的能力却非常的弱，这并不是我们希望看到的现象，它至少是对全民的素质提升没有好处的。研究生代表了这个民族的最高水平，我认为研究生都要注意提升自我这方面的能力，这是作为一名研究生需要承担的责任。

本文发表于《上海研究生教育》2015年第1期，是该刊对钟扬院长的一篇访谈文章，署名程诗婷、金鑫、包晓明，收入本书时略有修改。

原创的，就是世界的

——钟扬教授专访

从2012年起，《科学通报》邀请不同领域专家对当年度诺贝尔自然科学奖（以下简称"诺奖"），从专业研究和科学传播角度进行深度解读，并编辑出版了"解读诺贝尔自然科学奖"系列专题，受到读者的广泛关注。2014年末，本刊编委、西藏大学校长助理、复旦大学研究生院院长钟扬教授，带着对我国科学家早日获奖的期盼，围绕诺奖这一话题，与本刊记者作了深入交流，表达了他的独到见解，也为办好《科学通报》提出建议。

我们的话题是从日本的科学研究开始的——

对日本科学研究的关注使我开始认真思考诺奖

《科学通报》：请问您是从何时起开始关注和思考诺奖？

钟扬编委：我关注诺奖可以说是从关注日本的科学研究开始的。1989年我第一次迈出国门，到日本京都参加国际植物物种生物学会议，自此就一直关注日本科学家及其科学技术的发展。后来我在日本国立综合研究大学院大学（The Graduate University for Advanced Studies）获得博士学位，我所在的研究单位是文部科学省统计数理研究所。促使我对诺奖深入思考的是在2001年，日本出台了"第二个科学技术基本计划"，明确提出日本要在今后50年内获得30个诺贝尔奖，这在当时引起强烈反响。很多科学家认为，科学研究具有不确定性，不能像生产丰田汽车一样"生产"诺奖。连诺贝尔化学奖得主、日本名古屋大学教授野依良治都公开批评政府提出这样的目标"没有头脑"。但出乎大家意料的是，至2014年，日本在这些年间已收获了13个诺奖。据我了解，按照诺奖评选规则，最后有一个调查过程，而进入这个调查过程的日本科学家数量更大。因此，未来日本学者获得诺奖的前景仍是光明的。

  看来，我们的确需要正视日本的科研成就，了解他们成功的经验。然而，由于种种原因，我们对日本了解并不多。2010年冬天，我再次到文部科学省统计数理研究所做访问教授，其间用了大约3个月时间，着重思考日本科学家为何会获得诺奖。为此，还专门访谈了几位日本的诺奖获得者和一些重量级科学家，以及一位在日本科学理事会任职的官员型学者，了解日本获得诺奖的情况。

以诺奖为代表的科学奖励，所尊重的是原创

  《科学通报》：您对日本获得诺奖有了什么样的了解，又作了哪些思考呢？

  钟扬编委：在日本获得诺奖的科学家中，我注意到一个现象，他们发表在 *Nature*、*Science* 等国际顶尖期刊上的论文并不多，其中有些论文没有发表在所谓高影响因子的刊物上，甚至还有论文并未发表在英文刊物上。

  有一个典型的例子，汤川秀树是日本第一位诺贝尔奖获得者，也是一名完全由日本本土培养的理论物理学家。1932年，25岁的汤川秀树担任大阪大学讲师，他一方面从事科研教学活动，一方面也为日后申请博士学位做准备。1935年，他将研究成果写成《论基本粒子的作用》一文，发表在本国的专业学术期刊——《日本物理-数学会刊》，文中首次提出了著名的介子学说，预言了介子的存在。这一新理论使汤川秀树于1938年获得了大阪大学的物理学博士学位，也于1949年荣获了诺贝尔物理学奖。

  1946年，汤川秀树在京都大学基础物理研究所创办了一份日文本土刊物《理论物理学进展》，致力于向国外推介日本理论物理学研究成果，帮助日本科学家克服因国际竞争和语言障碍等对发表创新思想不利的因素。1973

年，两名年轻的助手（助理教授）小林诚和益川敏英合作在该刊发表了一篇题为《弱相互作用可重整化理论中的cp破坏》的论文，提出了著名的小林-益川模型，以解释弱相互作用中的电荷宇称对称性破缺。2008年，小林诚和益川敏英因此荣获了诺贝尔物理学奖。从这本日文刊物，国际学术界逐渐看到汤川秀树本人的高水平学术论文以及一批日本科学家的原创性研究。这本刊物最终培养出多位物理学诺奖得主。

另外一个故事发生在日本名古屋大学。阪田仓一是名古屋大学物理学教授，他立足于本土教育，虽然其本人没有获得诺奖，但为名古屋大学粒子物理专业培养未来的诺奖获得者奠定了基础。前面提到的益川敏英就是毕业于名古屋大学，益川敏英不仅不能用英文在国际刊物上发表文章，甚至很难用英文与国际同行交流。我们会认为，作为一位科学家，不懂英文是一件很不可思议的事情。而他却几乎不看国外刊物。他获得诺奖后提出的要求是领奖的时候可否讲日语，也是那时，他办了人生第一本护照出国了。这虽然是一个特例，但引发了我的疑问，为什么这样的科学家也能赢得科学的桂冠？

这其中有一个很重要的，但不太为国人所知的原因就是诺奖评选规则——如果某项研究进入诺奖评审程序，一定要了解当时的真实情况或原始记录，不管用什么语言记录，是否发表或发表在哪里，只要谁最先提出，当时记录了，就会被认定。这里，"原创"才是最重要的。

此外，我感觉诺奖评选注重原创这一原则也给我们一个新的启示。随着新媒体的发展，原创的机会应该是越来越多了，我们应该及时地记录各种科学发现和灵感。当你有了大数据，也许还可以开展数据驱动型科研或者说基于大数据的科研，而一些非常宝贵的数据可能形成原创性成果。以前写成论文是原创，现在分析数据也是原创。我们要抓住机会多做原创性的事情，多思考基本的科学问题。

我们要通过国际化提高自信，而不是通过国际化丧失自信

《科学通报》：那么，您是否认为发表在中文刊物上的原创性成果也有可能获得诺奖？

钟扬编委：当然，包括《科学通报》。刊物是科学研究中一个重要的交流平台。《科学通报》发表过一大批优秀的代表性成果，如袁隆平先生的水稻雄性不育理论等。我本人回国后的第一篇文章就发表在《科学通报》，近些年也为这份刊物做了一些服务性的工作。我觉得《科学通报》等一批中文学术刊物在让一名学生成为真正的科学研究者的过程中，起到了不可替代的作用。当年很多科学家都是先用中文发表研究论文，再翻译成英文，而不像现在直接用英文写作。尽管《科学通报》的中英文版已经分开了，但是坦率地说，就一篇论文本身的研究质量而言，用什么语言发表是没有差别的。比如，中英文论文都需要有新发现，在写作上都需要逻辑性，等等。在这个意义上，刊物是没有影响因子之分的。

日本诺奖获得者给我们的最大启示是要鼓励年轻学者去做原创性研究，这也可为我们办刊所借鉴。我们经常提到 Nature 和 Science 的栏目丰富，其实不光是栏目，我们还应该关注其内涵。浏览这些刊物的时候你也许会注意到有"Science in Russia"这样的内容，当期报道俄罗斯的科学。此外，还有阿拉伯世界的科学、伊朗的科学，等等。但在我们这本刊物中，还没有专门报道中国西部科学的栏目。比如，兰州大学的人才建设，西部民族地区的科学设施，以及西藏的科研，等等。也许我们需要去关注某些原来没有关注过的东西。

日本的本土科学刊物值得我们学习的是，要通过国际化提高自信，而不是通过国际化丧失自信。我不反对我国很多刊物走国际化道路。如果我们期刊国际化了，自信心增强了，大家反过来认为中文刊物也能够做得很好，就对了。现在也许正是时机，《科学通报》要保持自己的风格，保持自己的品牌，为年轻科学家的成长多出一份力。比如，为研究生办科技论文写作培训，一定会很受欢迎的。

我们的教育和科研活动过多地依赖考试和考核

《科学通报》：我国科学家对于诺奖有着特殊情结，长时间以来对其翘首以盼。您认为我国在科学领域至今未有诺奖产生的主要原因是在哪里？是否有一些妨碍因素呢？

钟扬编委：这个问题比较复杂，也许我们的原始积累还不够，时间未到；但我感觉，有一个因素我们要重点考虑，就是我们现阶段的教育、科研活动过多地依赖考试和考核。当我和大家一起探讨我国为啥没有获得诺奖的时候，我总是半开玩笑地说，可能我们并没错，错在诺奖——这么重要的奖励居然不考试？如果不吃不喝考上七天七夜，第一名肯定是中国人。大家听了都一笑而过。因为无论怎么改革，目前我国的科学教育工作基本上都是围绕考试和考核（甚至是"选秀"）进行的。我们太看中"选秀"之后的那些头衔。似乎没有这些头衔，别人就看不起你。像刚才那位不懂英文的日本科学家，我们很惊诧的是他竟然在名古屋大学获得了博士学位。而在我国，这样的人可能在幼儿园阶段就输在起跑线了。

我们一方面看到我国发表的论文越来越多，但另一方面原创的思想并不多。在全世界范围内，官员和管理者的影响面都是比较大的，他们会引导科研方向，决定科研经费的走向，但由于他们中的很多人已远离科研第一线，所以热衷于采用影响因子之类的"硬指标"来评价科研论文。事实上，只有科学家自己才最了解科学工作本身。我注意到有这样的评论，我国科学家很多文章，即使是发表在顶级刊物的，也可能具有某种跟风性质。当然，我们也应清醒地认识到，从数量到质量的转变需要有个过程。对刊物影响因子的过分依赖则在很大程度上反映了我们还没有足够的自信，因为我们不能通过自己的眼睛来判断论文的水准，而只能通过他人的标准来评价自己。

日本的科研规模总体比我们小，但在科研人才培养方面确有一些可借鉴的地方。一方面，他们大力倡导学习国外先进经验，比如他们很早就允许学生用英文撰写学位论文（在我国很多大学的学位管理条例中是不允许的）；另一方面，他们也为具有原创精神的学生大开绿灯。美国学者也常说，一个国家高等教育成熟的标志是，所谓一流的大学不可能在所有领域保持一流；也不可能容许一个二流的大学在所有的领域都是二流。看来，我国的教育和科学事业还需要时间继续成熟和完善。

第一位获诺奖的中国科学家，有可能是一位女性

《科学通报》：凭着您对诺奖的理解，可否猜测一下，我国将在何时、在何领域最早获得诺贝尔奖？

钟扬编委：大家都愿意相信20年内我国能够获得诺奖。中国的科研规模和投入持续增长，举世瞩目，因而很有可能达到 critical mass，即从量变到质变转化的临界量。不过，据我所知，目前能进入诺奖（自然科学奖）考察程序的中国科学家几乎没有，而我听说一位日本科学家已经被考察了20年之久。所以，梦想和耐心都是我们应该具备的。我与一位接待过诺奖评委的日本科学官员探讨过相关话题，她曾是日本最好的国立女子大学校长，卸任后成为日本科学理事会委员。她认为中国获诺奖是很有希望的，但中国科学家要沉下心来，多做原创性研究。我问她，根据她的了解，中国在哪些领域更有希望。她说，虽然她自己是生命科学领域的教授，但不能断言生命科学首先会获奖。交谈中，她的一个观点给我留下了极为深刻的印象。她猜测，如果中国科学家获诺奖，很可能最早出现在女性身上。因为中国女科学家的地位、从事科研的热情和规模超过了世界上任何一个国家，这令她非常惊讶。

仔细分析不难发现，我国首位获得拉斯克医学奖的屠呦呦确是女性。中国女性从事科学研究的规模不仅名列前茅，而且还有一批女科学家在默默地帮助别人。比如，我就认识一位国内很出色的学者，他的研究成果是世界级的，但他曾亲口对我说，他最重要工作的原始想法以及实验的起步都是他的太太做的。不妨设想一下，如果他真的有希望进入诺奖考察程序，查阅当时的研究记录，这个"原创"一定会被认定是他太太的。

但是，要让中国女科学家获奖的猜测成真，我们必须先改善许多不尽如人意的地方。我国的本科生、硕士甚至博士中女性的比例非常高，但高级科学家中女性的比例实在不高；要给女科学家们创造更好的条件。当然，也可能并不是条件不好，而是不少女性更多地顾及了其他社会角色，逐渐放弃了相对艰苦而成功率低的科研工作。因此，我们要重视我国的女科学家，

尤其是青年女科学家，要为她们加油鼓劲。

采 访 后 记

作为一名援藏13年的生物学家和教育工作者，钟扬教授对西藏有着深厚的、特殊的感情。采访即将结束时，他不忘借此机会，大声呼吁科研、教育界要关注西部，关注那些具有丰富的生物多样性和独特生态环境的区域。"梦想无论多么遥远，总驻守在我们心底。创新的心永远无法平静。只要心在不断飞翔，路就不断向前延伸。"正如他和学生们在微纪录片《播种未来》中描述的那样，他在坚守中追逐着梦想，在创新中播种着未来。

只要原创的，就是世界的！让我们一起祝愿我国科学家们在坚持创新中不断攀登世界科学的高峰！

本文原载于《科学通报》2015年第60卷第7期，是该刊记者安瑞对钟扬教授的一篇专访，收入本书时略有修改。

在我失联的日子里

题记：这是一个真实的故事。一些段落是我用左手（或右手）一字一句敲出来的，另一些则是我的学生卓雅帮我输入的。在过去的一周里，不期然的，我的人生一瞬间跌到了命运的关口，我在努力地回忆着每一个细节，希望能给各位村民提供真实的信息。所有材料和交流不宜到村外转发。
——钟扬

5月2日是我51岁的生日。前一天的夜里，我在疲惫交加之际，写了几段文字来描述我的出生。不知为何，我在句子开头提到了"我很累"，可能这就是内心真实的感受吧，我只希望能快快休息一下，不要再过每天睡眠三小时的日子了。

清晨七点，起床后吃了几块饼干权充早餐，就直奔学校1号楼去面试自主招生考试的高中生。今年复旦大学严格按照教育部自主招生规定，只同意了很少的学生来参加自主考试。换言之，这几乎算没有举行自主招生考试了。前几年，我们最多有四百个正教授花两到三天的时间来面试学生，每个学生十五分钟左右，那场面真是宏大。我曾做过理科组的一个组长，还到电视台讲过复旦自主招生是怎样录取学生的。今年倒好，理科一共只有2名学生，学校准备了5名考官，每位考生的面试时间是一小时左右。考试真是在相当愉快的环境中进行的。第一位考生，用优秀已经完全不足以来形容他了，你能想象，一位高中生已经发表了两篇SCI论文吗？而且都是通讯作者，就算把他放到研究生中也是相当难得的。我突然间感觉，他可能就是Bill Gates第二，因为Gates先生当年据说就是因为研究做得特别好，人生没有追求后才退学的，我上课讲的Gates退学就是这个版本，对于这样的人来说，做科学家对他们的挑战太小了。我不经意间问了他一个问题，你怎么看你的父母？这孩子告诉我，他的父母很有钱，但他不愿意过跟他们一样的生活，他们的知识面，他们的志趣，都跟他今后要当的科学家大相径庭。不知为何，我突然感到一阵悲哀，因为从35岁以后我渐渐体会到父母才是我最好的老师。也许，这么聪明的孩子从我的眼神中捕捉到一丝不快，他很快出现了少有的惶惑。我说："你不要紧张，我们就是聊聊天而已，今天是我的生日，现在是早上八点四十五分，我就是在这个时间出生的，所以我有点走神了，想起了我的父母。"这个孩子瞬间就恢复了自信，继续去回答那些漂亮而高深的科研工作。我给他打了高分，尽管我对他不了解，但无论如何，如果复旦大学能得到这样的学生来成为本科生，他也一定会是优秀的学生。临走时，他居然说了一句："老师生日愉快！"之后，我又花了一个小时去面试了一位才能略差，但说话直率的女生。也许这位女生是报考生命科学的缘故，我在她能否进入复旦大学学习的问题上也没有任何犹豫。因为我知道他们无论如何都比前几年数以百计的考生中的绝大部分要强得多。面试结束，天下起了倾盆大雨。我们吃了一盒盒饭就散了。

中午十二点整，一种从未有过的疲倦向我袭来，我只要在一点以前到银行给小儿子存入100块钱就可以上床睡一觉了。小儿子在上海的西藏中学读书，他所有的同学都是父母在西藏当地存进生活费，他们在上海每个月从卡里取来使用，我和儿子尽管在同一个城市，但也用这种方法来给他生活费。这是一件小得不能再小的事情了，我跑到银行，却发现今天是周末，这家银行没有开门，于是我就坐了半小时车去学校找他。今天是学校男生"放风"的日子，我在学校只见到了他班上的女生们，而学校外，一批批走过的都是高年级的男生们。我到他取钱的银行时，已经超过了规定时间十五分钟，我知道，说不定他已经失望地离开了。大雨之下，我打着伞，在银行边上的拉面馆（这是他和藏族同学最喜欢去的地方）来回走了几圈，最后还是决定退回银行等他。眼看着他们两小时的"放风"时间即将过去一半，我在银行门口透过浓重的雨帘目不转睛地盯着每一个路过的孩子。突然，我发现了小毛那瘦小的身影，他没有穿校服，他说校服太普通了，爸爸可能会看不见。他今天穿的衣服并不防水，但很鲜艳，他顶着大雨来到银行，没有见到爸爸，他就去吃了凉面，并给两位同学带了满满两碗六块钱的凉皮。我把他头上的水简单擦干净，像多数家长跟十几岁的孩子并没有过多的交流一样，我只是简单地问他："六块钱的凉皮就算周末的改善生活吗？"他回答："是的，同学们很喜欢。"办完了银行的事，我觉得还有时间，就问他是不是还想去外面吃点东西，他说今天的凉面吃得很饱了，不想吃了。我想也许是大雨和在雨中行走的不便冲淡了他的胃口吧。在一个屋檐下，一位老大妈在卖菠萝，我花20块钱买了两个菠萝，削好切好，在回学校的路上却没有找到有人卖小袋的盐巴。宿舍里有两个同学，其他几个还在补习功课。我和小毛拿着饭碗到小卖部，让售货员阿姨舀一勺盐来洗去菠萝的涩感，阿姨还很怜惜地说，这孩子很乖很懂事，但太瘦小。我把菠萝用盐水泡好，看着他兴高采烈地吃下，告诉他把另一个菠萝用盐水泡好，等下午三点的课后，全寝室同学分着吃下。他点了点头，说还有一段时间上课，他想去洗个澡。雨还在淅淅沥沥地下着，我没把他送到教室，只是看着他穿着短裤和我给他新买的拖鞋走进浴室，才转身回返。说实话，自从他独立生活以来，比孪生的哥哥足足矮了10公分，他要我买的最多的东西是部队用的压缩饼干，因为课间唯一能偷偷拿出来吃的就是压缩饼干了。至于我给他从拉萨带回的两箱尼泊尔方便面（类似我国的干脆面），更被他慷慨地送给了全班的每一个同学。他没有像我一样15岁离开父母，而是12岁不到就离开了我们去过集体生活。作为全校唯一的汉族学生，他所遭受的文化冲击，肯定不比我们当年去美国来的小。我相信有一天，他和哥哥能在不同的人生道路上做得比我强。

下午三点多，我才回到家中，睡意已经全无。我开始拿出草拟中的复旦大学与西藏大学合作协议，一字一句地校对。我已答应学校书记，在"五一"假期结束后，便提交修改稿。从初稿到修改稿已经历时两周，而这两周被北京和其他劳模活动弄得支离破碎。还好，我在大约五点四十五分左右初步完成了这一工作。

下午六点三十分,我没有跟任何人谈及生日晚宴的事。事实上,我也不准备弄什么生日晚宴。在众多的朋友邀请中,我挑选答应了一个非常小型的、类似家宴的活动,因为我跟这家人还比较熟识,也知道今天并不是一个喝酒的日子,天又下着雨,我只想安安静静的在九点前结束,回家早早上床睡觉。等我赶到浦东吃饭的地方时,已经六点四十五分了。五人寒暄了几句,马上就开始吃饭了。酒是我最爱喝的酒,但今天备的量并不大。到七点,刚刚酒过三巡。我告诉大家,七点整,东方财经将有一档采访我的节目。我还清楚地记得在这档节目录制完后,一位博士生在看我录完节目后对我说:"钟老师,您怎么看上去这么累?"我笑了笑,心想这样的日子很快就会过去了。我甚至可以利用到香港的三天时间好好休息下,尽管我是带队的,但其他同志已做了周密的准备,把事情安排得井井有条。说着,讲着,七点零五分,电视台开始播放我的节目,历时也就五分钟。席间的一位朋友反复用手机频频拍照,说了些鼓励的话。

傍晚七点二十分,情况突然发生了变化。不知为什么,我的右腿像灌了铅一样沉重,夹菜的右手只握住了一只筷子,而另一只筷子却掉在了地上,服务员给我拿来新的筷子后,我试图去抓住新的筷子,但却发现再也握不紧了。当时我没顾得上那么多,只想快点结束这短暂的饭局回去休息。于是我换左手夹菜,并去厕所收拾了一下,回来的时候,我一个趔趄,差点扑倒在地上。我看到了大家惊愕的目光,但我嘴里还一直说:"没事没事,今天太累了。"我的微信以很低的声音叫个不停,但是我听不见,我索性把微信打开放在桌子上。七点二十七分,汉波发来了一封短信,她说:"钟扬,此次回国我感触良多,希望能在今后为西藏多做些事……"我想回复"收到,谢谢"这几个字,但右手已经完全不听使唤,激动之中还把手机带到了地上,左手也再次把筷子带到了地上。一切都明朗了,这绝不是累的缘故,也不是一般的眩晕,这肯定是脑部出血的征兆。医学上,脑部出血常常用脑溢血来表示,与民间中风是一回事。

窗外,雨还在下着,而我一下子僵硬在那里,不知所措。短暂的沉默后,我的朋友们没有去捡地上的筷子，而是马上将我送往第二军医大学长海医院。朋友中，有一位解放军大校，也是我国药学领域的专家，今天的晚饭可以说是上天的安排，他用自己的车，和太太一起，马不停蹄地直奔长海医院急诊科室。节日的夜晚雨还在下着，窗外霓虹闪烁，穿过繁华的街道，医院大厅人来人往，也许其中很多人只是出来散步，只是会朋友，只是喝茶，只是做着再正常不过的事情，但我看起来都有一种异样感。我的脑子已经不大转动，但只是纠结一件事，我为什么当时不坚持把那四个字的回复发给汉波呢？今天，汉波看了这条消息，她也许会认为，钟扬为什么要这么傲慢呢？为什么一封私信居然七天不回？而她想过没有，她的这条短信，我可能永远不会回复，那将会是一种什么样的情景。谢谢汉波，你七点二十七分的来信，让我记下了试图回复的时间，给我提供了一个医学上最简单的坐标，尽管这一点都不是你发信来的本意吧。当我被手推车推进急救室，再送进CT检测仪的一瞬间，我开始感受到身体内密密流淌的鲜血，这当然是幻觉。只有当CT的片子出来，我才看到从大脑破裂的血管中流出的殷红血迹化作的CT片上一块块惊人的白斑。

当我被送进急诊病房时，我还在思考着5号下午去香港的飞机。说来也巧，5月8号下午有个会议，所以我决定不跟复旦代表团一起行动。负责修改行程的外事秘书倩洁问我："您的票已经改好，真的不要再改了吗？您的行程安排得太紧了。要是再改，可能买不着票了。"我跟倩洁并不太熟，就回了一句："不用再改了，补票费我出。"倩洁急忙辩解说："钟老师您误会我的意思了，我只是希望您不要再改了，您作为领队在香港已经没有多少休息时间了，钱肯定不要您出。"我回了一句："谢谢你！我当然也是开玩笑。"对香港访问事宜最后的讨论定格于七点差一分。二十分钟以后，这件事却发生了根本的转变。

现在回想起来，在这样一个平凡的节日之夜，参与此间的每一个人，我，我的朋友，随后赶来的一位同事，我们都做到了最好，几乎毫无偏差地实施了最佳的医学抢救——最快的时间，最准确的判断，离我们最近、质量最好的医院，而其他的医疗行动，只能留给明天了。时钟很快拨到了九点，在长海医院的急诊室里，我寄居于一个墙角的床上，那位大校整整陪了我一夜。这一夜是我的内心极度狂乱的一夜，我没有做好任何思想准备，没有对工作上留下的那么多报告，要参加的会议，出发前要见的学生，等等这一切做好交代。就像一条在海中不知疲倦地畅游的鱼儿，一下子被抛到了沙滩。这夜我一晚没睡，血压在200左右，护士也没有办法帮助我，只能希望出血点能止住，不要再扩大了。我右边的手脚已经不听使唤了，我唠唠叨叨地跟边上的人说着一些话，但由于口齿不清，事后我才知道，当时没人能听懂我在说什么。

九点左右，大儿子来了，他的身高已接近一米七，是一个阳光、健壮的小伙子。但不知为什么，我老是觉得，他是一个真正的小孩儿，没有任何人生的阅历。在今天，他显然吓坏了，甚至大人们在走廊外议论我的病情时，他也守着我，默默地不肯离开。他的手已经像大人一样大，但是没有力量，我的右手已经完全离开了我的身体，只能用左手摸着他的头顶，就这样不说话地待着。在这个浑浑噩噩的夜晚，我仿佛又看见小毛拎着凉粉，穿过街道的瘦弱身影。据说，那夜，小毛打了一夜电话，也没有听见我的声音。他无法想象，五个小时以前见过的父亲，已经到了一个陌生的医院，当了一个他出生以来就没见过的病人。他也许不得不要开始走自己的人生道路了。想到这，泪水禁不住浮上了我的眼眶。

未完待续。

本文写作于2015年5月，未正式发表。

复旦博学文库（第一辑）总序

为了进一步提高复旦大学人文社会科学高层次人才培养的影响力，传承中国文化和社会科学研究精神，展示我校博士研究生培养成果，复旦大学研究生院、党委宣传部、复旦大学出版社决定从人文社科类博士研究生学位论文中挑选一批优秀作品，以专著形式出版。首批入选的六篇博士学位论文，就是其中的代表。

总体看来，入选第一辑"复旦博学文库"的论文不仅涵盖面较广，涉及哲学、新闻、历史地理、国际关系、社会发展以及管理科学等领域，研究成果也体现出作者独特的学术视野和研究的深入程度。例如，李甜博士的《明清宁国府区域格局与社会变迁》，注重乡土文献的收集以及材料的准确释读，使其结论建立在坚实的文献及详细考证基础之上，将历史人类学方法引入历史地理学研究中，具有一定的创新性。又如，林青博士的《阿尔都塞激进政治话语研究》，围绕阿尔都塞思想中意识形态理论，将其置于"五月风暴"背景下考察其思想转变，全面剖析了阿尔都塞新政治逻辑的方法论和哲学基础，在讨论阿尔都塞理论的学术效应及其遗产方面取得了突破。此外，我们很高兴地看到，赵清俊博士的《纳米生物制药领域的创新绩效评价与机理研究》在交叉学科研究方面开展了有益的尝试，成为本辑文库的亮点之一。

需要说明的是，入选本辑文库论文的指导老师们也都具有较高的学术造诣。尽管每篇论文都是各位博士的独立之作，但这些成果与其导师的精心指导亦是分不开的。

编辑和出版"复旦博学文库",对我们探索中国现阶段如何培养高质量的人文社科类博士研究生具有促进作用。近年来,我国所培养的文科博士研究生数量在全世界名列前茅,这一方面反映了我国人文社会科学研究的繁荣,另一方面也让我们不免担忧所培养的博士研究生质量是否存在问题。从国家和上海市教育管理部门的要求以及社会对高层次人才的需求来看,在控制招生数量的同时,抓好培养过程的关键环节,做好学位质控工作业已成为目前博士研究生教育的"重中之重"。我们的博士研究生们也应当清楚地意识到,博士研究生阶段的学习与研究是一个十分艰苦的探索过程。每一项具有一定深度的研究成果,均是师生们反复斟酌选题、认真设计方案、仔细分析结果后所获得的,是他们的智慧和努力的结晶,也是随时间而积累的产物。事实上,博士研究生们为修改和完善论文而延长培养期限的情况也日趋普遍。尽管此次入选的论文还存在一些写作仓促的痕迹,但从总体质量上可以作为我校人文社会科学类博士论文的标杆。毋庸讳言,在当前较为浮躁的社会风气影响之下,许多科学研究中充斥着浮光掠影式的所谓"成果",甚至学位论文造假、抄袭等学术不轨行为也时有发生。出版"复旦博学文库"的初衷就是希望扭转这一现象,对提高我校博士研究生论文质量真正起到引领作用。

衷心祝愿我校研究生教育工作不断发展,收获越来越多高质量的博士学位论文,也期望"复旦博学文库"越办越好。

本文是钟扬教授为复旦大学出版社出版的"复旦博学文库"(第一辑)所写的总序,写于2015年10月,收入本书时略有修改。

研究生培养质量提升的解决之道

研究生的能力和素质培养,现在已经成为一个越来越受关注的话题。为此,教育部曾多次下发文件,要求高校重视研究生的培养质量。

事实上,研究生质量问题,主要是研究生的能力。这并非一所、几所大学或者哪一个部门能够完全办到,这是一个庞大的系统工程。

大学并非无限责任公司

研究生的能力培养贯穿整个教育过程,并非只在研究生阶段。只不过研究生阶段已是"教育接力赛"的最后一棒,前面积累下来的问题也许此时集中体现。我们的上级教育部门由于管得太多、管得太细,实际上已经将大学变成了一种无限责任公司——希望每个毕业的研究生都要成为一个各方面优秀的人。但一个研究生学习阶段也就只有两三年时间,而且每个人还有很多不同的想法,研究生教育只能在有限时间达到有限目标。人才培养问题都要等到研究生阶段解决,为时已晚。

比如,现在强调要关心研究生的心理问题。其实,有些研究生的心理问题在童年时期就已造成。仅仅因为他考取了研究生,似乎这些问题也要研究生阶段解决,这不太现实。只能在研究生招生过程中增加心理健康考查内容。

在我看来,我们的研究生教育首先要正确地认识到研究生有哪些能力是我们必须培养的,有哪些能力是我们培养不了的。对于不能在研究生阶段培养的能力，我们不应该浪费太多时间。而对于某些研究生，他们已具备很好的能力，甚至超越了导师，学校就要善于发现这种学生并且为他们提供最好的条件，让他们得到最好的发展。这样，研究生教育才会有意义。

不过，根据世界各国研究生教育发展的经验来看，中国在最近30年的研究生教育中所遇到的培养质量问题，也是西方国家200年来已经遭遇过的，可以说是成长中的烦恼和痛苦。

数据显示，现在我国有超过400所大学和研究所有博士学位授予权，2012年全国招收博士生67 216名。而1982年左右，我国第一批博士学位获得者仅17名，1984年全国博士生招生人数大约是1 000名。也正是因此，1985年一些大学开始成立研究生院。但谁能料到，30年后就会变成6万多名博士。记得2012年底我去国外参加研究生培养研讨会时，当我说出这个数据时，把国外的大学都吓着了。我们的解释是我们国家发展的迫切需求使得我们必须有这么大的量。

但有目共睹的是，研究生数量上升却伴随着质量的下降，这已成为目前研究生教育遇到的最大挑战。最直观的例子是，过去博士并不那么容易遇到，但现在要遇到一个博士很容易。而且现在博士一般不那么爱讨论专业，不管做什么研究的都是如此。甚至你与他交谈很久也很难感受到他是一名博士，也许就是文凭之外的所谓气质不像吧。

研究生质量提升必须直面问题

为了检查研究生培养质量，复旦大学做了两次"问题驱动型"调查，来查找研究生培养中存在的问题。之所以是"问题驱动型"，是因为我们必须直面研究生培养中的真实问题，只有发现问题，才可能解决问题。

我看过很多质量调查报告，都是谈到大部分情况（如95%）是好的，少部分（如5%）是有问题的。但我感觉未必如此。为此，我们在全校范围内访谈研究生，调查受教育者的感受，他们如何看待研究生学位，希望得到什么，哪些部分没有得到满足……从被教育者的角度来审视我们研究生培养中的问题。

这个调查针对全校二年级研究生。之所以选择二年级是因为，一年级学生充满了梦想。二年级往往是"纯粹的梦想"开始破灭，这时研究生自己也会意识到问题很多，他们也有解决问题的愿望。而到三年级，学生就开始为工作或者深造烦恼，无暇思考这些问题了。

过去的调查是看打分和综合，但缺点很多。比如，一个专业25个研究生，三年后发表25篇论文，人均一篇论文。但仔细一看会发现，这25篇是15个同学写的，剩下10个学生一篇论文都写不出。所以一定要一个个谈，每次调查都有上千人。

部分教授也参加调查。调查中有一个很能说明问题的例子，文科的学生反映师生关系不太密切，很多情况下学生可能两三个月没见过导师。而理科学生反映导师一天见他三遍，甚至把办公室搬到实验室门口，老问学生实验做得怎样，学生"压力山大"。那么，如何了解文科的老师和学生是不是在研究指导上配合密切？其实很简单。调查学生，在过去三个月中，导师和他之间讨论最多的科学问题是什么？调查好学生后，可以给老师打电话：在过去三个月中这个学生和你讨论的问题是什么。有时候，老师和学生说的就是牛头不对马嘴。老师有时还替学生打马虎眼，但学生非常大胆，他们如果对导师不满，往往会直接告诉我们，这个老师我真的不喜欢，他总是忙自己的事情之类。所以这样的调查还是很真实的。

那么，院系对研究生培养的态度是什么呢？凡是要研究生指标的多是院长来，软磨硬泡，希望增加指标。谈研究生质量，来的一般则是副院长。不过调查过两轮后，现在情况有所改观了。

研究生和老师其实是学术共同体，研究生培养的问题解决了，学科和科研的质量也会上升；否则，教授再好、硬件再好，学科质量也很难上升。最明显的例子是，有一个院系虽然有国家重点学科，但在985高校中排名不理想。一所比我们学校排名高的大学的院长决定加入本校。引进过来若干年，还是问题多多。调查发现，我们可以购买硬件，可以引进团队，但我们无法引进学生,因为学生和教授是一个环境内的共同体。这个共同体变了，结果也会变。

上　编

研究生培养质量四大问题

我们的质量调查发现以下四个问题，我相信这四个问题在很多学校研究生中都存在，只是比例多少。

第一，缺乏远大的理想和人生目标。远大理想和创新人才、行业颠覆性人才培养密切相关。退而求其次，很多学生来读研究生却连自己想干什么都不知道。很多人说，我妈觉得我现在工作不好，希望我读个研究生。这算什么目标？因为找工作难，所以来读一个研究生。这不是把研究生事业给毁了吗？

第二，缺乏从事科学研究的热情。很多人来问我，你们专业好不好找工作，我小孩要不要读博士。以前我觉得这个问题很难回答。现在，看了那么多研究生访谈报告后，我终于找到了一句回答的话——他不要读研究生，尤其是博士。

如果你想读博士，那么最好问一问自己的内心，你曾经心头燃烧的科学研究之火是不是已经熄灭。如果这团火熄灭了，那无论是好导师还是好大学都无法让它燃烧。如果火不大，那么我们可以把它放大，但如果已经熄灭了，那还读下去，基本上是没救了。这也是导致科学研究水平低下的原因。

麦可思教育调查发现，目前学术学位研究生当中只有四分之一真正有热情，我个人认为这个估计还有一点偏高。我经常问导师，你们实验室招的研究生有没有一个特别爱科学？如果没有我深表同情，因为低于这个比例。如果真的有一个就恭喜你。要是每个学生经过你的教育都特别爱科学，那你应该买彩票，一般的导师都没有如此好运。

第三，缺乏科学论文写作的训练。这件事太普遍了。我们今年开始下大力气解决，就是想让研究生写出来的东西像一个研究生写的。我们要成立研究生论文写作服务中心。为什么？因为我坚信论文写作是可以训练的，不像上面两个问题是不能训练的，只能在招生时多一双眼睛，同时去碰运气，后面这个是可以训练的。

第四，缺乏必要的时间和精力投入。这是管理问题。我们扪心自问，到底师生双方有多少时间和精力的投入？有的人虽然在职，但想着学习的事。

有的人虽然是脱产,但灵魂完全离开了这个校园,这样的人在以前是不应该读研究生的,但现在,在我们队伍中有很多。

在解决我们的问题之前,可以先学习一下国外的先进经验。当然,研究生培养还要坚持特色,要有信心。其实,国外高等教育成熟的一个标志是绝不会有任何一个一流的大学在所有的领域都保持一流。因此,大家也不会看到一个二流的大学在所有的领域都是二流。

研究生培养问题解决之道

问题说了那么多,有些在短时间内还很难解决。但无论如何,我们的教育工作者不能放弃自己的责任。即使知道研究生的有些能力我们不能直接去培养,但我们要尽可能地提升和改造他们的素质。

第一是发现问题的能力。很多研究生之所以不像研究生就是因为缺少发现问题的能力。

现在很多研究生做论文,总需要老师先提出问题。如果研究生自己不能发现问题,以后到单位如何开展工作?对一个理工科学生来说,如果能够准确地发现问题,那问题已经解决一半了。不能发现问题的学生往往文章也写不好。我看过很多研究生的论文,发现他们最写不好的就是引言部分。如果你问学生为什么研究这个问题,学生第一句就说是老师让我做的。

而且不能解决的问题不是问题。所有要研究的问题不是漫无边际的,必须是已经重新定义并可以尝试解决的问题。

第二是选择课题的能力。做课题就是让研究生从发现问题逐渐走上解决问题之路。

一些老教授的话很有道理:硕士是刚刚从通识教育走向科学研究的第一步,因此他所挑选的课题应该让他在从事科学研究的第一天起就尝到科研的甜头,发现科研的乐趣。这不同于将未经训练的学生当成某种形式的劳动力并直接投入高强度的科研攻关工作,两者甚至有天壤之别。

第三是交流与写作的能力。

很多学生在写作上达不到研究生的标准。这里说的写作,中英文是一致的。我发现,写作不好的学生往往也不喜欢阅读,这在很大程度上限制了他们长远的发展。甚至一些学生缺乏逻辑，平时说话做事都是颠三倒四的。

我们刚成立的写作服务中心正在联系一批教授，他们愿意花时间去帮助自己和别人的学生提高写作水平。我们期望学生匿名来问诊，看自己论文写作水平达到什么级别，再有针对性地进行训练。

如何诊断呢？比方逻辑问题。我们挑一个真实的话题，拿关键词到百度上去查几段话。每一段话在200字以上。为什么200个字？我们发现现在学生的思维不连贯和手机100多个字的片段化有关。而一个研究生如果每次还只会按100多个字思考和论述问题，那肯定不够格。

我们把几段话顺序打乱后，请同学们来重新理顺。我们很多硕士生和博士生都理不顺。这与专业无关，这是逻辑问题。从这一点看，我们的研究生培养问题有多严重——即便准许学生抄论文，很多人都抄不对！

第四是团队组织和协作的能力。研究生阶段若不培养，以后也许再也无法培养了。

我经常抽时间到中学和小学做报告。我始终认为，目前教育上忽视得最多的是团队协作能力的培养。我们的研究生甚至发自内心地认为，从幼儿园起，他的成功都是靠个人取得的。事实上，我们的应试教育不断强化了这种意识——你的每一次考试成功都是你一个人努力的结果，不需要和别人合作。高考如此，考研还是如此。所以他一点都没有想过，在他今后人生道路上需要跟人合作。但事实并非如此。

对大多数科学家来说，团队组织和协作应该是他所有能力中最强的能力。为什么？因为它在一定程度上违背了人（包括科学家）的本性——能不与人合作就不与人合作。可是，现代社会的很多工作的确需要我们去克服这个人性的障碍。

所以，我们一定要在研究生阶段为学生补上这一课。否则的话，如果他在研究生阶段没有在这方面受过训练，今后在工作单位很快就会暴露出这一能力的缺乏并尝到苦果。

本文发表于《文汇报》2015年11月20日第7版，收入本书时略有修改。

弘扬劳模精神　培育创新人才

一、发扬劳模精神和劳动精神，全面建成小康社会

在我的身边，有一大批同行刻苦工作、锐意创新，特别是在我所研究的生命科学领域，我国的研究水平与世界先进水平的差距在不断缩小，这与一代又一代科研工作者的艰苦奋斗是分不开的。习近平总书记说："做研究，就要甘于寂寞，或是皓首穷经，或是扎根实验室。"夜晚，实验室那一盏盏不熄的灯光就是这种精神的真实写照。在条件相对艰苦的西部学校，在堪称"世界第三极"的青藏高原，我们更是要弘扬劳模精神，克服常人所不能克服的困难，发挥劳动创造的巨大能量，才有可能改变落后面貌，实现西部经济建设和社会进步的跨越式发展。

对年轻一代而言，劳动精神更需要大力提倡。无论在哪个领域工作，也无论收入多少，劳动创造财富的理念不能改变。目前，高校学生普遍追求丰富的个人生活和优越的物质条件，这是社会发展使然，无可非议，但必须通过劳动来创造，而不是坐享其成。事实上，不劳而获的思想现在还大有市场，比如我们大力纠正但屡禁不绝的学术不端行为和论文抄袭现象，就是部分人想走捷径，不劳而获思想作祟的具体表现。

二、提升创新能力，培育创新人才

当前，国内外、校内外学习深造的机会越来越多，为广大青年人的成长提供了诸多选择，但是提高学历还是提升能力是有一定差别的。换句话说，学历高并不等于能力强。我们曾经对本科生甚至研究生毕业找工作难的问题进行过较为细致的调查，发现有些专业的学生就业难的一个重要原因是由于学校在学生能力培养方面存在缺陷，理论和技术脱离实践的弊病十分明显。例如，有些专业的导师和研究生热衷于做"高大上"的研究项目，把发表论文作为科技创新的唯一指标，而该专业在我国工业制造领域的重要性主要集中在工艺和集成创新方面，这就需要我们的师生深入到企业和经济建设的第一线，学习和研究书本上缺乏的实际问题。另一方面，我们的确有许多工作在生产第一线的劳动模范，他们虽然学历不高，但经过长期学习和刻苦钻研，已逐渐成为技术革新能手和发明专家，成为我们创新创业和经济建设不可缺少的中坚力量，他们的发明和创新对我们社会发展所产生的巨大价值是许多并不解决实际问题的论文所不能比的。作为新时期的教育工作者，我们不只是向青年学生灌输知识，而是应当把提升青年学生的创新能力放到更为重要的地位，把培育创新人才作为高等教育改革必须下大力气实现的目标。

三、充分发挥教科文卫体工会的特殊作用

在产业工会中，教科文卫体工会是党政与知识分子沟通联系的桥梁和纽带，广大教职工对工会发挥的作用寄予了厚望。在新的形势下，如何发挥工会的作用值得我们去研究和尝试。除了发扬我国各级工会密切联系群众的优良传统外，还可以借鉴一些发达国家工会在保护职工权益方面的经验。例如，近年来我们从国外引进了大批优秀人才，也难免会遇到一些与国外原单位或竞争者的纠纷。国外通常以法律诉讼方式解决这类问题，而我国习惯于由所在单位出面进行行政调解。我们发现，在充分保护引进人才的

权益并与外方进行有效沟通和协商方面，工会在很多情况下可以发挥比校方更为积极的作用。因为在国际事务中，不同国家的工会组织为自己的工会会员争取权益是一种更高效、更人性化的方式，并能有效地规避行政风险。我们发现国际上已有相对成熟的做法，而国内高校和科研单位在这方面才刚刚起步。总之，教科文卫工会在维护职工权益方面还有许多工作可做。

本文发表于《中国教工》2016年第6期，收入本书时略有修改。

一个招办主任儿子的高考

这是我所经历的1979年的高考：全省录取率不到4%，我所在班级80%的同学是农村户口，一半考上了北大、清华和科大。而除了我，参加高考的省地市招办主任的孩子，竟无一人上大学……

在当今中国，1 000个人就至少有1 000个高考故事。

一

我的大学梦源于38年前的早春二月。1977年恢复高考后的第一届大学生即将入学，我所在的中学在大操场上举行了隆重的欢送仪式。锣鼓声、鞭炮声、欢呼声震耳欲聋，4名考上大学的同学胸佩大红花，精神抖擞地站在高台上，接受学校的表彰和师生的夸赞。他们还不到我校应届高中毕业生总数的1%，却成了全校2 000多名学生心目中真正的英雄。作为一名即将毕业的初中生，我仿佛看到了人生的希望和前进的榜样，那从未走进过的大学校园对我而言似乎也不再遥远了。

当时，有一句话叫"路线对了头，一步一层楼"，用在我和我的小伙伴们的学习上还算贴切——寒暑假再不东跑西窜了，而是自己找一个空无一人的教室，一边手抄文革前的各种习题，一边查教科书上的答案，反复做练习。不用父母催促，没有补习班，也没有补课教师，更没有花冤枉钱，一个假期下来，一门不及格的课程就"补"上了。自己偷偷一盘算，照这样再努力两年，上大学还是很有希望的。

不料，新学期一到学校，我顿时傻了眼。班上涌进了一大批家境贫寒、拼命读书的农村孩子。记得全班年龄最小的那位同学，来校住读时连被褥都没有，是班主任老师动员同学们四处找报纸和稻草铺在他的床上，才使他艰难地度过了寒冬。但一考试，这个身材瘦小、木讷寡言的同学立马就像变了一个人，成绩总是遥遥领先。就是这批农村同学，把我一下子从班上第二名挤到十名开外。更令我喘不过气来的是，他们决定提前一年参加高考。对当时急于"跳出农门"的一代人而言，高考确实是改变命运的唯一出路。在他们的激励和"裹挟"下，我不得不奋力追赶这支奔跑的队伍，期盼早日和他们一起实现自己的大学梦。

转眼就到了1979年高考报名的日子。与班上成绩拔尖同学（就是今天所谓的"学霸"）固有的差距，加上因祖母去世随父母去料理后事耽搁了一些课程，使我对自己信心不足。尽管如此，我还是想和同学们一起去搏一搏机会，提前参加高考。

二

从1974年起，我的父亲开始担任地区教育局高等院校招生办公室主任。在这个中部省份的贫困地区，出生农家而又刻苦攻读的寒门子弟比比皆是。1977年前，父亲负责的招生工作完全是推荐工农兵学员之类的政治任务。恢复高考后，高校招生成为了全社会关注的敏感工作，父亲不仅经常通宵加班，而且每年高考录取期间都要被借调到省招生办公室"全封闭工作"一两个月。说实话，那时的高考实在单纯，不仅没有如今这么多花样的免试推荐和加分，而且全国统一试卷、全省统一录取分数线。

不过，当时也碰到过所谓"公平性"的问题。由于国家没有明文限制提前高考，后来居上的低年级同学似乎对应届毕业生和往届毕业生形成了威胁。我所在的中学就有大概80名非毕业班同学想报名参考，这引发了部分应届毕业班同学和家长的担忧，他们向省招办反映了此事。省招办迅即责成我父亲去了解情况,并给考生和家长们一个满意的答复。

几天后,父亲找我谈话。从他严肃的表情,我就知道大事不妙。他告诉我,提前高考的情况组织上已讨论过,形成了一个还算公平的解决方案,明天会到学校与师生见面。但无论如何,你必须放弃这次高考的机会。我问父亲:为什么?父亲说,作为招办负责人来处理几万考生关心的问题,他的儿子就非避嫌不可。我无话可说,只是觉得委屈,一摔门就走了。

第二天,父亲到我所在中学来宣布组织意见:凡提前高考的同学必须办理提前毕业手续,在校学生不能违背"机会均等"的原则,也就是说在读生只能参加一次高考。也许是这个方案确实公平,抑或是有人听说我放弃了高考机会,这次协商会气氛出乎意料地好。最后,全校仅21人决定提前毕业后参加高考,应届生和往届生也都表示满意。省里得知此事,迅速面向全省数十万考生推广我们地区的处理方案并将此原则沿用多年。

我却一直想不通,拒绝和父母谈话,一个人生着闷气。班主任老师来家访时对我说,我们都知道这件事对你是不公平的,但你父亲决心很大,大家对他的做法也很服气。你不要再惹他生气了,不能参加正规的高考,去考科大少年班怎么样?

我知道科大少年班是一个举国瞩目的英才计划,一年在全国就招收20多个人吧。头一年,我们地区成绩最好的一位同学都没考上,我能有什么希望?学校建议我去报考科大少年班明显是为了宽慰我失落的心情而已。

不料,父亲竟然同意了。他说,这的确是一场成功机会不大的考试,对你也不太公平,因为科大单独考试、改卷和录取,如果考不上少年班,你也不能像正常高考一样可以被其他大学录取,但你还不到15岁,今后的机会还很多。今年你去试一下,锻炼自己。明年起我退出招生工作吧,直到你能考上,行吗?父亲的话打动了我,我知道他的工作是十年浩劫后一个知识分子难得的机会,放弃招办工作对他不也是一种不公平吗?

接下来的日子异常煎熬,我只能鼓足勇气,迈向了通往科大少年班的"独木桥"。在通过高淘汰率的初审、复试和面试后,我考上了科大第三期少年班。也许是这场考试太过严酷吧,我差一点对自己失去了信心。当我接到录取通知后，我告诉母亲，先不要通知父亲。她知道我的内心深处还是无法谅解父亲。

三

几天后，还在省城招办参加全封闭招生的父亲托人叫我去一个招待所见他。见到父亲后，他告诉我今年的高校招生录取已全部结束，各大学都已离开招待所了，招办明天也要撤出。从母亲那里得知我考上科大少年班的消息，父亲非常高兴，他的同事们也纷纷表示祝贺，都想见见我。父亲叮嘱我说，高考是一件"几家欢乐几家愁"的事，不要太得意忘形，刺激别人。省招办主任孙伯伯的儿子今年已是第二次高考，这次又是只差几分，落榜了。

孙伯伯是一个温文尔雅的上海人，文革前的研究生，但大家都说他做事公正，没有架子。听说他曾为我们地区一个身体条件略差（可能是色弱）的农村考生和几所大学都拍了桌子，最终使这个考生圆了大学梦。当年像这样有能力、敢担当的领导确实都能声名远扬。

见到孙伯伯，他拍着我的肩膀说：小伙子，祝贺你。你给你父亲争了光，也给我们招办的子弟带了个好头。我们在做"为人作嫁"的工作，其实并不想耽误自己的孩子，但有时不好兼顾啊。你父亲跟我谈过你想提前高考的事，本来我们是同意的，但他还是要避嫌，因为我们省的升学压力太大，你父亲所在的地区压力更大，他不得不这样做，你能理解他吗？

四

37年过去了，我的父亲早已退休，但那年高考前后的点点滴滴令我们父子俩至今难以忘怀。孙伯伯任省招生办公室主任和省教委主任十多年，儿子却一直未能上成大学。是的，这就是我所经历的1979年的高考：全省录取率不到4%，我所在班级80%的同学是农村户口，一半考上了北大、清华和科大。而除我之外，参加高考的省地市招办主任的孩子竟无一人上大学。现在回想起来，那时的高考纪律也没有到三令五申的地步，但违法乱纪和徇私舞弊之事并不多。我自己后来当上了大学教授，也多次参与过学校本科和研究生招生工作。与当年相比，如今招生的多元化、个性化、信息化程度更高，领导越来越重视，程序越来越规范，公示越来越详细，但为什么民众越来越紧张，质疑也越来越多呢？

说实话，我极不愿意参加各种名目繁多的高等教育改革会议，尤其是涉及所谓教育公平性的讨论。大量的会议几乎耗尽了我们这个民族宝贵的时间、智慧和耐心。其实，我们每个人心里都清楚，任何时候任何地方都不存在绝对的公平，教育如此，高考亦然。但是，倘若决策者和管理者的子女总能规避不公平并且又总是让普通人承受不公平的话，我们的制度肯定有什么不对劲的地方。不信？甭管多么重要的会议和多么贴心的沟通，究竟还有多少侃侃而谈的领导和专家们能同大家分享一下自己孩子真实的教育故事呢？如果真能这样做的话，无休止的会议一定会少掉许多。即使非开会不可，讨论的时间也将大大缩减。

本文发表于2016年5月28日《文汇报》第六版，后授权"知识分子"微信号转载，署名索顿，收入本书时略有修改。

复旦博学文库（第二辑）总序

2013年7月，复旦大学研究生院、党委宣传部和复旦大学出版社决定联合策划出版"复旦博学文库"，计划从每年毕业的人文社会科学类博士研究生的学位论文中评选若干篇优秀论文，以学术专著的形式结辑出版。2015年出版的"复旦博学文库"第一辑即收录了六篇博士论文。

今年，又有七篇博士学位论文入选"复旦博学文库"第二辑，论文作者分别来自哲学学院、马克思主义学院、外文学院、管理学院和历史地理研究中心，研究方向则涉及马克思主义哲学、思想政治教育、英语语言文学、传播学、历史地理和企业管理等学科。这些论文是在各学位评定分委员会初评后推荐的二十余篇论文基础上，经学校评选委员会遴选产生的。无疑，它们代表了我校目前人文社会科学优秀博士学位论文的水准，反映了我校博士研究生们在这些学科领域中创新性研究的广度与深度。

尽管我国的博士研究生教育已取得了长足的进步，但博士论文质量不高的问题依然十分突出。从近年来各级论文盲审和抽检等渠道所获得的信息看，存在严重缺陷的博士论文比例依然居高不下。这些问题大致涉及以下几个方面：一是缺乏独到的见解和细致的分析，许多研究工作虽然不乏某些闪光点，但只对一些表面现象和统计数据进行浮光掠影式的探究，缺乏有洞察力的认识；二是投入研究的时间不够，人文社会科学研究通常需要较长时间的积累，但一些研究生原有的学术功底不太扎实，攻读博士学位研究生期间在学习和科研上投入时间又不够，达不到博士学位论文的要求；三是写作不规范，缺乏缜密的逻辑和推理。更有甚者，学术违规行为和抄袭现象屡禁不止。因而，进一步端正学风，提高博士研究生培养质量，成为了我校导师、研究生和管理者需要共同承担的艰巨任务。

我们衷心期望"复旦博学文库"第二辑的出版不仅能促进各相关学科领域的发展，而且能为人文社会科学方向在读博士生们的学习和科研提供可资借鉴的范例，促进我校博士研究生学位论文整体质量的提高。

本文是钟扬教授为复旦大学出版社出版的"复旦博学文库"（第二辑）所写的总序，写于2016年6月，收入本书时略有修改。

钟扬老师给家长的一封信

你好,我是"植物家族历险记"板块的主讲人,复旦大学教授,钟扬。

在《我和"科学队长"》那篇文章中,我提到了自己小时候的一些经历。在我小学二年级的时候,为了验证我在书上读到的有关电池的知识,我用家里的盐和醋去改造家里手电筒的电池,结果这些电池统统报废了。我的母亲是一位中学化学老师,当她看到这一堆废电池时,并没有责骂我。而且从那以后,她常常带我到中学化学实验室去观摩实验课,还带我去蓄电池用硫酸厂参加生产。这对我以后的化学学习起到了意想不到的作用。

这个真实的故事引起了很多家长的兴趣。当然,也有些家长更加焦虑,因为他们既不是中学化学老师,也不知道应该把孩子带到哪里才能接触到科学知识。我理解他们的心情,想跟他们说几句话。关于如何让小孩子们学好科学知识,我在这里很想表达两层意思:

第一,对不同的人而言,学习科学的意义可能不一样。换句话说,并不是每个孩子都要当科学家,而不同的科学家其实特点也不一样,没有统一的培养模式。我认为,在目前的科学教育中,最缺乏的还是实践能力和批判性思维。至于是学好化学还是学好数学,哪门科目在中考或者高考中所占的比例更高,其实并没有那么重要。就实践能力而言,学习物理、化学和生物中所接触到的实验比单纯做练习题要有趣得多,也更适合动手能力的培养。遗憾的是,目前,实验反而成为了我们基础教育的薄弱环节。就我所知，其中一个原因是这种教育对教师的要求比较高。至于批判性思维，最重要的是"怀疑一切"的态度，包括质疑书本知识的正确性。前段时间，我在北京史家胡同小学讲了一次课，当我提到《十万个为什么》上的"错误"，以及怎样通过实验来检验这些错误时，孩子们感到非常惊讶——这些孩子即将小学毕业，在此之前，它们从未听一位老师说过书上居然还有错误。

我要表达的第二层意思来自我的观察。我发现有些家长的教育方式应该调整，他们特别喜欢帮孩子们制订时间规划。例如，孩子周末的什么时间段该看什么书，做什么事，都由家长一一安排，有些家长甚至把时间安排精确到了分钟。我觉得孩子这种时时刻刻都在家长掌控之中的成长，不一定会达到预期的效果。家长的基本理由是孩子太小，自己不会科学地规划时间。其实，绝大多数生物个体都可以通过对环境的适应来形成有利于自身发展的节奏和节律。对于什么时间应该学习什么科目，什么时间身体需要休息，孩子们自己应该逐渐体会到自然的适应过程，而家长们过度的"好心干预"，有时会干扰了其学习节奏和生活节律。作为一名生物学家，我认为，孩子们保持自己适合的节奏和节律非常重要，但目前很多父母并没有认识到其中的重要性。

尽管我们不主张父母在孩子的科学学习过程中过度地预设目标或者干扰其节奏，但父母的陪伴还是应该得到鼓励。如果在收听"科学队长"音频时，父母能够和他们一起分享科学知识，孩子可能更容易体验到学习的乐趣。我们祝愿每一位孩子都能够健康成长，快乐学习，和父母共同收听"科学队长"。

本文据"科学队长"微信公众号音频节目"植物家族历险记"音频内容整理而成。

达尔文进化论过时了吗？

达尔文进化论是我们用科学思维去系统理解生物进化之谜的起点。不过，学习和研究进化论的人常常要面对两大疑问：一是达尔文进化论正确吗？二是达尔文进化论过时了吗？

过去100多年来，围绕第一大疑问的争论虽从未停歇，但主战场已逐渐从宗教和信仰之争转向科学证据之争。以古生物学领域为例，不断涌现的科学事实在弥补达尔文当年未曾发现的"缺失环节（missing link）"的同时，也在检验着达尔文进化论的预测能力——这是科学理论的一个重要标志。例如，2004年在加拿大发现的"大淡水鱼（Tiktaalik）"化石揭示了从鱼类（鳍）到陆生动物（腿）之间的过渡状态，被公认是"种系渐变论（phyletic evolution）"的一个极好例证，因为其发掘工作是以渐变论为依据，在潘氏鱼（Panderichthys）和棘螈（Acanthostega）化石地质时代之间的地层完成的。当然，达尔文进化论并非完美无缺，它确实存在"可证伪"之处——这恰好是科学理论的又一个重要标志。以自然选择理论为例，它在孟德尔遗传学再发现后和分子进化（molecular evolution）的中性学说（neutral theory）建立之初就受到了强烈挑战，但各种不能用自然选择理论简单解释的新证据最终还是拓展了人们对进化动力和机制的认识，而不是摒弃该理论。

第二大疑问则涉及达尔文进化论在今天是否依然"合用"。相关争论虽局限在科学界内部，但也异常激烈。逻辑上，任何一种科学理论随着时间的推移和时代的进步都会面临"过时"的问题——只要这个理论被后来的学者批判过或修正过。显然,这一标准未免太过绝对。就实际的科学研究工作而言,不妨采取一些相对标准或者"程度"指标来判断一种科学理论是否"过时"。这些指标可以包括:该理论的逻辑框架现在是否已经动摇?该理论与后来的新理论差异程度有多大?该理论与后来获得的科学事实之间的冲突程度有多大?等等。然而,对于众多研究者而言,即使按如此"粗略"的标准对达尔文进化论与其他进化理论进行比较与分析也绝非易事。一方面,这与我国现阶段的科学教育理念有关,绝大多数人对科学理论还是秉持"黑白分明"或者"非对即错"式的看法,对达尔文进化论更是如此。实际上,仅从今天的认识视角和科学体系去理解前人的各种学说,通常都很难作出公允的判断。另一方面,能搜寻到的研究资料浩若烟海,而有价值的可靠信息又十分有限。的确,在过去100多年间,进化研究被认为是科学界最热门也是最富有争议的领域,各种学说和思想层出不穷,鱼龙混杂,使得我国学者很难在短期内全面了解这一领域的发展状况,厘清其中纷乱的头绪。

所幸的是,最近上海科学技术出版社出版的《进化着的进化学——达尔文之后的发展》(以下简称"进化着的进化学")一书能帮助读者较快了解达尔文之后进化科学不平凡的发展历程。该书详细梳理了一个半世纪以来进化论的"进化轨迹":首先,19世纪的科学哲学思潮对进化论的理论框架产生了重大影响,尤其是魏斯曼(A. Weismann)和海克尔(E. Haeckel)为此做出了巨大贡献;随后,19世纪下半叶依据古生物学的新发现而发展起来的新拉马克主义(neo-Lamarckism)和定向进化理论(orthogenesis),以及20世纪初孟德尔遗传规律被重新发现后兴起的"孟德尔学派(Mendelism)"对达尔文进化论形成了巨大的挑战;其后,数量遗传学(quantitative genetics)和(实验)群体进化学(population evolution theory)的出现又使进化论"柳暗花明",尤其是以杜布赞斯基(T. Dobzhansky)为核心人物提出的综合进化论(synthetic theory of evolution)巩固了解释生物进化动力的自然选择理论;最后,20世纪60年代由日本学者木村资生(M. Kimura)等人提出的分子进化的中性学说,与综合进化论进行了长达20年的争论,最终也获得了分子证据所支持的"半壁江山"。

通读完《进化着的进化学》,读者可以尝试从不同角度将达尔文进化论与其他进化学说进行比较,来判断达尔文进化论是否已经过时。这里,不妨按前述三条标准作一些简要说明。在达尔文进化论的理论构架方面,美国学者迈尔(E. Mayr)在其晚年进行了最为系统的研究。他将达尔文进化论从逻辑上归并为五个理论:物种可变理论、共同祖先理论、渐变理论、物种增殖理论和自然选择理论。其中自然选择理论作为达尔文进化论的核心原理,至今仍是进化生物学不可动摇的科学基础。可以说,没有自然选择理论,整个进化生物学也就无法发展。

在达尔文进化论与后人建立的新理论之间的差异程度方面,可以看看美国古生物学家辛普森(G. G. Simpson)在1982年纪念达尔文逝世100周年的文章中所举的一个例子。达尔文将古生物学作为生物进化的证据,但他在世时所能掌握的古生物材料少之又少,这曾使他困惑不已。达尔文去世后,生物化石的数量、种类及相关知识有了令人惊讶的增长,人们逐渐积累了很多可以验证进化速率与方式的标本。基于此,越来越多的争论涉及"渐变论"和"间断平衡论(punctuated equilibrium)"之间的比较。初看起来,达尔文所建立的渐变论认为生物跨越地质年代的变化是渐进而缓慢的,而间断平衡论则认为这些变化是快速进化与进化速率几乎为零的不间断反复,因而它被人们看成是一种新的理论。然而,辛普森认为,间断平衡论不过是通过挑选出两个极端而削除许多中间的实例而已。换言之,达尔文已经注意到古生物学事实,但缺乏足够的材料,而间断平衡论并非是革命性的想法,它只是从构成系列的古生物证据中挑出了两个极端,再构造了一个新的名词(术语)。

在达尔文进化论与后来获得的科学事实相冲突方面,最好的例子莫过于分子证据了。诚然,达尔文本人不可能对100多年后兴起的分子生物学有任何见解,而当木村资生等人将核苷酸和蛋白质数据置于群体遗传学的数学理论(尤其是基于扩散模型的理论)中时,发现了用达尔文进化论中的自然选择理论无法解释的情形,于是提出了分子进化的中性学说——分子水平的进化与变异，与其说是自然选择起作用，还不如说是突变和遗传漂变在起主要作用。大量分子数据的观察表明，进化过程中不使氨基酸发生变化的置换（同义置换，synonymous substitution）比使氨基酸发生变化的置换（非同义的置换，nonsynonymous substitution），其发生频率高得多，而为各种不同蛋白质编码的同义置换速率几乎相同，具有明显的"分子钟（molecular clock）"特征。因此，中性学说认为随着功能限制减少（即功能重要性降低），核苷酸进化速率将收敛到由突变率所决定的上限。换言之，分子水平上认为越是难以与自然选择发生关系的性状，其进化速率越快，这就是导致自然选择理论不能用于解释分子进化的原因所在。

有趣的是，木村资生本人在坚持分子进化就是随机固定中性（或近中性）突变的同时，也认为表型进化依然遵循达尔文的自然选择原理。根井正利（M. Nei）进一步认为表型进化最终可以由分子进化来解释，而自然选择无论在分子进化还是表型进化中只有保留有利突变和消除有害突变的作用。而随着分子生物学的发展，突变的含义已得以极大延伸，甚至涉及调控网络变异、基因重复与丧失、不同基因重组和基因组重构等复杂进化现象。越来越多的证据表明，这些新的突变类型与表型进化相关，但在很大程度上已不同于中性学说创立之初所提出的由随机遗传漂变所固定的模式。因而，中性突变和自然选择在分子进化中的相对重要性都需要接受新数据的检验。同时，近年来通过对分子水平的突变进行统计分析以检验自然选择存在与否的方法论研究也取得了长足进步。总之，在分子进化研究领域亦不可轻言自然选择理论已经过时。

《进化着的进化学》作者庚镇城先生几乎是我最早认识的复旦遗传系教授。20多年前，我还在中国科学院系统工作，与大学教授交往并不多。在一次进化生物学研讨会上首次遇见庚先生，他对分子进化学说的介绍给我留下了深刻印象。几年后，我成为复旦大学生命科学学院教授，开始与庚先生有了较多接触。他虽已退休多年，身患眼疾，但笔耕不辍，尤其是他在2009年为纪念达尔文诞辰200周年撰写出版的《达尔文新考》一书，为全面了解达尔文进化论的诞生提供了新的资料和视角。例如，史料表明达尔文确实收到过孟德尔寄来的论文，但进化论巨匠究竟为何与遗传学创世之作失之交臂，值得思考。2014年，庚先生又撰写出版了《李森科时代前俄罗斯遗传学者的成就》一书，详尽记载了从1918年到1945年苏联特殊时期遗传学非同寻常的发展历程。

今天，呈现在读者面前的这本《进化着的进化学》则以广阔的视野和详细的考证，展现了达尔文进化论一个半世纪以来各种理论思潮的演变以及不同学科证据的融通对进化思想的影响。能完成如此广度和深度的研究与考证工作，与庚先生不平凡的求学经历和不间断的勤奋思考是分不开的。作为改革开放后的第一批留日学者，他曾在东京大学和庆应义塾大学从事合作研究，并与日本国立遗传所和神经生物所等研究机构的学者建立了密切联系。其时，正值日本进化生物学处于国际学术前沿的鼎盛阶段，庚先生如饥似渴地学习国外先进经验和研究成果，并及时介绍给复旦大学师生及国内学术界同仁，对加快我国青年学者的成长，促进我国进化生物学的发展发挥了极为重要的作用。值得一提的是，庚先生在研究和推介国外学者文献的同时，也提出了自己的观点。例如，他曾建议用"中立理论"来替代"中性学说"，尽管由于约定俗成的原因未被广泛采用，但其观点本身是完全正确的。

掩卷而思，我折服于本书作者对进化学史抽丝剥茧般的梳理，也感怀于"结束语"中谈及多年来的著述艰辛。让我们衷心祝愿庚镇城先生身体健康，老当益壮，为我国的进化生物学教育与研究做出新的贡献。

本文发表于《科学》2017年3月第69卷第2期"科学书屋"栏目，收入本书时略有修改。

基因密码

——生命与哲学

科学是一种什么样的游戏

今天要讲的"生命与哲学"只是我在第一线做科研的一点感悟,供大家参考。我想介绍一些新知识和新思维。新思维就是批判性思维。"批判性思维"这个词很时髦,最近,我到上海实验学校给初中学生讲课,标题中也明确用了"批判性思维"。批判性思维实际就是质疑,平时我们要求孩子们不仅要学习,而且要学习正确的知识。可是,科学里面都是"正确的"知识吗?

我给中国科学院上海生命科学研究院的研究生作过的报告有一个视频,标题叫"科学是一种什么样的游戏",介绍了波普尔的科学哲学——科学的基本特征是其中包含错误。最近,我在《科学》杂志上还讨论"达尔文进化论"是否过时的问题。什么叫科学理论过时了?无非就是错了,但错误居然是科学的基本特征。科学的对立面是宗教(信仰),信仰需要证实,而科学则需要证伪。这样就确定了我们科学家的职责。科学家的基本素质之一就是研究科学(甚至读一本书)并找出它的错误。而我们的教育培养的大多数人,却是读了一本书赶紧说这本书是对的,然后背下来参加考试。考试有标准答案,把答案一记就算完了。就这样一直读,读成了博士。我对他们说:"现在写一篇论文,论述前人是错的。"他们就傻了,因为从没考虑过前人的错误。不喜欢批判性思维并不是我们民族天然有缺陷。可是抱歉,科学就是要用批判性思维,没别的道路可走。

举个例子吧,牛顿提出力学定律后,很多人开始不相信,但经过实验后很快就信了。再后来的上百年时间内,牛顿力学都快变成"放之四海而皆准"的学说了。但一个只会背诵牛顿力学的物理学者,在科学界一点地位都没有,因为他违背了科学的游戏规则。他是个会讲解牛顿力学的教师,但不是好的物理学家。

后来,出了一位伟大的物理学家,他的名字叫爱因斯坦。他说牛顿错了,至少在物体运动速度接近光速时牛顿力学是错的。他这么做使得众人都承认他是伟大的物理学家。为什么?第一,他说牛顿错了;第二,牛顿力学因为有错误倒成了真正的科学。换言之,爱因斯坦不仅通过说"牛顿错了"证明了自己是世界上最伟大的科学家,而且也帮助牛顿力学成为科学。

对这一问题的讨论如果要上升到哲学层面,那么仅凭刚才说的故事就不行了。我们要经常性地提一些问题,然后大家一起思考,哪些地方对,哪些地方错。重申一遍,我们的判断有两个基础:一是批判性思维,就是怀疑一切;二是只要是科学的,就一定包含错误。

基　因　科　学

基因是什么?虽然现在"基因"这个词到处都是,比如"文化基因""材料基因"等,但都是借用的。基因是一门学科的基础,这门学科叫作"Genetics"(遗传学),许多以"gene"开头的英语单词都与"生殖"和"生产"有关。

孟德尔

遗传学是由格里格·孟德尔(Gregor Johann Mendel,1822—1884)提出来的,他本来是学自然科学的,但大学没有读完,后来的正式职业是传教士。他离开维也纳大学后去了今天的捷克共和国。他的人生目标是当修道院院长,空余时间管理修道院前面的小院子,他就用这片院子建立了遗传学。

孟德尔种了一批豌豆,然后进行观察和分析。一种豌豆的豆荚是绿色的,另一种是黄色的,如果这两种豌豆杂交会发生什么事？他拿一支毛笔沾上一种豌豆花粉，再送给另一种。当然，蜜蜂也可以做这个事，但这样做实验不准确。无论是自然还是非自然的授粉，绿色和黄色豆荚的豌豆进行杂交的结果可能是黄色，也可能是绿色的。顺便提一句，生物学家就把结果叫"杂（交）种"，杂种有可能产生杂交优势。

通过杂交实验，孟德尔发现了一个规律，假定绿色和黄色的豆荚都是纯种的，它们杂交后肯定是绿色的，因为绿色是显性的，而黄色是隐性的。但是，如果杂交出来的两个绿色再产生的后代就会产生性状分离现象——既有黄色，又有绿色。所以，孟德尔发现，就算杂交再有优势，到了下一代也必须分离。

孟德尔用统计学方法进行分析，最后得出显性和隐性的三比一规律。孟德尔并不知道是什么东西控制遗传变成这个样子，但他猜测一定有一种我们不知道的因素控制着遗传，因此他用了一个词"Factor"，猜测遗传应该存在哲学上的内因。直到1914年左右，丹麦植物学家约翰逊（W. Johannsen, 1859—1927）才发现了这个内因——基因（gene）。

有意思的是，我们今天使用的科学词汇很多来自日语，但这个词不是。"gene"在日语中翻译成"遗传子"，中国科学家翻译成基因，算是充分理解了它是一种基本因子。

基因在什么地方呢？后来，人们发现它就是染色体上的DNA，也就是脱氧核糖核酸。最早发现的脱氧核糖核酸来自于果蝇的唾液腺，肉眼都可以看到。当然，我们还需要看清它的内部结构。一位聪慧而美丽的女科学家罗莎琳德·富兰克林（Rosalind Elsie Franklin，1920—1958）拍摄了世界上第一张DNA（B型）的X射线衍射照片。她已经感觉到它可能是螺旋状的，但无法确定是单螺旋、双螺旋还是三螺旋。当时，还有两个聪明的年轻人弗朗西斯·克里克（Francis Harry Compton Crick）和詹姆斯·沃森（James Watson），他们也在思考DNA的结

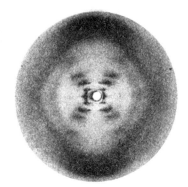

富兰克林拍摄的DNA（B型）X射线衍射照片

詹姆斯·沃森（左）与弗朗西斯·克里克

构。沃森去访问的时候，富兰克林给他看了那张衍射照片，沃森看了几分钟（一个历史性的时刻）后把照片还给了富兰克林。他回去对克里克说："我已经找到了结构，但拿不到照片，请你太太（一个画家）用笔画下这个结构。"接下来，两人连夜撰写论文。这篇不足千个英文单词的论文在 Nature 杂志发表后，改变了整个生物学的面貌（两个人后来也因此获得了诺贝尔奖）。富兰克林非常郁闷，几年后得乳腺癌去世了，真是天妒英才。的确，这个世界上只有极少数人在特定情况下能成为所谓的成功者。

下面，讲一讲批判性思维。当人们发现基因的基本模样后，任何一种质疑都有可能获得诺贝尔奖。举一个例子：基因是整整齐齐排列在染色体上的吗？芭芭拉·麦克林托克（Barbara McClintock，1902—1992）一直研究玉米，发现她的玉米没有一根是纯紫色的，也没有一根是纯黄色的，而是两种颜色镶嵌的。所以，她猜测基因排列不是整齐的，而是跳跃的。此言一出，所有人都认为她是疯子，嘲笑了她几十年。后来科学家们才发现基因真的有"跳跃"现象，或者称为"转座子"。麦克林托克也终于在81岁获得了诺贝尔奖。

一个基因能决定一个功能吗？1993年，美国加州大学的一位女教授发现了一个基因，叫做"Brca1"。1995年她又发现了"Brca2"，这两个基因合在一起大约能预测乳腺癌遗传风险的25%。或者说，如果你检查出这两个基因的致病突变，就有四分之一的可能罹患乳腺癌。当然，还要看家族的其他成员，比如外婆、姨妈、妈妈是否患乳腺癌。如果她们发病，你就属于真正的高危人群，风险可高达75%，这就是好莱坞明星安吉丽

芭芭拉·麦克林托克

娜·朱莉要去动手术的原因。

这是一个通过单基因预测癌症风险的例子。说到癌症，它和基因的关系很密切。世界上第一个能够完全早期检测和控制的癌症是妇女的宫颈癌，其元凶是德国科学家哈拉尔德·楚尔·豪森（Harald zur Hausen）发现的HPV病毒（2008年他也被授予了诺贝尔奖）。现在，我们仅凭疫苗就可能让高达60%的妇女终身免受宫颈癌的困扰。有可能获得诺贝尔奖的还有乳腺癌和前列腺癌研究。人们在寻找许多癌症的相关基因，但遗憾的是大多数癌症基因还不靠谱，比如肺癌和肝癌等，都太复杂。这些发现都说明了单基因是有作用的，但多基因共同作用情况更多。于是，就需要基因组计划。

我们现在能确认有功能的人类基因有30 000多个，估计有功能的有50 000多个。我们将人类基因和小鼠基因对比，大概80%左右是相同的，令人震惊。很多宗教人士反对现代生命科学研究，说它亵渎了我们人类的尊严，其实当人类基因组和小鼠基因组摆在一起时，我们真的没有什么尊严了——两者的基因太相像。给人类尊严最后一击的是黑猩猩，我们与黑猩猩的基因更相像。在哲学意义上，我们何以为人？还好，我们与黑猩猩的微小差异几乎全部在大脑上。我们之所以能够统治世界，用的是脑子。

人类掌握了基因组测序技术以后，基因组就相当于一个探测器，可以知道我们是由什么组成的。艾伦·威尔森（Allan Wilson）率先在加州大学伯克利分校研究了147个人的线粒体基因组。线粒体基因的特殊之处在于它百分之百由母亲贡献，父亲一点贡献也没有。艾伦·威尔森用线粒体基因序列构建了一棵进化树，结果发现树的底端全是黑种人，暗示人类来自非洲。当时，奥斯卡金像奖最佳影片授予了电影《走出非洲》，所以人们也将威尔森的假说叫做"走出非洲假说"，或者"夏娃假说"。

我的同事金力院士（现任复旦大学副校长）在2004年用父系基因资料分析东亚人是哪来的。他用了上万个样品，结果表明我们是非洲来的，而且来的时间还不长（十万年内）。

生命与哲学

既然讲"生命哲学",首要问题就是"生命"是什么呢?它不能全靠生物学家回答。病毒就是一道难题。病毒到底是不是生命呢?有人说是,有人说不是。为什么呢?因为它有DNA(RNA),但也可以像无机物一样。它并不具备生命的全部特征。我们只能在一般意义上认为这些特征包括能量、遗传、意识等。

第二个问题是繁殖。达尔文进化论中有趣的一点就是关于性的进化。性在生命中有什么意义?它决定了繁殖。这个世界为什么是两性世界?两性是不是必需的?可以明确地说,不是必需的,很多生物可以无性繁殖。比方说克隆,原意是指以幼苗或嫩枝扦插(无性繁殖或营养繁殖)的方式培育植物,现在扩展到动物克隆了。对两性繁殖认识的重要补充来自蜜蜂,因为它是三性的——除雌性和雄性外,还有一种是工蜂,它不承担繁殖任务。不过,两性世界的确在自然界中有平衡优势,也是最有效率的:无性繁殖的最大的优点是抛弃了"需要找对象"这样一条限制,可以自己为自己传宗接代,但缺点是多样性越来越低,就好比是一种复制(copy);有性繁殖的优点是充满了新奇,每一次都可能有不同的组合。例如:科学家曾研究过一个问题,全世界的家长都宣称对子女是一碗水端平的,但孩子们都感觉父母是偏爱的。父亲到底偏爱谁?科学家调查的结果非常明显,父亲偏爱长得像自己的孩子。理由很简单,它来源于一个古老的进化根源——雄性动物无法确定哪个真的是自己的孩子。所以,在多子女情况下,父亲会投资那个可能继承自己最多基因的孩子。

第三个问题是关于人类胚胎的。目前,科学家公认人类的新生命是从胚胎第十四天开始的。原因是,研究发现胚胎发育到第十四天开始有了痛感。什么是疼痛?人类知道自己的疼痛,但为何能够由己推人?或者由人推己?这就是伦理道德。如果你用针去扎胚胎,它感到疼痛就如同你扎在自己身上一样。

最后,再谈几个有哲学含义的问题。第一个问题是有没有与天赋相关的基因呢？回答是可能有，但我认为不要过于追求它。曾有一个科学家（Nelson Freimer）想研究天赋基因。有一天，他让助手加班，他的助手说我要去教人弹钢琴。一个洗瓶子的人为什么会教钢琴呢？助手说他家很多人会弹钢琴，科学家领悟到这可能是一个音乐世家，而且很快发现他们家族很多人能用自己的耳朵辨别绝对音高。科学家从该家族中分离出"辨别绝对音高"的基因，并到纽约的茱莉亚音乐学院去测试，发现茱莉亚音乐学院有高达三分之一的孩子能辨别绝对音高，远远高出世界上其他任何地方。这不就是天赋基因吗？且慢，还有三分之二的人并没有这个所谓的天赋，说明这个天赋对成为音乐家并不是必需的。后来还发现这个基因和某些疾病有关联。例如：一直困扰这个家族的失眠症就是常人承受不了的。你是否愿意自己的孩子拥有这样的基因？

另一个吸引人但也可能不靠谱的是行为基因研究。有些人行为怪异，该不该由基因负责？迪恩·哈默（Dean Hamer）先生最先捅了这个马蜂窝，在 *Science* 杂志发表了同性恋基因研究。有人指出，这类研究的方法中有一些缺陷，如剩下的人是否有非同性恋基因？还有一些申请，要研究犯罪基因，是否杀人犯都是遗传和基因决定的。所以，整个社会对基因说明一切的情况产生了巨大的疑问。如今，我们讲得最多的还是疾病基因，其他像追求天赋、行为或者把人类分成三六九等的基因，我们都必须慎之又慎。

最后一个是长寿问题。目前，世界上的长寿研究大部分都不靠谱，其中归纳法占了主要部分。从方法论来说，找不出更科学的方式，一般很难做实验。当然，科学家也有办法，比方经常用来做长寿实验的动物叫做线虫。这个动物是自由生活的多细胞生物，它从出生到死的细胞数目大致不变；它能活五至七天，这意味着科学家们可以做很多次实验。实际上，用线虫做研究材料是敲除个别基因来看它是否影响长寿。换言之，虽然这样很难发现延长寿命的基因，但可以找出妨碍寿命的基因。科学家经过长期研究终于取得了突破，现在公认的有好几个决定长寿的基因，因为如果把这些基因逐一敲除后，线虫能活三十天，这是一件很了不起的事。

这类文章发表后当然成为热门话题，但也很快展现出哲学意义。为什么?一个好消息是人也有同源的基因。这是一种强烈的暗示,即可能它也与人的寿命有关,但至今为止并没有人做过人的相关实验。坏消息是这个基因具有生殖功能。科学家们意识到长命百岁和断子绝孙可能是同一件事。更重要的是,线虫实验表明妨碍长寿的基因要敲得早敲,生育之后再敲就没用了,真可谓"鱼和熊掌不可兼得"。这就是长寿的两难境地,它是一个严肃的哲学命题,值得深入探讨。

　　本文发表于《书城》2017年第8期"知本读书会"栏目,收入本书时略有修改。

教育最重要的是释放学生的学习力

复旦大学研究生院院长、生命科学学院教授钟扬在2017年9月25日因为车祸离开人世，怀念他的不仅仅有高校和科学界的同行，还有很多基础教育界的老师和校长们，甚至中学生和小学生，他们或多或少都曾经受到过钟扬的指导。

很多人提到钟扬，都会说到，他在孩子们的心中播下了科学的种子。不仅如此，中小学邀请他去进行科普讲座或者是请他去和老师交流，他从来都是来者不拒。

2017年2月，他在接受本报记者专访时，详细阐述了他关注中小学基础教育的缘由，从中可以窥见他对教育的思考——

身为一名大学教授和研究生院院长，我为什么关注基础教育？因为高等教育与基础教育不打通，就无法培养真正高质量的人才。

当我们读书时，大家都有着差不多相同的信念，学生的成绩25%取决于自己努力，75%取决于学校老师的教学。但是不知从什么时候开始，大家不再相信那75%了，都觉得家长和课外培训才是决定孩子是否优秀的因素。

我在中科院系统和高校已经工作了32年，基础教育培养的人才最终都送到了这里，但是在过去那么多年，我没有感受到送到我这里的学生在沉重的负担中完善了知识结构。

钟扬教授为小朋友们做科普。

恰恰相反，送到我手中的学生，他们的知识体系显得割裂，一些本该在基础教育阶段养成的能力没有得到很好的培养。这一切都使得我去思考，这是为什么？

我认为，很重要的一点是，基础教育和高等教育这一整个连贯的教育体系被割裂了，甚至连基础教育和高等教育的话语体系都是割裂的。

在过去十多年甚至更长时间中，教育有了快速的发展，甚至可以说，已经进入到和发达国家的教育对标的阶段了，这意味着民众对高质量教育的需求也在改变。

近年来，教育减负的呼声很高，特别是在大城市，但是不是大城市的学生负担最重呢？那为什么20世纪70年代在上海读书时，一周上六天课，星期天还要打扫卫生，都没有觉得负担重呢？所以减负不是笼统地说，学生的负担重，学习内容多，减少就能够解决问题了。

钟扬教授带领学生采集标本。

教育是要培养聪明人的。那么世界上什么样的人最聪明?

我曾经在一次教育论坛上讲了一个故事,有一个小孩跟着父亲到山里去,在那里他遇到了一位最聪明的老者。小孩问老者,我手掌里有一个蝴蝶,蝴蝶究竟是活是死?老者说,蝴蝶就是你想要的样子。

其实这就是教育的本质。

这个寓言有着多重含义。首先,不能预设答案;第二,压力在第三方。

究竟小孩是那只蝴蝶,我们学校是那只蝴蝶,还是我们自己就是那只蝴蝶,抑或是教育就是那只蝴蝶?

还有一个著名的寓言故事——笼中的金丝雀。这是西方著名的隐喻。金丝雀是对瓦斯最敏感的动物。矿工下井时常常带着金丝雀的笼子,一旦金丝雀不叫,就说明瓦斯泄露了,矿工必须立即撤离。

其实,我们的孩子就是这些金丝雀,对于我们这个年龄的人来说,外部的世界很难对我们产生什么很大的震撼,但是孩子们不一样,他们来到这个世界时,大多是白纸一张。

教育的问题往往折射的是社会问题,现在我们常常说,孩子很脆弱。其实不是现在的孩子太脆弱了,而是社会的问题最终反映在孩子身上,而孩子们没有那么强大的承受力。

"不能让孩子输在起跑线上"这句话很有名，但这并不意味着能在终点取胜。教育是一场接力赛，要培养聪明人的话，需要每一棒之间都有策略，否则，一个学生从第一棒跑到最后一棒，每一棒都在冲刺。

一个孩子出生后，接受的教育可以说是一场接力赛。教育的第一棒曾经是小学教育，现在也许被提早到了学前、胎教阶段。"不输在起跑线上"很有名，但这并不意味着在终点取胜。

众所周知，接力赛要获胜必须有策略，要有松有紧。而且，接力赛最好是一个教练，他会告诉你，接力赛为了整体的成功，四棒的力量会有均衡的安排。

但是，在教育的接力赛中，却是四个教练，四个指标体系，每一个教练都指挥自己这棒的运动员拼命冲刺，只为这一棒的指标达到最优。落实到每一个孩子，他一个人跑四棒，但要接受四个教练的考评，得冲刺四次。

可以说，这样是用局部的最优来替代整体的成功。我们的中学有可能特别厉害，甚至可能被评为全世界最好的第二棒。但，我们有没有想过，到第三棒时，我们的"运动员"可能连棒都接不住了？

遗憾的是，我们第二棒的考评只考评第二棒，第三棒能不能接住，谁关心？这也导致很多"运动员"到了本科开始睡大觉，因为，他的人生已经冲刺三四次了，甚至在第一棒前就开始冲刺了，还怎么持续地冲刺呢？

也有人说，孩子的学习热情在冲刺中被消耗光了，这我也不认同。因为学习的热情未必总量固定，而是可以被激发出来的，就像科学早就发现，一个人的大脑容量也未必是固定的，也许可以越用越多。

但是，排除孩子个体之间的差异，仅从教育的规律来看，安排他冲刺四次，那也是错误的。

何况在冲刺的过程中，按照一般规律，我们要集中精力做好一件事情，势必会放弃其他的一些事。但，如果我们放弃的恰恰是下一棒最需要的，那到底该怎么办？

而且，由于我们每一棒都在冲刺，每一个局部冲刺的过程中，下一棒往往不知道上一棒用的什么策略来交棒。就像我是大学教授，我并不知道中小学送来的是什么样的学生。

坦率地说，大学对于中小学的改革究竟在改些什么，可以说是一无所知。而中小学对于大学的要求，也可以说是一无所知。最简单的例子，大学认为中小学有些课程内容不该减，减了以后，读大学乃至读研究生会影响他们的思维习惯。但中小学的校长可能会说，学生负担重，就该减。

如果教育在中学就结束了，情况也许没那么糟糕，但教育到研究生才结束，那就糟糕了。因为中小学根本不知道学生如果读到研究生需要什么。这也是中小学叫我去做讲座，我从来不拒绝，不遗余力地去讲的一个重要原因。

再打个比方，减负从本质上来说，是一种权衡，如果精力时间无穷多，那根本不存在负担。但有限时间精力的情况下，减负的优先域在哪里？比如我是从事生物多样性研究的，要保护一个城市的生物多样性，最简单的做法就是，什么都不要建。但是城市要发展，那必然要在发展和生物保护之间进行权衡并选择相对好的策略。

减负也是如此，对比知识和方法，显然是先减知识后减方法，可是从现实来看，我们的中小学教育内容中，恰恰缺少科学的方法。

当我们传授知识时，老师很少告诉孩子知识发现的过程，这样，孩子们就只是死记硬背了这个知识点。但是，如果学会了举一反三的本领，孩子自己会学到很多的知识点。到了研究生阶段，如果还没掌握科学的方法，那么孩子就需要记住无穷多的知识点。这就使我们处于两难，一方面知识点减少，学习内容就空无一物，如果增加知识点，孩子就不堪重负，如果教会他们方法，他的负担不就轻了吗？

所以，我们的减负现在陷入了怪圈，简单的减负以量为指标，可是因为质的不够，家长想提高质，只好通过其他方式，无形中又增加了量。

但是，我不认为大学教师直接参与基础教育就能解决问题。比如热火朝天的科学、技术、工程、数学综合教育（STEM），现有情况下，如果我参与过多，势必会增加孩子们的负担，但如果以减负为由不去做，我又于心不忍。

减负需要像转化医学一样,把高等教育和今后社会需求的教育理念,转化成中小学能够实施的内容。这需要时间,需要慢慢推进。否则,如果过时的知识太多,知识点太破碎,那么,要想在短时间内提高质量,势必会增加大量的负担。任何好的实验都要以时间为代价,如果心急,那就很可能失败。

更新我们的教育理念,用科学的方法重新审视我们的教育,根本上来说,是用科学的方法提高学习的质量,释放学生的学习力。

在孩子的时间和空间有限,而各种信息越来越丰富的今天,孩子们的负担为何越来越重?

20世纪六七十年代出生的一代变成了焦虑的家长,他们认为今天小孩要达到他们当年的程度,才能取得一点成绩,这无形中成为孩子负担的来源。

从表观遗传学来看,一些并非由基因控制,而是由于环境造成的压力,可以造成上一代DNA修饰上的改变,这些改变能够对下一代产生影响的痕迹。

同样,当下,在知识越来越多的情况下,还用以前的方法教育孩子,那么孩子的负担必然会加重。所以减少学习时间,提高学习质量就成了减负的精髓。

学习就像艺术,必须要有留白,可惜的是,大多数家长,包括老师都并不知道这个道理,而总是希望填满学生的时间,通过无穷无尽的操练来提升学生的成绩。

以写作为例,从我所教的学生来看,学生总体写作水平不容乐观是明显的事实。从作文写作中就可以看出当下基础教育的现状。

我们都知道,即便是作家,当堂写作文,也未必能写出高分作文。写作文需要深思熟虑。但是,现在我们的中小学,当堂作文的量太大,使得学生学会了一些在45分钟之内写作的套路,但却没有学会写作文的科学方法,到了真正需要写作时,没有能力完成。

在寒假里,我指导一位中学生写作。这名学生拿了他写的五篇日记来给我看,我只挑出一篇来修改。他大吃一惊,因为学校规定,他们一个假期要写十篇日记。但是我认为,连一篇都写不好,干嘛要写十篇?

我给中学生上课时也告诉他们，好文章是改出来的。我自己的文章不改上七八遍，都不会拿给老师看。现在让他们一个寒假写七八篇作文，学生怎么会改？只写一篇改十遍比写十篇一遍不改要强。

反观美国对小学四年级的作文要求，写作动机、方法、逻辑都包含在内，甚至还有数据事实和观点的要求。这才是写作培养的重要能力，但是这些如果要加入中小学教学，势必要减少量的要求。

因此，可以说，原先的教学要求不仅让学生的负担很重，还掩盖了落后的训练方式。

说到底，教育最重要的是要释放学生的学习力。

孩子们的生理成熟度和100多年前不一样，社会节奏也快了，如果我们的中小学教育还在沿用以往的模式，负担一定会重。

所以应该借减负这个机会，促进教育全面深化改革，而且减负应该是提升教育质量的突破口。

全世界的教育都面临着巨大的难题——在今天的课堂上传授昨天的知识，这能否应对明天的挑战？可以说，谁先认识到未来的教育，谁就可能抢占先机，在这个意义上来讲，我们的教育发展还有很大空间。

本文发表于《文汇报》2018年3月30日"文汇教育"栏目，是该报于2017年2月对钟扬教授所作的一次专访，署名姜澎，收入本书时略有修改。

我和科学队长

1970年初春，因幼儿园搬家，我意外地成为一名小学插班生。从那时起，一套残缺不全的《十万个为什么》开始成为我小学阶段最喜欢的课外读物。在反复阅读和独自思索之后，我渐渐明白了一些条目，其中不少还牢记至今。这样一段自学科学的经历究竟给我的一生带来了什么呢？

我相信科学能深入儿童的心灵。在既没有学校科学课程，也没有家长和课外辅导的情况下，一个小孩子其实很难靠几本书来准确地了解科学道理。不过，书中那些遥远的故事及其承载的有趣知识会随着时间的推移，慢慢渗透进脑海。在某个不经意的时刻或场合，这些似是而非的知识也许会与新的思想活动碰撞出火花，并以独特的科学气质展现出来。只有此时，长大的孩子才能真正体会到"春风化雨，润物无声"的人生意境。科学教师也会有相同的感受。几年前，有档电视节目中请来一个近十年前听过我科普讲座课的学生来谈体会，其中谈到那次讲座对他选择人生道路的影响一段对我触动最大。

我相信科学能培养人们的探索精神。自然科学与社会科学的差异促使年幼的学习者也要将"动脑"和"动手"结合起来，才能摆脱"书呆子"的桎梏。记得我小学二年级时从书中读到了一点电池的知识，立即就将家里手电筒中的大电池倒出来，用铁钉打出小洞，再往洞内灌各种各样能找到的酸性液体。当我的母亲——一位中学化学老师看到那一堆废电池时，不但没有责怪我，还将我带到化学实验室去观摩实验课。最后,她带我去校办工厂参与汽车蓄电池用浓硫酸的稀释工作,彻底打消了我对剧烈化学品的恐惧,也激发了我学习化学的兴趣。至于以后高考化学成绩名列前茅,反倒成了一项副产品。

在参与"科学队长"活动之前,我已热衷于青少年科普活动,包括为中小学生举行科普讲座,撰写和翻译科普著作,以及承担上海科技馆和自然博物馆的中英文图文版工作,等等。青少年科普是一种令人愉悦但费时费力的工作,对科学家本身其实也是一种挑战,绝非"没有时间"和"不感兴趣"那么简单。今天,我高兴地看到"科学队长"之类的科普活动大大增强了单一努力的效果,更乐于投身其中。

不过,我想告诉家长和孩子们,"科学队长"节目只是提供了学习科学的一个入口。听别人讲故事是快速了解科学概况的途径,但只有勤于思考和动手实验才是真正了解科学的不二法门。希望你们能在听完"科学队长"后,对身边的事物产生更大的好奇心,去思考更多,学习更多,这才是我做科普的初衷。

本文由"科学队长"微信公众号提供,推送日期为2017年3月17日,收入本书时略有修改。

世界屋脊上的种子收集者

"植物猎人"与"奇异果"

我是钟扬,植物学家,来自复旦大学和西藏大学,非常高兴有机会跟大家讲一讲种子的故事。

作为植物学家,我们经常说,一个基因可以拯救一个国家,一粒种子可以造福万千苍生。

举个例子,1984年,我大学毕业以后到中国科学院武汉植物研究所工作。我们所里面最重要的一种植物是猕猴桃,它是1904年由英联邦国家的传教士和那些专门来寻找各种各样奇花异草的人在湖北宜昌农村意外发现的,我们叫那些人"植物猎人"。"植物猎人"们找到这种植物以后,从树上剪下来20多根枝条带到了英联邦国家。最后这种植物传到了新西兰,通过杂交,成了新西兰非常重要的植物品种。

在我们现在的植物学家看来,这简直是个意外。为什么?我们今天知道了,这种植物是雌雄异株,所以如果当时他光把雄的剪回去,或者光把雌的剪回去,是无论如何都不能繁殖的。而当时全世界的生物学家没有一个人知道植物雌雄异株的机制。

猕猴桃在新西兰的种植取得成功这件事让中国植物学者们很心酸,但是新西兰的植物学家也不轻松，为什么？因为他们国家最重要的农业产业，居然建立在中国的三棵植物身上。假如他们当时取的这三棵植物并不是整个猕猴桃种群里最好的呢？如果这个猕猴桃有一种病害，有一种虫害，有一种特殊的东西能够对它进行毁灭性的打击呢？所以新西兰的植物学家心里非常明白，真正的遗传宝库在中国。

而仅仅在湖北，在武汉，我们大概收集了70多个猕猴桃品种。这70多个品种有的并不好吃，有的长得并不好看，但是它们是真正的种子。

猕猴桃只是一个例子，更重要的例子是粮食作物的"绿色革命"，那就是矮秆基因HYV的发现。科学家罗曼·保尔从野生资源中筛选出了矮秆基因，从此植物不再需要长那么高，特别是农作物，长矮一点，让它多结一些种子。发现矮秆基因的这位先生获得了诺贝尔和平奖。在我们国家，大家也知道袁隆平先生在海南岛发现了一种叫野稗的野生稻子资源，然后通过反复的选育，终于得到了杂交水稻，带来我们农业上的一场革命。

可见，获得种子对我们的未来是一件非常好的事。这些种子可以为我们提供水果，可以为我们提供花卉，改善我们的生活，更重要的是，可以为我们提供粮食。还有比它更重要的吗？有，那就是医药。

但是非常糟糕的是，由于全球环境的破坏，人类活动频繁，很多植物在人类了解和知道它们能否被利用之前，就已经灭绝了。

怎么办？很多科学家提出了各种各样的方案。世界上目前最引人注目的种子库是斯瓦尔巴特种子库，我们称之为"种子方舟"或"末日种子库"。它设置在离北极1 000公里左右的永久冰川冻土层里面。这个种子库不仅在工程上、科学上设计得非常精妙，而且还特地考虑了人类在遭受核打击和停电的情况下，这些种子能保存多久。

斯瓦尔巴特种子库不仅能够保护一批种子，更重要的是有非常强烈的警示意义。它建在北极这个地方，让我们清楚地了解到生物多样性的现状并不乐观，我们应该行动起来。

邱园里竟然没有一颗来自西藏的种子

我从武汉调到复旦大学工作后,发现上海在我国的生物多样性方面排名倒数第一,北京排名倒数第二。在这两个生物多样性相对贫乏的地区,集中了我国生物多样性研究差不多一半的人才。后来我申请了援藏。

青藏高原是生物多样性的热点地区,到这样的地方去收集种子有着特殊的意义。它是全世界第一批确定的二十几个生物多样性的热点地区之一。

青藏高原的植物有多少呢?目前我们按科算,是212个科,占到了我国植物品种总数的32.9%。

青藏高原一共有将近6 000个高等植物物种,就是说能够结种子的,占到全国总数的18%。更为重要的是,其中1 000个左右的植物物种是西藏特有的,我们称之为"特有种"。这些植物不仅数量很大,而且质量非常好。

顺便提一句,即使是这样一个庞大的数字,我们认为也是非常保守的。因为一些地区交通不便,譬如我们在日喀则地区发现了一个嘎玛沟。我们第一次进去的时候,单程需要七天的时间,所以100年来植物学家没有在这个地区留下记录。我们估计的样本数目也显然不包括这些地区所特有的稀奇古怪的奇花异草。

在青藏高原寻找植物的工作需要坚持比较长的时间,我大概坚持了十多年,其间我和我的学生,包括我的第一个藏族博士,我们花了大概三年的时间收集西藏巨柏的种子。巨柏的种子非常难收集,但我们把这个种子给收集齐了。

学植物的都知道英国皇家植物园邱园,那里的种子收集覆盖面非常广泛,因此人们可以据此分析全世界种子的现状以及评估环境变化以后种子的状况。而且,邱园的科学权威特别多。在那里,我们找到了蔡杰先生。当我们要他为中国植物学家提供植物种子研究的资料的时候,他马上同意加入我们的团队。

我们分析了当时邱园所有的植物,发现里面居然没有一颗来自中国西藏的种子。因此,它关于全球变化的预测,在这个版图上少了一块。所以我们写了一篇小文章，阐述全世界气候变化的形势下，必须采集和保存西藏的种子。Nature杂志发表了我们这封来信，同时也呼吁世界科学家重视西藏的种子。

大家都来尝尝光核桃

在中国的植物中，我们最讨厌收集的种子大家知道是什么吗？椰子。一颗那么大，而我们需要收集8 000颗。我们大概需要两辆卡车才能把它们拖回来，然后这才算收集到了一科的种子。

光核桃呢，实际上是光核的桃，不是光的核桃。就是说，我们一般吃到的桃子中间那个核是皱皱巴巴的，而光核桃的核是光滑的。这个桃子有什么用呢？没什么用。目前查来查去，大概它最大的用处就是在藏药里面有少许的用途吧。但是我们需要它，也许它就像猕猴桃一样，多少年以后可能就跟我们的水蜜桃杂交了。杂交品种有什么优点呢？水蜜桃很好吃，那光核桃呢？它抗虫、抗病、抗旱、抗寒。利用这些优点，我们可以通过非转基因的方式，经过杂交，再加上自然选择，培育出一种新型的桃子品种。所以我们知道它潜在的意义是很大的，得先把它收集起来。

一个桃子里面有一颗种子。所以我们先收集8 000个桃子，装了两大麻袋，把它们运回拉萨，放在我的实验室里面。

如何把里面的种子取出来成了一个难题。如果有自动化的取种方法就非常好，但是没有，所以我就把这些桃子摆在门口，铺满一个台子，所有路过的老师、学生，每个人都必须尝一尝。尝多少呢？七颗。我们认为如果超过了十颗，很多同志一辈子都不想再见我了。他们都非常淳朴，特别是藏族朋友，一边吃一边呸呸呸，然后告诉我这个东西不能吃。

确实，我也知道不好吃，因为它是猴子吃的。可是我们必须这样把它吃完，用牙刷好好地把种子给刷干净，刷完以后用布擦干，擦干以后再晾干，因为不能暴晒，暴晒会降低种子的质量。

我们把这收集到的8 000粒种子送到中国科学院昆明植物所的种子库里面。科学家们看有没有裂的、被虫蛀的，把外观检查做完了，就抽样进行发芽试验。发芽试验结束后进行登记，最后筛选5 000颗种子，把它们封到瓶子里面。这样可以保存80年到120年，算作一个样本。那一年夏天，我做了500个样本。

我们做遗传资源的很辛苦。第一，要沿着海拔2 000多米到3 000米这条线，慢慢地搜寻种子。由于不同的个体之间有可能发生杂交，我们规定两个样本之间的空间距离不得小于50公里。这样我们一天要走800公里，每走过50公里看见一个物种的种子，就赶紧收集几个，装上麻袋，然后开车去另外一个点。

第二，在整个西藏境内，任何一个物种不得收集超过5个群体的种子。这样算来，大约7年时间，我们收集了大约1 000个物种，占到了西藏物种的1/5。今年起，我们要在墨脱开始新一轮的收集。

照这样进行下去，在未来的20年，我们有可能收集到西藏高等植物的75%。

100年后的种子罐

肯定有人会问，你收集这么多种子，对我们普通人来讲意义何在呢？

答案是：为了应对全球的变化。假设一百多年以后还有癌症，假设那时候大家发现一种植物有抗癌作用，然而由于地球气候的变化，这种植物已经灭绝了，幸好一百多年前有个姓钟的教授采过它的种子。

都一百多年了，姓不姓钟有什么关系呢，是不是教授又有什么关系呢？但是届时如果发现了那5 000粒种子，拿出来一种，哪怕只有500粒能活，最后只有50颗能结种子，那个植物不就复活了吗？

当然也有人问，如果一百年以后这个种子没有用了怎么办呢？我期待看到种子没有用的那一天。说明什么？说明那个植物还在。我们连这样的尝试都不用做该多好。

还有人问我，从国内一流的复旦大学，转移到世界"最高学府"的西藏大学，有什么体会。

我说那个地方是高原，什么都慢，培养人也特别慢。今年我的第五位博士德吉毕业了,是藏族人,也是我培养的七个少数民族博士中新近毕业的一名。

我在复旦大学可以培养很多博士,但是他们不一定对我们刚才所说的种子收集或者说高劳动强度而低回报的工作真的有那么大的兴趣。但是在西藏,我培养的藏族博士毕业以后,至少这五个里面有四个都留在了西藏大学,都在西藏工作。

曾经有一位著名导演要给我们拍一部宣传片。那个片子最后起名叫《播种未来》。这个五分钟的微电影被送到国际电影节去参展,拿了微电影的纪录片金奖。很高兴,这个片子有了这样的结局,我想得奖原因一方面是西藏美丽的风光,一方面也来自种子赐予我们的灵感和一切。

本文发表于《中国青年报》2017年9月12日,后被《课外阅读》杂志第22期转载,收入本书时略有修改。

下编

钟扬教授各学术领域代表性论文[*]

I. 生物数据模型与信息系统

钟扬教授早期主要从事生物数据模型及信息系统的设计与实现工作，为实验生物学研究提供新的手段与技术平台。在中国科学院武汉植物所期间，利用计算机对包括睡莲目在内的多种资源植物的科目进行数量分类研究，其中关于荷花品种的数量分类研究的相关论文（《荷花品种的数量分类研究》）荣获1988年湖北省自然科学优秀学术论文。

1996年，钟扬教授在美国加州大学伯克利分校和密歇根州立大学合作研究期间，在国际上率先提出了一种新的交互分类数据模型（UNIC）和检测系统树差异的新测度，并据此建立了一个基于分类本体论思想的交互分类信息系统（HICLAS），为不同生物分类系统的整合提供了技术保障。相关论文在国际植物分类学会会刊 *Taxon* (Data model and comparison and query methods for interacting classifications in a taxonomic database)和生物信息学刊物 *Bioinformatics* (HICLAS: a taxonomic database system for displaying and comparing biological classification and phylogenetic trees)等发表后引起广泛关

---

[*] 因各期刊格式、体例略有不同，选文原则上遵循原文体例，除个别明显错误核正外，其他体例细节未作统一。

注。国际生物多样性信息学权威F. A. Bisby教授在其为*Science*撰写的专题综述中介绍了该模型和系统。其后，美国和欧洲学者发表的一些生物多样性相关的重要模型和系统（如TaxLink和IOPI等）均参考引用了UNIC模型和HICLAS系统，并肯定了其优点和前瞻性。

近十年来，钟扬教授在国内建立了一批分子数据库，如通过整合水稻和拟南芥等基因表达和数量性状位点（QTL）信息，建立了鉴定植物抗逆相关候选基因的数据库系统PlantQTL-GE (PlantQTL-GE: a database system for identifying candidate genes in rice and *Arabidopsis* by gene expression and QTL information)；建立了PB转座子及小鼠插入突变的数据库系统PBmice (PBmice: an integrated database system of *piggyBac* (PB) insertional mutations and their characterizations in mice)，该数据库系统能储存、分析分子遗传与发育生物学的相关数据，支持了该领域的实验生物学发展；此外，还建立了多蛋白查询整合信息系统MPSS，可通过ID正规化工具批量提取蛋白质信息(MPSS: an integrated database system for surveying a set of proteins)。上述工作均发表于*Nucleic Acids Research*和*Bioinformatics*等生物信息学权威杂志。

荷花品种的数量分类研究

莲（*Nelumbo*），也称荷花，在我国长期的人工栽培过程中，形成了众多优良的品种。近年来，引入美洲黄莲（*N. lutea*）与亚洲莲（*N. nucifera*）的杂交品系，更增添了不少新的内容。但迄今为止，关于品种的分类方法尚存在争论。事实上，由于荷花品种间的差异，仅仅在少数性状的基础上进行定性描述难以阐明其间的亲缘关系。本文则以部分花莲品种为材料，运用电子计算机进行数学分类的定量分析，为合理解决荷花品种分类系统问题提供科学的依据：

1. Q型聚类分析有助于揭示品种间的亲缘关系。
2. R型聚类分析说明了性状间的生物学距离。
3. 主成分分析初步反映了各性状所占的信息比例。

钟扬，张晓艳.荷花品种的数量分类研究.《武汉植物学研究》1987年第5卷第1期，49-58.

荷花品种的数量分类研究[*]

钟 扬　张晓艳

（中国科学院武汉植物研究所）

提　要　本文运用电子计算机进行了荷花品种的数量分类研究。结果表明：

1. Q型聚类分析有助于揭示品种间的亲缘关系。
2. R型聚类分析说明了性状间的生物学距离。
3. 主成分分析初步反映了各性状所占的信息比例。

关键词　荷花；数量分类；聚类分析；主成分分析

莲（*Nelumbo*），也称荷花，在我国长期的人工栽培过程中，形成了众多优良的品种。近年来，引入美洲黄莲（*N.lutea*）与亚洲莲（*N.nucifera*）的杂交品系，更增添了不少新的内容。但迄今为止，关于品种的分类方法尚存在争论[1,7]。事实上，由于荷花品种间的差异，仅仅在少数性状的基础上进行定性描述难以阐明其间的亲缘关系。本文则以部分花莲品种为材料，运用电子计算机进行数学分类的定量分析，为合理解决荷花品种分类系统问题提供科学的依据。

---

[*] 承蒙邓惠勤、熊治廷等同志提出宝贵意见，唐佩华老师、张玲、张平等同志给予了帮助，特此致谢。

下 编

材料与方法

1. 以花莲71个品种作为分类运算单位（Operational taxonomic unit, OTU）。详见表1。

2. 选取34个性状（Character），这些性状数据是1985年6月—8月间在中国科学院武汉植物研究所莲藕组实验地的观测纪录，并经过统计学处理及筛选得到。在考虑性状的稳定性和加权等问题时，参考了过去的实验记录及部分参考文献[2,8]。

性状的编码分三种情况：二元性状（简称"二"）6个，有序多态性状（简称"多"）13个，连续数值性状（简称"数"）15个。详见表2。

本文的全部数值运算使用BASIC语言编写程序，在我所IBM PC/XT和APPLE Ⅱ型电子计算机上运行通过。

表1　71个花莲品种及亲本组合

Table 1　A list of name of 71 lotus varieties and their parents

序号 No.	中　名 Chinese name	学　名 Scientific name	亲本组合 Parents
1	美玉	N. nucifera×lutea cv. 'Meiyu'	（绍兴红莲 × 红碗莲）$_{F1}$ × 美洲黄莲
2	满江红	N. nucifera cv. 'ManJianghong'	红碗莲 ⊗
3	春光	N. nucifera cv. 'Chunguang'	红千叶 × 白碗莲
4	天女散花	N. nucifera×lutea cv. 'Tiannusanhua'	小舞妃 × 美洲黄莲
5	童艳	N. nucifera cv. 'Tongyan'	白碗莲 × 白千叶
6	龙飞	N. nucifera×lutea cv. 'Longfei'	红碗莲 × 美洲黄莲
7	白鸽	N. nucifera cv. 'Baige'	解放公园红莲 × 白千叶
8	红牡丹莲	N. nucifera cv. 'Hongmudanlian'	绍兴红莲 × 红碗莲
9	冰娇	N. nucifera cv. 'Bingjiao'	妃翠莲自然授粉
10	小樱红	N. nucifera cv.'Xiaoyinghong'	*
11	翠菊	N. nucifera×lutea cv. Cuiju'	红碗莲 × 美洲黄莲
12	小公主	N. nucifera×lutea cv. 'Xiaogongzhu'	碗儿红 × 美洲黄莲
13	小佛手	N. nucifera cv. 'Xiaofoshou'	妃翠莲自然授粉
14	朝霞	N. nucifera cv. 'Zhaoxia'	娇容醉杯 × 白碗莲
15	彩蝶	N. nucifera cv. 'Caidie'	红碗莲自然授粉

续表

序号 No.	中 名 Chinese name	学 名 Scientific name	亲本组合 Parents
16	小桃红	*N. nucifera* cv. 'Xiaotaohong'	红千叶 × 白千叶
17	万寿红	*N. nucifera*×*lutea* cv. 'Wanshouhong'	红碗莲 × 美洲黄莲
18	冰心	*N. nucifera* cv. 'Bingxin'	白碗莲$_{F3}$⊗
19	友谊红2号	*N. nucifera* cv. 'Youyihong 2'	中日友谊莲 × 红碗莲
20	飞菊	*N. nucifera*×*lutea* cv. 'Feiju'	红碗莲 × 美洲黄莲
21	黄舞妃	*N. lutea*×*nucifera* cv. 'Hangwufei'	美洲黄莲 × 舞妃莲
22	红双喜	*N. nucifera* cv. 'Hougshuangxi'	唐招提寺青莲 ⊗
23	点绛唇	*N. nucifera* cv. 'Dianjiangchun'	寿星桃 × 白花建莲
24	小舞妃	*N. lutea*×*nucifera* cv. 'Xiaowufei'	美洲黄莲 × 亚洲莲
25	碧玉莲	*N. nucifera* cv. 'Biyulian'	满江红 ⊗
26	娇阳	*N. nucifera* cv. 'Jiaoyang'	*
27	君子莲	*N. nucifera* cv. 'Jiunzilian'	（中日友谊莲 × 红碗莲）$_{F1}$× 红千叶
28	白菊	*N. nucifera* cv. 'Baiju'	（中日友谊莲 × 红碗莲）$_{F1}$× 红千叶
29	江南红	*N. nucifera* cv. 'Jiangnanhong'	满江红 ⊗
30	飞虹	*N. nucifera* cv. 'Feihong'	红碗莲 × 寿星桃
31	翠玉莲	*N. nucifera* cv. 'Cuiyulian'	寿星桃 × 满江红
32	喷云	*N. nucifera* cv. 'Penyun'	*
33	小寿星	*N. nucifera* cv. 'Xiaoshouxing'	红碗莲 × 寿星桃
34	小碧台	*N. nucifera* cv. 'Xiaobitai'	白孩莲 ⊗
35	红杏	*N. nucifera* cv. 'Hongxing'	*
36	银芍药	*N. nucifera* cv. 'Yingshaoyao'	（红碗莲 × 美洲黄莲）$_{F1}$⊗
37	长瓣小桃红	*N. nucifera* cv. 'Changbanxiaotaohong'	白孩莲 × 美洲黄莲
38	重瓣粉妆	*N. nucifera* cv. 'Chongbanfenzhuang'	绍兴红莲 × 红碗莲
39	吉祥莲	*N. nucifera* cv. 'Jixianglian'	（中日友谊莲 × 红碗莲）$_{F1}$× 红千叶
40	粉青碗莲	*N. nucifera* cv. 'Fenqingwanlian'	红碗莲 × 唐招提寺青莲
41	美中红	*N. nucifera*×*lutea* cv. 'Meizhonghong'	美洲黄莲 × 亚洲莲
42	娇容碗莲	*N. nucifera* cv. 'Jiaorongwanlian'	红千叶 × 红碗莲
43	白碗莲	*N. nucifera* cv. 'Baiwanlian'	*
44	娇容醉杯	*N. nucifera* cv. 'Jiaorongzuibai'	绍兴红莲 × 粉碗莲
45	红樱莲	*N. nucifera* cv. 'Hongyinglian'	白碗莲自然授粉
46	醉杯	*N. nucifera*×*lutea* cv. 'Zuibei'	中日友谊莲 × 美洲黄莲

续表

序号 No.	中 名 Chinese name	学 名 Scientific name	亲本组合 Parents
47	睡芙蓉	N. nucifera cv. 'Shuifurrong'	红边玉蝶 ⊗
48	露月	N. nucifera cv. 'Luyue'	白花建莲 ⊗
49	玉锦	N. nucifera cv. 'Yujin'	红碗莲自然授粉
50	樱桃小口	N. nucifera cv. 'Yingtaoxiaokou'	小舞妃 × 满江红
51	霓碗	N. nucifera cv. 'Niwan'	红碗莲自然授粉
52	赛菊	N. nucifera cv. 'Saiju'	白碗莲自然授粉
53	紫云	N. nucifera cv. 'Ziyun'	满江红 × 小满江红
54	积翠	N. nucifera cv. 'Jicui'	白孩莲 × 碧降雪
55	红棉	N. nucifera cv. 'Hongmian'	红碗莲 × 牡丹莲
56	太真出浴	N. nucifera cv. 'Taizhenchuyu'	白碗莲自然授粉
57	玉玲珑	N. nucifera cv. 'Yulinglong'	*
58	含笑	N. nucifera cv. 'Hanxiao'	白孩莲 ⊗
59	出水黄鹂	N. nucifera cv. 'Chushuihuangli'	*
60	碧莲	N. nucifera cv. 'Bilian'	*
61	大洒金	N. nucifera cv. 'Dasajin'	*
62	大满江红	N. nucifera cv. 'Damanjianghong'	*
63	红霞	N. nucifera cv. 'Hongxia'	*
64	秋水长天	N. nucifera cv. 'Qushuichangtian'	古代莲
65	红芍药莲	N. nucifera cv. 'Hongshaoyaolian'	*
66	天娇	N. nucifera cv. 'Tianjiao'	*
67	红蕊	N. nucifera cv. 'Hongrui'	*
68	白千叶	N. nucifera cv. 'Baiqianye'	*
69	并蒂莲	N. nucifera cv. 'Bingdilian'	*
70	水红	N. nucifera cv. 'Shuihong'	红边玉蝶 ⊗
71	红边玉蝶	N. nucifera cv. 'Hongbianyudie'	红碗莲 × 白碗莲

* 亲本组合不明。The parents are unknown.

Q分析与讨论

1. Q型聚类分析的数学运算步骤：[9]

对原始数据进行标准化（Standardization）；对标准化数据计算分类运算单位之间的相关系数（Correlation coefficient）；在相关系数矩阵的基础

上，进行Q型聚类，采用效果较好的UPGMA（Unweigted pairgroup method using arithmetic averages)法，作出OTU分类结果的树系图（Dendrogram)（见图1）。

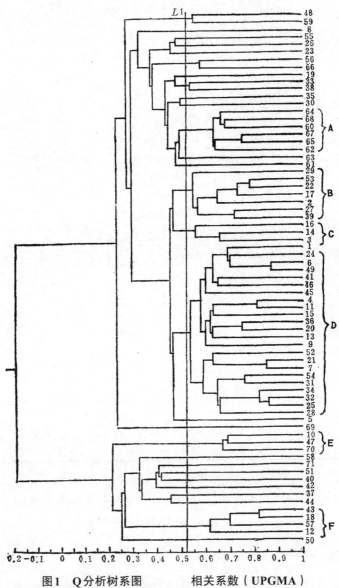

图1 Q分析树系图　　相关系数（UPGMA）
Fig. 1　The dendrogram of Q cluster analysis (71 OTU's)
Correlation coefficient (UPGMA)

聚类分析后，需要确定一条分类等级划分界线。在此，以类群聚合的水平为纵坐标，聚合次序为横坐标，将全部聚合过程描画成一条阶梯式的折线，称为聚类的结合线[6]（见图2）。从结合线看，类群的聚合在纵轴上的分布是不均匀的，从而构成系统分类的等级性。等级分界值取在结合线中跳跃位置的中点。从图2中可以得到：

$$L_1=\frac{1}{2}(0.5377+0.5042)=0.5210$$

将L_1标在树系图（图1）中。

表2 荷花品种的分类性状

Table 2 The characters of lotus varieties

编号 No.	性状 Characters	编码类型 Code types	编号 No.	性状 Characters	编码类型 Code types
1	花蕾长/花蕾宽[1)] Bud length/bud breadth	数 N[3)]	18	花瓣尖部色 Petal tip colour	多 M
2	花蕾长/花蕾尖距[2)] Bud length/bud tip distance	数 N	19	花瓣中部色 Petal middle colour	多 M
3	花瓣数 Petal number	数 N	20	花瓣基部色 Petal lower colour	多 M
4	雄变瓣数 Petaloid anther number	数 N	21	是否洒金 Blade maculate	二 T
5	雄蕊数 Stamen number	数 N	22	洒金颜色 Maculate colour	多 M
6	心皮数 Carpel number	数 N	23	外瓣情况 Petal droop	多 M
7	花径 Floral diameter	数 N	24	附属物颜色 Subsidiary colour	多 M
8	花高 Floral altitude	数 N	25	瓣脉明显否 Nerved clearly	二 T
9	花梗高 Pedicel altitude	数 N	26	叶梗粗弱 Petal thick-slalked	多 M
10	外瓣长/外瓣宽 Petal length/petal breadth	数 N	27	叶片光滑否 Blade smooth	多 M
11	外瓣长/外瓣尖距 Petal length/petal tip distance	数 N	28	叶片平展否 Blade level	多 M

续表

编号 No.	性状 Characters	编码类型 Code types	编号 No.	性状 Characters	编码类型 Code types
12	立叶长/立叶宽 Lamina length/lamina breadth	数 N	29	叶绿浓淡 Blade green deep	多 M
13	立叶梗高 Petiole altitude	数 N	30	立叶前开花否 Bloom before vertical leaf	二 T
14	花梗高/叶梗高 Pedicel altitude/petiole altitude	数 N	31	结实否 Fruit-bron	二 T
15	雌变形式 Pistil change	多 M	32	叶梗刺明显否 Petiole thorn clearly	多 M
16	花托变化 Receptacle change	二 T	33	重瓣系数[8] Perianth multiplezation coefficient	数 N
17	花蕾颜色 Bud colour	多 M	34	内瓣尖绿否 Inside petal tip green	二 T

1) 通过计算原始数据获得的间接性状。下同。Indirect Characters.

2) 尖距:最宽处到尖部的距离。下同。Tip distance bistance between the most breadth and tip.

3) N: Numerical characters; T: Two-state characters; M: Multistate characters.

2. 参照树系图,在等级分界线L_1内,选择A、B、C、D、E、F六组进行讨论。

A组中全部为大型品种。这些品种花型、花色虽有所区别,但在该水平上仍聚为一类。由此说明植株大小在品种分类中占有重要的地位(在主成分分析时还将继续讨论),这与传统分类的结果相一致。同时,指出了某些特殊情况。例如,OTU69(并蒂莲)和OTU61(大洒金)同属大品种,但花型及花色甚为特殊,使两者在这一分界线内还不能聚合。

以下5组均为小型品种。其中B、C、F三组均体现出花型与花色的一致性。B组为红色、重瓣花型;C组为红色、单瓣花型;F组为白色、单瓣花型,且植株最为矮小。

具有美洲黄莲亲本的品种大都聚合在D组内。该组花的颜色以白色为主，花型既有单瓣又有重瓣。而传统的分类结果没有将其归为一类。

E组中全部是立叶前开花的品种。这三个品种在0.67的水平上聚为一类，而在很远的水平（0.22）才与其他品种相聚，表明这三个品种具有相对的独立性。

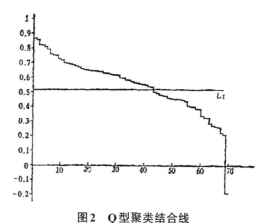

图2　Q型聚类结合线
Fig. 2　The combined-line of Q cluster analysis

同时，说明立叶前开花这一性状与植株的其他性状相关，从而表现出较大的相似性。而在传统分类系统中没有对这一现象引起足够的重视。

上述讨论说明，在相关系数和等级分界线基础上的定量分析比较适合于分类学的研究，具有较为合理的生物学意义[5]。

R分析与讨论

1. R型聚类分析的数学运算步骤：[9]

对标准化数据计算分类性状间的距离系数（Distance coefficient）；在距离系数基础上，进行R型聚类分析，仍采用UPGMA法，作出性状分类的树系图（见图3），作出结合线与等级分界线（见图4）。从图4中可以得出：

$L_2 = \dfrac{1}{2}$（9.30+8.61）=8.96

将L_2标在树系图（图3）中。

2. 参照树系图，在等级分界线L_2内，选择A、B、C、D、E、F六组进行讨论。

A组性状集中反映了植株的大小。

B组性状均与花色相关。由花蕾颜色可初步推断出其他性状。瓣脉的明显程度也与花色显著相关，如深色花的瓣脉较为明显。

C组性状反映了花的重瓣性。一般情况下，重瓣性愈强，雌蕊及雄蕊的瓣化程度愈高，重瓣系数也愈大。

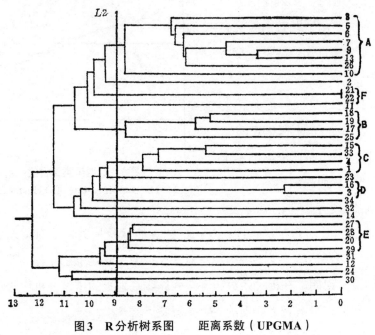

图3　R分析树系图　　距离系数（UPGMA）
Fig. 3　The dendrogram of R cluster analysis (34 characters)
Distance coefficient (UPGMA)

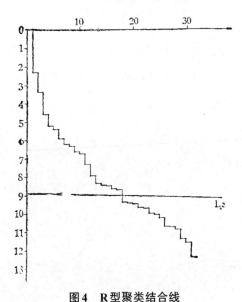

图4　R型聚类结合线
Fig. 4　The combined-line of R cluster analysis

D组中性状16与性状3的距离很小。已经证明，这一结果主要来自于OTU69（并蒂莲）的特殊情况。花托的全部瓣化，使花瓣数无限增加，因而在远离其他品种的同时，增大了两个性状的相关性。在消除OTU69的影响后，我们发现与性状3距离较小的是性状33，说明重瓣系数在很大程度上也取决于花瓣数。这一结论与重瓣系数的计算公式并不矛盾[8]。

E组性状反映了立叶的情况。

直观来看，如立叶浓绿的品种，叶表面较为光滑且平展。

F组仅有花瓣洒金性状。说明这种花瓣嵌合现象的遗传性并不对其他性状产生影响。

以上结果表明，用距离系数进行性状的聚合是可行的。性状分组的研究对于今后进一步的工作（如染色体定位，连锁等遗传现象）具有一定的价值。

主成分分析与讨论

1. 主成分分析（Principal component analysis, PCA）的数学运算步骤[3][注]：

对标准化后的原始数据计算性状间的内积矩阵S_N；计算S的特征根λ_i和特征向量U_{ij}[5]；将34个特征根依大小排列，计算相应每一坐标轴所占有的信息比例，结果见表3；计算各性状对主成分的作用，即求出34个性状对各主成分的负荷量（Loading）。前三个主成分的结果见表4。

表3　34个特征值及占有信息比

Table 3　The 34 eigenvalues and their information per cent

序号 No.	特征值 Eigenvalues	信息百分比 Information per cent	序号 No.	特征值 Eigenvalues	信息百分比 Information per cent
1	466.79	19.61	18	26.96	1.13
2	391.13	16.43	19	25.80	1.07
3	272.87	11.47	20	24.44	1.02
4	186.06	7.82	21	19.59	0.81
5	144.03	6.05	22	16.16	0.68
6	119.78	5.03	23	14.60	0.60
7	97.87	4.11	24	12.67	0.53
8	83.43	3.51	25	11.04	0.46
9	77.93	3.27	26	8.53	0.35
10	58.69	2.47	27	7.63	0.31
11	55.19	2.32	28	7.11	0.30

［注］　用品种去排序性状应采用PCA的逆分析，它与正分析在运算步骤上没有本质区别。

续表

序号 No.	特征值 Eigenvalues	信息百分比 Information Per cent	序号 No.	特征值 Eigenvalues	信息百分比 Information Per cent
12	49.15	2.07	29	3.95	0.17
13	45.52	1.97	30	3.45	0.14
14	42.74	1.80	31	2.00	0.08
15	40.20	1.69	32	0.36	0.02
16	33.94	1.43	33	0.16	0.01
17	30.17	1.27	34	0.03	0.00

2. 从主成分分析的意义出发，结合表3、表4可以看到：

对第一主成分影响最大的是性状13（立叶梗高）、性状9（花梗高）、性状26（叶梗粗弱）、性状7（花径）等（用*标出，下同）。这几个性状在前面的分析中已经显示了明显的相关，证明第一主成分是反映植株大小的。

表4 34个性状对前四个主成分负荷量表

Table 4 The loadings of characters

性状 Characters	第一主成分 First principal	第二主成分 Second principal	第三主成分 Third principal	第四主成分 Fourth principal	性状 Characters	第一主成分 First principal	第二主成分 Second principal	第三主成分 Third principal	第四主成分 Fourth principal
1	-0.37	4.37	-0.03	-1.77	18	-3.64	0.65	-5.72*	2.56
2	-3.50	3.28	-0.73	-1.21	19	-3.60	0.57	-5.67*	1.84
3	-2.95	4.35	3.87	4.71*	20	1.45	-5.13	1.22	-1.36
4	-0.18	4.49	1.77	-4.38*	21	-4.56	-0.27	-1.35	-0.63
5	-5.00	-2.45	-2.55	-2.32	22	-4.56	-0.27	-1.35	-0.63
6	-6.13	-2.29	-0.41	-3.95	23	3.11	4.61	-2.56	-1.95
7	-6.63*	-3.24	2.75	-0.51	24	2.06	1.13	2.49	3.04
8	-4.91	-3.44	3.27	2.63	25	-1.88	2.27	-4.55	2.21
9	-7.50*	-0.55	2.43	0.19	26	-6.80*	1.74	0.62	-0.81
10	-3.34	2.39	-1.01	-2.70	27	3.89	-1.96	-0.42	2.93
11	-1.86	1.29	0.59	-0.42	28	3.49	-4.91	-0.79	2.03
12	-1.49	-2.37	4.34	2.37	29	-1.07	-6.01	1.19	1.07
13	-7.54*	-1.36	1.87	0.47	30	1.69	0.54	-1.26	0.24
14	-1.33	3.91	-0.85	-0.37	31	-1.08	-5.09	-2.36	-0.03
15	0.67	6.48*	2.33	0.70	32	1.06	1.82	-1.84	1.20
16	-2.56	2.80	3.74	5.84*	33	0.86	-6.93*	2.38	0.19
17	-2.54	2.18	-5.92*	3.05	34	1.64	0.85	5.11	-3.50

对第二主成分影响较大的是性状33和性状15，第二主成分体现了荷花的重瓣性。

第三主成分中花色因素影响较大，依次为花蕾颜色，花尖部色和花中部色等。

对第四主成分影响最大的是性状16，其次是性状3。显而易见，花托完全瓣化导致花瓣数无限增加，但同时发现，对该主成分影响较大的还有性状4。R分析已表明，性状4与性状33密切相关。可见，第四主成分仍是反映花型的重瓣性。但R分析结果中，性状4、性状16和性状3距离较远，这一问题在后面的主成分中更为突出。因此，我们认为对第四主成分以后的分析意义不大。

需要指出，第二主成分与第四主成分实际上是从两个不同角度阐明了重瓣性在分类中的地位。第二与第四主成分所占的信息比之和超过了第一主成分的信息量。这似乎表明，在各种情况中，人为地把植株大小定为第一分级标准的问题尚需讨论。[7]

通过主成分分析还可以发现，仅取前15个主成分就可保留总信息量的90%左右，将这一结果与R分析结果相对照更说明问题。这给我们今后在进行性状观察、取舍和记录过程中，分清主次、强调重点，提供了定量化依据。

目前，关于荷花品种分类系统的定量分析仅仅是一个开端，还有待于今后进一步的研究。作者通过初步探索，认识到在深入开展荷花种质资源的生物学研究基础上，广泛应用现代数学方法和电子计算机技术建立"荷花品种资源管理系统"的必要性和可能性，这将成为我们努力的目标。

参 考 文 献

1. 王其超，1982：荷花品种整理及品种分类研究初报。武汉园林科技，(1)，1—24。

2. 王其超等，1985：荷花新品种选育研究初报。武汉植物学研究，3(1)：81—87。

3. 阳含熙等，1981：植物生态学的数量分类方法。科学出版社，232—252。

4. 杨汉碧等，1983：中国杜鹃花属高山杜鹃亚组的数量分类研究。植物研究，3（3）：75—86。

5. 张巨洪等，1983：BASIC语言程序库。清华大学出版社，233—238。

6. 徐克学等，1983：我国人参属数量分类研究初试。植物分类学报，21（1），34—43。

7. 倪学明，1983：莲的品种分类研究。园艺学报，10（3），207—210。

8. 黄国振，1983：荷花重瓣化及其遗传基础的初步探讨。武汉植物学研究，1（2）：139—142。

9. Sncath, P. *et al*., 1973（赵铁桥译，1984）：数值分类学。科学出版社。

Data model and comparison and query methods for interacting classifications in a taxonomic database

分类学数据库中交互分类的数据模型、比较和查询方法

　　将信息论的观点应用于生物分类，来获取分类学的概念作为分类数据实体，并开发一个系统来管理这些概念以及它们之间的谱系关系。为了开发数据模型，有必要将明确的定义应用于一些通常没有精确定义的分类术语，并创造和定义一些新的术语和概念。我们概述了一些比较交互分类和查询层级分类数据库的方法。同时在互联网上公开了一个称为HICLAS（HIerarchical CLAssification System，层级分类系统）的程序/数据库系统，它提供了一个X-Window（Linux系统上的可视化窗口）界面用于查询分类的数据。

　　Zhong, Y., Jung, S. W., Pramanik, S., Beaman, J. H. Data model and comparison and query methods for interacting classifications in a taxonomic database. 1996. *Taxon*, 45(2): 223-241.

　　本文获国际植物分类学协会（International Association for Plant Taxonomy, IAPT）的授权收入《钟扬文选》。

Data model and comparison and query methods for interacting classifications in a taxonomic database

Yang Zhong[1,3], Sungwon Jung[2], Sakti Pramanik[2] & John H. Beaman[1]

Summary

Zhong, Y., Jung, S., Pramanik, S. & Beaman, J. H.: Data model and comparison and query methods for interacting classifications in a taxonomic database. — Taxon 45: 223-241. 1996 — ISSN 0040-0262.

An information-theoretic view has been applied to biological classification to capture taxonomic concepts as taxonomic data entities and to develop a system for managing these concepts and the lineage relationships among them. In order to develop the data model, it has been necessary to apply explicit definitions to several taxonomic terms that generally have not been precisely defined and to coin and define several new terms and concepts. Methods are outlined for comparing interacting classifications and querying hierarchical taxonomic databases. A

---

1 Department of Botany and Plant Pathology, Michigan State University, East Lansing, MI 48824, USA.
2 Department of Computer Science, Michigan State University, East Lansing, MI 48824, USA.
3 Current address: Wuhan Institute of Botany, Academia Sinica, Wuhan 430074, Hubei, People's Republic of China.

program/database system called HICLAS, which provides an X-Window interface to query classification data, is available on the Internet.

Introduction

An area of biological database management, discussed previously by Beach *et al.* (1993), that is only now being effectively optimized is the processing of specimen data and taxon-level infonnation within the context of the taxonomic hierarchy, allowing for multiple alternative classifications and nomenclatural data processing at all hierarchical levels. New classifications are usually built by sharing, changing, and tuning taxonomic concepts of existing classifications. Developing an infonnation model for taxon infonnation should be based on the important criterion of sharing and adjusting taxon concepts of existing classifications when new classifications are proposed. Any infonnation-processing model for classifications must capture both the taxonomic concepts and the relationships among them as data entities.

Various taxonomic database management systems (DBMS) have been designed and implemented for processing the taxonomic data elements that are based on classifications [cf. Allkin, 1988; Allkin & Winfield, 1989; Beaman & Regalado, 1989; Crosby & Magill, 1988; Gibbs Russell & Arnold, 1989; Gomez-Pompa & Nevling, 1988; Humphries *et al.*, 1990; Pankhurst, 1991, 1993; Skov, 1989]. These systems employ a simple approach to managing the taxonomic hierarchy and associated taxon data entities, principally by tracking nomenclatural changes and the resulting synonymic links as indices to the underlying classifications.

Some commercial database systems allow direct representation of hierarchical data. IMS of IBM Corp. and System 2000 of Intel Corp. are two such hierarchical database systems available on large computers. Although these systems are able to represent parent-child relationships through hierarchical schemata, they lack the flexibility to represent a variable number of levels in the hierarchy. They are not suited for multiple, overlapping hierarchies. A requirement for taxonomic database systems is the ability to modify parent-child relationships, dynamically including changes that would link new levels. Such requirements are not easily implementable with traditional hierarchical database systems.

We are using directed graphs to represent the structure for biological classifications and interactions between classifications. A directed graph is a finite non-empty set of vertices (nodes) and a set of edges, such that the edges are ordered pairs of vertices. A set of directed graphs represents the taxonomic hierarchy as well as the relationships between hierarchies of related classifications. The main idea is to combine various related classification hierarchies into a unified structure that we call UNIC (union of classifications), which is a directed graph with each node representing a taxon concept based on a distinguishable set of character data. A node (taxon view) includes the taxon name, author or authority, date, and optionally a numerical reference to the publication.

Managing and tracking formally published names yields only part of the history and relationships of taxonomic concepts, because the rule systems of biological nomenclature were not designed to uniquely identify all taxonomic concepts. The rules of nomenclature specify rigorous and conservative protocols for adjusting the names of taxa after changes are made in the classification of a group, but they do not respond to many modifications to taxon circumscriptions (i.e., changes to taxon definitions or logical boundaries). A DBMS that works with names and their associated nomenclatural type concepts as indices to taxonomic concepts is constrained in two major ways: (1) it is unable to record any changes to taxonomic concepts that do not involve the reassignment of a name; and (2) it is unable to perform cross-classification compansons to any extent.

In the case of the first constraint, most taxonomic database systems record names and establish the logical link between a particular taxon and its synonyms. This kind of nomenclature, however, does not uniquely identify the new taxon concept, because of the reuse of the oldest acceptable name for the redefined taxon. Taxonomic database systems that use names as indices to taxa must use additional devices to distinguish between two concepts with the same name.

With respect to the second constraint, cross-classification comparisons of taxonomic concepts are essentially impossible when taxon names and not concepts are the basis of a taxonomic database system. This is true for comparisons between classifications of those concepts that lose their unique identity (i.e., when one and the same name is used for different concepts), but it is also true for comparisons of concepts that are uniquely indexed by the nomenclatural system. The difficulty for automated comparisons of classifications is largely due to a relatively static data model resulting in inflexible data structures.

Processing and analysing data in a DBMS can be effected in several modes. It is theoretically possible to analyse classification information in any type of file or database system, given enough time and programming effort. A distinction must be realized between developing one-time programs for specific queries and a more general approach to managing classification information that would permit ad hoc queries over any number of taxonomic and nomenclatural data elements. The distinction is important because taxonomic DBMS that require repeated, customized programming to access data are not practical for taxonomic research. In this paper we define the terminology necessary to manage classification data from a computer-systems view, present a new data model for the management of hierarchical classification data, and outline a methodology for querying classification data and comparing classifications.

Terminology

This section does not consider all the terminology used in traditional taxonomy. Rather, it redefines some taxonomic terms and provides some new concepts for constructing the data model outlined in the next section.

Classification. — A classification is a taxonomic treatment for a particular taxon or more taxa with certain taxonomic rank ranges based on character data and associated with related taxa. Normally, a classification is published in an official scientific publication distributed worldwide. Simple checklists or enumerations are not considered to be classifications.

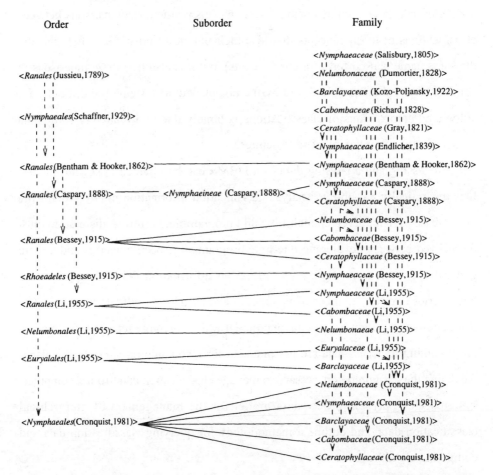

Fig. 1 A partial UNIC structure for classifications of the order *Nymphaeales*, and examples for explaining the operations "origination", "move", "merge" and "partition". The dotted lines with arrows represent operations for taxon views.

Taxon view. — The taxon view is a basic and central concept of this paper. Beach *et al.* (1993) have discussed the principal classification data elements and proposed using directed graphs, in which every node not only includes a taxon name but also publication information such as a bibliographic reference, to represent the structure for classifications and interactions among classifications. Here a taxon view is considered to be a triplet of the following elements (an interrogation mark [?] can be inserted after any element, and a publication number added — see below): taxon name, author or authority, and year. For example, <*Nymphaeaceae* [Salisbury, 1805]> and <*Nymphaeaceae* [Caspary, 1888]> are two of the taxon views concerning the family Nymphaeaceae as shown in Fig. 1. The elements of the taxon view are defined as follows.

— *Taxon name:* A scientific name for a particular taxon in accordance with the *International code of botanical nomenclature* (the *Code*; Greuter, et al., 1994) but not including author citation. For example, the taxon names in the two taxon views indicated in the previous paragraph are the same, i.e., *Nymphaeaceae*.

— *Author and authority:* An author is the person or persons who validly published a name of a taxon. An authority, in contrast, is the person or persons who published a classification that may or may not affect the name used for the taxon or the taxonomic placement of that group. For example, Salisbury is both the author of the name *Nymphaeaceae* and authority of the taxon view <*Nymphaeaceae* [Salisbury, 1805]> but Caspary is the authority of the taxon view <*Nymphaeaceae* [Caspary, 1888]> (see Beach, *et al.*, 1993:250).

— *Year:* The year(s) in which the classification was published.

— *Publication number:* An arbitrary index number of the publication, which permits to distinguish different classifications for the same taxon by the same author or authority in the same year. (Publication numbers are here omitted, or set to zero, and references are included among the "Literature cited" at the end, in the usual way.)

A taxon-view group is defined as a complete set of taxon views that have the same taxon name. They can be written as <Taxon name, author>. For example, <*Nymphaeaceae*, Salisbury> represents a taxon-view group concerning the family *Nymphaeaceae*.

Taxon-view types. — Every taxon view can be classified as either a primary taxon view or a secondary taxon view. It also can be classified as a parent taxon view and/or a child taxon view, according to the hierarchical relationship between a particular taxon view and related ones.

— *Primary taxon view:* A taxon view that was published by the author of the taxon name. In a taxon-view group there is only one primary taxon view, the one published first. For example, <*Nymphaeaceae* [Salisbury, 1805]> is a primary taxon view in the taxon-view group < *Nymphaeaceae*, Salisbury>.

— *Secondary taxon view:* A taxon view that was published by a later authority using the same taxon name as for the primary taxon view. In a taxon-view group all of the taxon views except the primary taxon view are secondary taxon views. If necessary, they could be ordered as first secondary taxon view, second secondary taxon view, etc., e.g., according to the publication year. For example, <*Nymphaeaceae* [Caspary, 1888]> is a secondary taxon view in the taxon-view group <*Nymphaeaceae*, Salisbury>.

— *Parent-child taxon view:* In a narrow sense a parent taxon view is a taxon view which includes one or more views at a lower taxonomic rank. Views one rank lower are called child taxon views, i.e., direct child taxon views for the parent taxon view. Views two (or three...) ranks higher or lower are referred to as grandparent-grandchild (great grandparent-great grandchild...) taxon views. They are also considered as ancestor-descendant taxon views in a broad sense. For example, <*Nymphaeaceae* [Caspary, 1888]> is a child taxon view of <*Nymphaeineae* [Caspary, 1888]> and a grandchild taxon view of <*Ranales* [Caspary, 1888]>. <*Nymphaeineae* [Caspary, 1888]> is both a parent taxon view of <*Nymphaeaceae* [Caspary, 1888]> and a child taxon view of <*Ranales* [Caspary, 1888]>. Other relationships may be inferred by analogy. The relationships between parent taxon views and child taxon views, i.e., parent-child relationships, can be explicit or implicit. This means that the parent-child relationships may be explicitly stated in the publication or may be inferred from a publication when they are not explicitly stated.

Operations for taxon views.— In order to explain more complex lineage relationships among taxon views, seven operations for one or more taxon views are proposed as follows.

— *Origination:* The creation of a primary taxon view through the description of new material. In other words, an author establishes a new taxon name and its concept. This operation relates to a single taxon view. The following operations affect two or more taxon views.

— *Move:* The movement of a taxon view from one classification to another classification without changing its taxonomic rank. Usually when a child taxon view is moved from its parent taxon view to another existing parent taxon view at the same taxonomic rank, the parent-child relationship is changed but the taxon name in the child taxon view is not changed (infrageneric taxa are often exceptions, see examples in Fig. 4). For example, <*Cabombaceae* [Li, 1955]>, a child taxon view of <*Ranales* [Li, 1955]>, was moved to <*Cabombaceae* [Cronquist, 1981]>, whose parent taxon view is <*Nymphaeales* [Cronquist, 1981]>, by the authority Cronquist (1981).

— *Merge:* The creation of a secondary taxon view by combining two or more earlier taxon views. For example, <*Nymphaeaceae* [Li, 1955]> is merged with <*Nymphaeaceae* [Salisbury, 1805]> and <*Euryalaceae* [Li, 1955]> to create a new secondary taxon view, <*Nymphaeaceae* [Cronquist, 1981]>. Note that if a taxon view is merged with other taxon views to a new secondary taxon view, the merged taxon views are no longer recognized by the authority of the new taxon view.

— *Partition:* The creation of two or more taxon views by decomposing an earlier taxon view. This operation is the reverse of a merge. For example, <*Nymphaeaceae* [Bentham & Hooker, 1862]> was partioned into three taxon views < *Nelumbonaceae* (Bessey, 1915)>, <*Cabombaceae* [Bessey, 1915]>, and <*Nymphaeaceae* [Bessey, 1915]>.

— *Promotion:* The promotion of a taxon view from a lower rank to a higher rank. It is obvious that the taxon name and taxonomic rank are changed so that the earlier taxon view is different from the resulting one. Fig. 2 shows two taxon views at the tribe rank that are promoted to family rank.

— *Demotion:* The demotion of a taxon view from a higher rank to a lower rank. This is the reverse operation from a promotion. The demotion of the taxon view <*Nymphaea pubescens* [Willdenow, 1799]> (at species rank) to <*Nymphaea lotus* var. *pubescens* [Hooker & Thomson, 1872]> (at variety rank) is shown in Fig. 3. The important difference between the two operations promotion and demotion, on the one hand, and move, on the other, is that promotions and demotions are effected at different taxonomic ranks and moves at the same rank.

— *Recognition:* The continued use of a particular taxon view in a new classification with the retention of the concept and parent-child relationships of this taxon view. Examples of the recognition operation for taxon views are shown in Fig. 4.

The seven operations described above can represent the methodology for most taxonomic treatments involving taxonomic revision. However, there are a few constraints for taxon views and operations. These will be discussed in the next section.

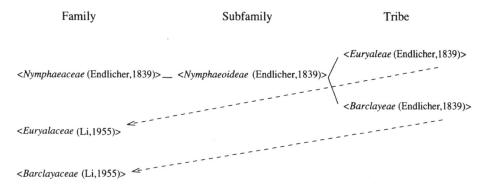

Fig. 2 A partial UNIC structure for classifications of the family *Nymphaeaceae*, and an example for explaining the operation "promotion". The dotted lines with arrows represent promotions from tribe rank to family rank.

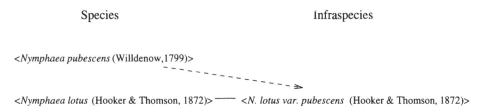

Fig. 3 A partial UNIC structure for classifications in the genus *Nymphaea*, and an example for explaining the operation "demotion". The dotted line with arrow represents the demotion from rank of species to form.

Data model

A new data model suitable for storing and comparing interacting classifications is outlined below. This data model supports a set of graph-oriented data structuring tools. These are classification trees and operational trees. A classification tree is a tree that captures a particular view of hierarchical classification. Operational trees are the trees created by the six lineage operations for taxon views defined in the previous section. They consist of "move trees", "merge trees", "partition trees", "promotion trees", "demotion trees", and "recognition trees", which are fundamental to providing information-theoretic views of interacting classifications. These operational trees represent the semantic interactions between classification trees. In the classification and operational trees, nodes correspond to taxon views. These views are partitioned and stored as groups in the database, based on the idea of an ordered meta entity set (OMES), as defined below.

Ordered meta entity set. — An OMES consists of entity sets in which each set consists of zero or more taxon views at a particular taxonomic rank. Thus, we use taxon views and taxonomic ranks (or simply ranks) to denote the entities and entity sets, respectively, in the rest of this paper. The following is the formal definition of OMES:

$$OMES = \{rank\ 1, rank\ 2, ...\ rank\ k\}$$

where all ranks must be ordered hierarchically so that rank 1 is always the highest and rank k is always the lowest. Each rank is either mandatory or optional. There is no limit on the number of ranks in an OMES so long as all of them are

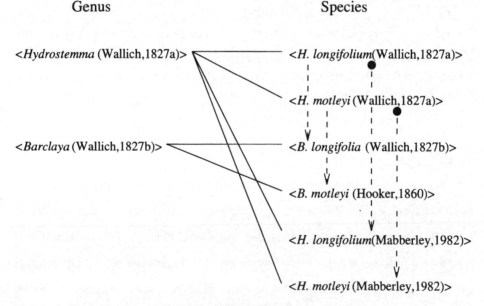

Fig. 4 A partial UNIC structure for classification of the genus *Hydrostemma*, and an example for explaining the operation "move" at the species rank (the taxon name usually is changed) and the operation "recognition ". The two dotted lines with only an arrow represent the move, and the two dotted lines with both an arrow and a solid circle represent the recognition.

Table 1 Taxonomic hierarchy provided for in the Code. — Use categories: M=mandatory (the 'principal ranks' of the *Code*, Art. 3.1); O=optional (ranks that must not necessarily be present in the classification of a plant taxon).

No.	Rank	Use	Abbreviation
1.	kingdom	M	K
2.	subkingdom	O	SK
3.	division (phylum)	M	D
4.	subdivision (subphylum)	O	SD
5.	class	M	C
6.	subclass	O	SC
7.	superorder	O	SPO
8.	order	M	O
9.	suborder	O	SO
10.	superfamily	O	SPF
11.	family	M	F
12.	subfamily	O	SF
13.	tribe	O	T
14.	subtribe	O	ST
15.	genus	M	G
16.	subgenus	O	SG
17.	section	O	SE
18.	subsection	O	SSE
19.	series	O	SR
20.	subseries	O	SSR
21.	species	M	SP
22.	subspecies	O	SSP
23.	variety	O	V
24.	subvariety	O	SV
25.	form	O	FO
26.	subform	O	SFO

hierarchically ordered. The taxonomic hierarchy provided for in the *Code* (Art. 4.2) is shown in Table 1. The OMES corresponding to the taxonomic hierarchy of Table 1 is {kingdom, subkingdom, ... form}. The number of elements of this OMES is 26. Usually an OMES for a particular classification is a subset of the OMES given in Table 1. For example, the OMES between family and genus ranks is {family, subfamily, tribe, subtribe, genus}.

Classification tree. — Tree constructions have been widely used to represent hierarchical classifications, especially in phenetic and cladistic taxonomy. However, there are many differences between such trees due to the different aims and definitions. Examples of classification trees in which every node is a taxon view are shown in Fig. 5. The tree structure representing the family-genus-species relationships in Fig. 5A for the species V_{SP4} and V_{SP5} is different from that of the species V_{SP1}, V_{SP2}, and V_{SP3}. This is because the second structure has the additional rank, subgenus. Note that the schema for these data contains a species rank having two parent ranks, namely, subgenus and genus. This violates the constraints of a tree structure with children having a unique parent. Most traditional hierarchic data models allow tree structures in their schema definitions, but interactions between trees requiring structural adaptations are not allowed. Thus it is difficult to handle classification data in the traditional hierarchical database systems such as the Information Management System (IMS) of IBM. Our new data model supports dynamic hierarchical relationships through binary recursive relations. All classification trees are stored in a single relation (table).

Operational trees. — Six operational trees have been defined. These correspond to the operations for the taxon views defined in the previous section. The operation "origination" for taxon views is not considered to create an operational tree because it takes place only at a single node. All operational trees are stored in a single relation in the same manner as are those for classification trees.

— *Move tree:* A move tree is a tree that captures the history of the movement of position of a taxon view at a particular rank among the different classification trees. Examples of move trees are shown in Fig. 6. Note that the move tree is available only for taxon views at the same rank because the move operation takes place only at the same rank.

— *Merge tree:* A merge tree is a tree that captures the history of the aggregation of the concepts of a taxon view at a particular rank. Examples of merge trees are shown in Fig. 7.

— *Partition tree:* A partition tree is a tree which captures the history of the decomposition of the concept of a taxon view at a particular rank. Examples of partition trees are shown in Fig. 8. On the basis of form, a partition tree is the opposite of a merge tree.

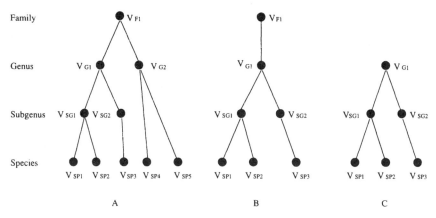

Fig. 5 Examples of classification trees. Each node with letters and number represents a taxon view. *A*, a complete classification tree of V_{F1} between family and species ranks; *B*, a subtree of V_{F1} between family and species ranks; *C*, a complete classification tree of V_{G1} between genus and species ranks, also a subtree of V_{F1} between genus and species ranks.

— *Promotion tree:* A promotion tree is a tree which captures the history of the promotion of the concept of a taxon view from a lower rank to a higher rank. It is very much like a move tree, but the difference is that the promotion tree is only for taxon views at different ranks, not for the same rank as are move trees.

— *Demotion tree:* A demotion tree is a tree which captures the history of the demotion of the concept of a taxon view from a higher rank to a lower rank. It is the opposite of a promotion tree.

— *Recognition tree:* A recognition tree is a tree which captures the history of the recognition of the concept of a taxon view at a particular rank.

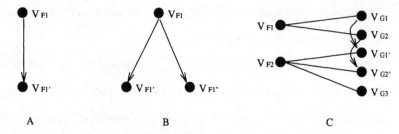

Fig. 6 Examples of move trees. Each node with letters and number represents a taxon view. Taxon views with apostrophes represent secondary taxon views. The arrows indicate the direction of the moves. *A*, the simplest move tree with two single nodes (one to one); *B*, a move tree with more than two single nodes (one to two or more than two); *C*, the combination (union) of classification trees and move trees.

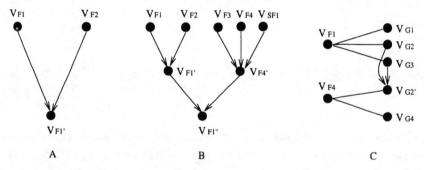

Fig. 7 Examples of merge trees. Each node with letters and number represents a taxon view. Taxon views with apostrophes represent secondary taxon views. The arrows indicate the directions of the merges. *A*, the simplest merge tree (two to one); *B*, the dual merge tree (note: the merge may take place at different ranks, e.g., $V_{SF1} \rightarrow V_{F4'}$); *C*, the union of classification trees and a merge tree.

Constraints for operational trees.— Various combinations of classification trees and three operational trees, i.e., unions of the trees, are shown in Figs. 6-8. Three other operational trees, i.e., promotion, demotion, and recognition, also can be combined with classification trees. In fact, a UNIC structure representing comprehensively the hierarchical relationships within each classification and lineage relationships between interacting classifications is a union of all classification trees and operational trees for particular related taxa at a particular taxonomic rank range. Several constraints, however, have been recognized for these operational trees.

These are described below.

— *Constraint 1:* Direct parent-child relationships between taxon views at two distinct ranks are not allowed to be created if there is a mandatory rank between the two ranks. For any optional rank, if it is used in a particular classification in which a parent taxon view is at a mandatory parent rank, then the number of child taxon views at the optional rank must be two or more and the mandatory parent taxon view can only have direct parent-child relationships with all the child optional ranks. For example, if a particular genus has subgenera and sections, the only direct parent-child relationship

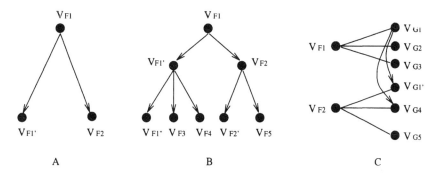

Fig. 8 Examples of partition trees. Each node with letters and number represents a taxon view. Taxon views with apostrophes represent secondary taxon views. The arrows indicate the direction of the partition. *A*, the simplest partition tree (one to two); *B*, the dual partition tree; *C*, the combination of classification trees and a partition tree.

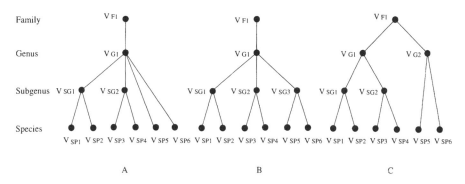

Fig. 9 An example for explaining constraint 1. *A*, illegitimate; *B*, legitimate by adding V_{SG3}; *C*, legitimate by adding V_{G2}.

below the genus would be with a subgenus and not with a section subordinate to the subgenus. This is a structural constraint for hierarchical type relations of a classification tree. In the example shown in Fig. 9A it is not allowed that V_{G1} have direct relationship with V_{SP6} and V_{SP7} with the skipping over of any taxon view at the subgenus rank. Figs. 9B-C show two legitimate cases in which V_{SP6} and V_{SP7} belong to the newly created subgenus or genus.

— *Constraint 2:* Any operational trees cannot be created if their taxon views skip a mandatory rank between their ranks. Especially, the move tree and recognition tree cannot contain taxon views at different ranks. This structural constraint for the operational trees is due to the nature of operations for taxon views.

— *Constraint 3:* For merge trees, a new tree cannot be created out of existing trees if they contain one or more common nodes. For other operational trees, a new tree cannot be created out of the same type of existing trees. The union of classification trees and operational trees is shown in Fig. 10. The interactional constraints (4 and 5) between different types of trees will apply.

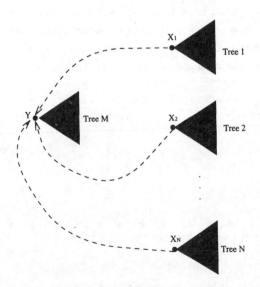

Fig. 10 Interactional constraints between the classification trees. The triangles represent classification trees; the arrows represent operation trees.

— *Constraint 4:* All the descendent taxon views of the taxon view Y in the tree M at the mandatory ranks must be related as operational trees with all the descendent taxon views of the taxon view X_1 in the Tree 1 to X_n in the Tree N at the corresponding mandatory ranks.

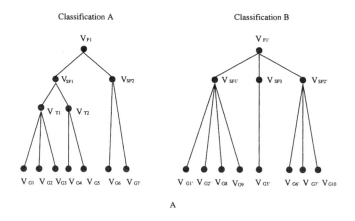

Fig. 11 An example of comparison of two classifications at the range of family to genus. The letter "V" with letters and number represents a taxon view. The letter "N" with letters and number represents a taxon name. *A*, two classification trees; taxon views with apostrophes represent secondary taxon views. *B*, results of the comparison ; "no comparison" means that the taxon view at a particular rank is absent in at least one classification.

— *Constraint 5:* Two or more taxon views cannot be merged if they are children of different branches of a partition tree.

Methodology for comparing interacting classifications

Two approaches can be used for comparing interacting classifications. The first involves taxon names (Fig. 11). Two taxon names in different taxon views may be the same or different. Because a taxon name plays an important role in a taxon view, it is meaningful that two or more taxon views should be compared with each other to find out the same or different names to tell the taxonomist what has happened in a new classification that replaces previous classifications. (Note that provided one precludes the use of descriptive names, such as *Dicotyledones*, at the suprafamilial ranks, it is impossible that there be the same taxon name in taxon views at different

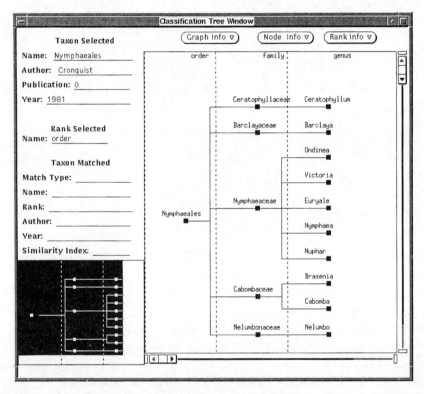

Fig. 12 An example of a classification-tree window. Note that a macro view of the entire tree is displayed at the lower left, whereas details are shown on the right. Various areas of the detailed tree can be accessed by vertical and horizontal scrolling. Menu buttons allow access to selected nodes or ranks for lineage information.

taxonomic ranks according to the *Code*; therefore, only taxon views at the same rank can be compared in this sense of comparison. Also note that when one classification has a particular rank and another lacks it, the classifications cannot be compared at this rank.) In our new data model, any classification has been considered as a tree within a certain range of ranks. Therefore, if all taxon views at a common range of ranks in two classifications are compared, the same and different taxon names at each rank of this range will be obtained and listed separately.

The second approach for comparing interacting classifications concerns more than two classifications. Thus, three kinds of comparisons are possible: (1) comparison of all classifications with each other for all taxon views; (2) comparison of one specific classification with others; (3) comparison of parts of a particular classification with others.

A mathematical model for comparing interacting classifications has been developed. This model provides an algorithmic framework for computer implementation. Details will be published separately elsewhere.

Implementation

We have implemented a hierarchical classification system called HICLAS, based on the methodologies described in the previous sections. HTCLAS is available on the Internet, and provides an Open Window^{TM}-interface on a workstation Unix platform. It can also be accessed with PC and Macintosh computers with X-Window. The system consists of an intuitive mouse-based graphical user interface connected to a Sybase® database server at Michigan State University. The interface allows a user to display and browse hierarchical taxonomic information. Fig. 12 is a sample view that could be presented for browsing. In this figure a macro view displays a full tree for a selected classification, and in the detailed view any part of the tree can be displayed by scrolling vertically or horizontally, allowing detailed access to a selected node or rank. The windows also have the capability to display lineage information.

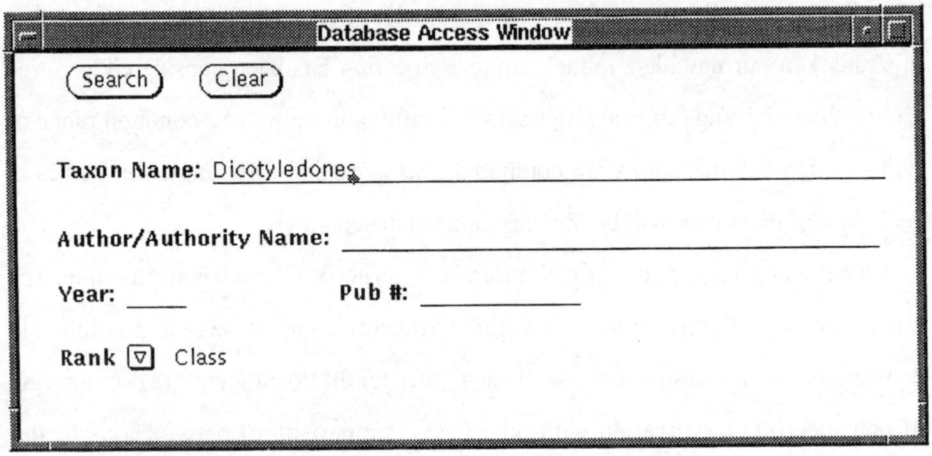

```
┌─────────────── Database Access Window ───────────────┐
│  ( Search )   ( Clear )                              │
│                                                      │
│  Taxon Name: Dicotyledones_____   │
│                                                      │
│  Author/Authority Name: _____   │
│                                                      │
│  Year: ____        Pub #: _____                 │
│                                                      │
│  Rank [▽]  Class                                     │
└──────────────────────────────────────────────────────┘
```

Fig. 13 An introductory search window that allows the user to select specified database entries.

To help users find specific information in large databases, the system contains a variety of search capabilities for pinpointing information of interest. Figs. 13 and 14 show a method for selecting a specific database entry for the class [*Magnoliopsida*]. The system contains a comparison tool that permits users to selectively compare taxonomic classifications. Built-in comparison operations and custom queries are implemented, based on the query algorithms outlined in the preceding section. A user is able to build, edit, and execute comparison expressions or use system-defined expressions. Fig. 15 shows the comparison-tool window along with a comparison expression that is ready to be evaluated.

Conclusions

The new data model and comparison and query methods described above employ the concept of "taxon views", which can carry more historical taxonomic information about a particular taxon and related taxa than is possible with regular taxon names. This model and the methods or algorithms could become powerful tools for storing, displaying and analysing taxonomic data in which hierarchical and lineage relationships are involved. The system has the potential to be applied and should be useful in the taxonomic revisionary process, especially when databases are being constructed by specialists in particular groups. Furthermore, the concept of taxon view could be expanded for describing any operational taxonomic unit (OTU) or hypothetical taxonomic unit (HTU) used in cladistic and phenetic taxonomy. Currently we are developing procedures that will allow the data model and comparison methods to represent cladograms and phenograms as well as taxonomic classifications.

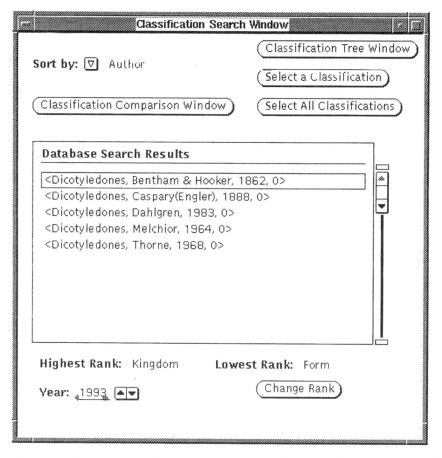

Fig. 14 A selection window that lists various previously selected classifications and provides options for further operations as indicated in the window.

The main data sources for this system are monographs, revisions, and indexes.

Completeness and accuracy of the data are very important for the model and methods to be most effective. It would be possible to associate other data, e.g., on type specimens, geographical distributions, descriptions and illustrations, with the skeleton classification data and taxon views that we have currently implemented.

We have assembled three data sets for testing the system. These are the order *Nymphaeales* (from class to infraspecific taxa), the genus *Nageia* (from family to infraspecific taxa), and the families *Verbenaceae* and *Lamiaceae* (from order to genus). Various commercial relational database management systems can be used to create data tables and manage the data, along with data coding and input procedures we have developed for this purpose.

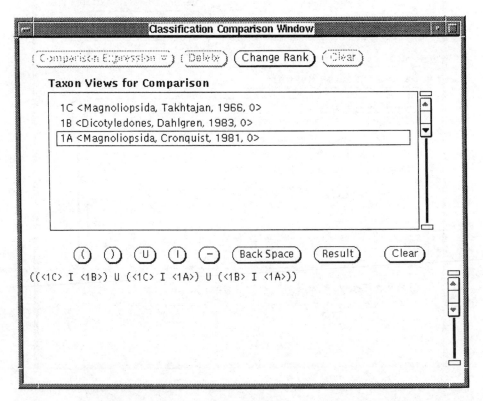

Fig. 15 A classification-comparison window providing comparison tools and a comparison expression that is ready to be evaluated.

Acknowledgements

The research was supported by National Science Foundation grants BSR-8822696 and DIR-9021656. We appreciate the computer programming by Michigan State University Computer Science students Stephen Perkins and David English, and thank Dr. Sibnath Das for collecting the input data set for HICLAS.

References

1. Aitkin, R. 1988. Taxonomically intelligent database programs. 315-331 *in*: Hawksworth, D. L. (ed.), *Prospects in systematics*. [Syst. Assoc. Vol. 36.] Oxford.
2. Aitkin, R. & Winfield, P. J. 1989. *ALICE user manuul, version 2*. Kew.
3. Beach, J. H., Pramanik, S. & Beaman, J. H. 1993. Hierarchic taxonomic databases. 241-256 *in*: Fortuner, R. (ed.), *Advances in computer methods for systematic biology: artificial intelligence, databases, computer vision*. Baltimore.
4. Beaman, J. H. & Regalado, J. C. 1989. Development and management of a microcomputer specimen-oriented database for the flora of Mount Kinabalu. *Taxon*. 38: 27-42.
5. Bentham, G. & Hooker, D. J. 1862. *Nymphaeaceae*. 45-48 *in*: Bentham, G. & Hooker, J. D. (ed.), *Genera plantarum*, 1(1). London.
6. Bessey, C. E. 1915. The phylogenetic taxonomy of flowering plants. *Ann. Missouri Bot. Gard.* 2: 109-164.
7. Caspary, J. X. R. 1888. *Nymphaeaceae*. 1-10 *in*: Engler, A. & Prantl, K. (ed.), *Die natürlichen Pflanzenfamilien*, 3(2). Leipzig.
8. Cronquist, A. 1981. *An integrated system of classification of flowering plants*. New York.
9. Crosby, M. R. & Magill, R. E. 1988. *TROPICOS: A botanical database system at the Missouri Botanical Garden*. St. Louis.

10. Dumortier, B C. 1828. *Analyse des families de plantes*. Tournay.

11. Endlicher, S. L. 1839. *Genera plantarum secundum ordines naturales disposita*, part 12. Wien.

12. Gibbs Russell, G. E. & Arnold, T. H. 1989. Fifteen years with the computer: Assessment of the "PRECIS" taxonomic system. *Taxon*, 38: 178-195.

13. Gómez-Pompa, A. & Nevling, L. I. 1988. Some reflections on floristic databases. *Taxon* 37: 764-775.

14. Gray, S. F. 1821. *A natural arrangement of British plants*. London.

15. Greuter, W., Barrie, F. R. Burdet, H. M., Chaloner, W. G., Demoulin, V., Hawksworth, D. L., Jorgensen, P. M., Nicolson, D. H., Silva, P. C., Trehane, P. & McNeill, J. 1994. International code of botanical nomenclature (Tokyo Code) adopted by the Fifteenth International Botanical Congress, Yokohama, August-September 1993. *Regnum Veg.* 131.

16. Hooker, J. D. 1860. [Illustrations of the floras of the Malayan Archipelago and of tropical Africa.] *Trans. Linn. Soc. London.* 23:155-172.

17. Hooker, J. D. & Thomson, T. 1872. *Nymphaeaceae*. 114 *in:* Hooker, J. D., *Flora of British India*, 1. London.

18. Humphries, J. M., Biolsi, D. & Beck, R. 1990. *MUSE tutorial and reference manual*. Ithaca, NY. Jussieu, A. L. de, 1789. *Genera plantarum*. Paris.

19. Kozo-Poljanskij, B. M. 1922. Vvedenie v filogenetičeskuju sitematiku vysših rastenij. Voronež.

20. Li, H. L. 1955. Classification and phylogeny of *Nymphaeaceae* and allied families. *Amer. Midl. Naturalist.* 54: 33-41.

21. Mabberley, D. J. 1982. William Roxburgh's "Botanical descriptions of a new species of *Swietenia* (Mahogany)" and other overlooked binomials in 36 vascular plant families. *Taxon*. 31: 65-73.

22. Pankhurst, R. J. 1991. *Practical taxonomic computing*. Cambridge.

23. Pankhurst, R. J. 1993. Taxonomic databases: the PANDORA system. 229-240

in, Fortuner, R. (ed.), *Advances in computer methods for systematic biology: artificial intelligence, databases, computer vision.* Baltimore.

24. Richard, A. 1828. *Monographie des orchideés des Iles de France et de Bourbon.* Paris.
25. Salisbury, R. A. 1805. *The paradisus londinensis.* London .
26. Schaffner, J. H. 1929. Principles of plant taxonomy, VII. *Ohio J Sci.* 29: 243-252.
27. Skov, F. 1989. Hypertaxonomy — a new computer tool for revisional work. *Taxon.* 38: 582- 590.
28. Wallich, N. 1827a. [Letter on *Hydrostemma*, in LXXXVIII. Proceedings of learned societies]. *Philos. Mag. Ann. Chem.* 1: 454-455.
29. Wallich, N. 1827b. Description of a new genus of plants belonging to the order *Nymphaeaceae*: in a letter to H. T. Colebrooke, Esq., F. R. S., F. L. S. *Trans. Linn. Soc. London* 15: 442-448.1827a.
30. Willdenow, C. L. 1799. *Caroli a Linné Species plantarum ... editio quarta,* 2(2). Berlin.

HICLAS: a taxonomic database system for displaying and comparing biological classification and phylogenetic trees（摘要）

HICLAS：一个用于展示和比较生物学分类和系统发育树的分类数据库系统

人们开发了众多数据库管理系统来处理生物学分类或系统发育信息的各种分类数据库。在本文中，我们提出了一个可以处理交互分类和系统发育有关的特定分类群的整合系统。

信息论观点（分类观点）已被用来捕捉分类概念作为分类数据实体。我们开发了一个适于支持语义交互动态观点的层级分类的数据模型，以及一个用于交互分类的查询方法。分类观点和数据模型的概念也可以被扩展到携带系统发育树中的系统发育信息。我们设计了一个基于分类观点的原型分类数据库系统,称为HICLAS（层级分类系统），并设计实现了数据模型和查询方法。该系统能有效地用于分类的修正过程，尤其当数据库是由特定群体的专业人员建立的时候，该系统还能用于比较分类和系统发育树。

免费访问网址：http://aims.cps.msu.edu/hiclas/

Zhong, Y., Luo, Y. N., Pramanik, S., Beaman, J. H. HICLAS: a taxonomic database system for displaying and comparing biological classification and phylogenetic trees. 1999. *Bioinformatics*, 15(2): 149-156.

PlantQTL-GE: a database system for identifying candidate genes in rice and *Arabidopsis* by gene expression and QTL information（摘要）

PlantQTL-GE：一个通过基因表达和数量性状位点信息来鉴定水稻和拟南芥中候选基因的数据库系统

我们设计实现了一个名为"PlantQTL-GE"的基于网络的数据库系统，来促进基于数量性状位点（QTL）的候选基因鉴定和基因功能分析的进行。我们收集了大量的基因、微阵列数据和表达序列标签（EST）形式的基因表达信息，以及来自多个来源的水稻和拟南芥的遗传标记。该系统整合了这些多样化数据来源并有一个易于使用的统一网络界面。它支持针对QTL标记区间或基因组位点的QTL查询，并在水稻或拟南芥的基因组上展示已知基因、微阵列数据、EST，以及在其他植物中的候选基因和相似的假定基因。QTL区间内的候选基因将基于匹配的EST、微阵列基因表达数据和调控序列的顺式元件而被进一步注释。该系统可在http://www.scbit.org/qtl2gene/new/ 免费访问。

Zeng, H. Z., Luo, L. J., Zhang, W. X., Zhou, J., Li, Z. F., Liu, H. Y., Zhu, T. S., Feng, X. Q., **Zhong, Y.** PlantQTL-GE: a database system for identifying candidate genes in rice and *Arabidopsis* by gene expression and QTL information. 2007. *Nucleic Acids Research*, 35: D879-D882.

PBmice: an integrated database system of *piggyBac* (PB) insertional mutations and their characterizations in mice（摘要）

PBmice：小鼠中*piggyBac* (PB) 转座子插入变异及其特征的整合数据库系统

DNA转座子*piggyBac*（PB）是一种近期被采用的在小鼠中进行大规模突变诱导的诱导剂。我们设计并实现了一个整合数据库系统，称为PBmice（PB Mutagenesis Information CEnter，PB突变诱导信息中心），用于存储、检索和显示从小鼠基因组的PB转座子插入中提取的信息。本系统围绕这些插入的信息，包括基因组定位，侧翼基因组序列，靶基因的表达水平和用捕获载体得到的捕获基因的表达模式。系统同时还存档了PB转座子插入相关的小鼠表型数据，允许用户通过快速或高级查询检索与单个或一组插入相关的基因型和表型信息。基于序列的信息可以通过交叉检索其他基因组数据库（如Ensembl等）。PBmice使用的BLAST 和GBrowse 工具可为插入相关的附加信息提供增强版的搜索和显示。用户还可以通过人性化的界面查看PB插入的总数和在基因组的分布，以及每一个PB插入小鼠品系是否可以获得。PBmice可通过如下网址免费访问http://www.idmshanghai.cn/PBmice 或http://www.scbit.org/PBmice/。

Sun, L., Jin, K., Liu, Y. M., Yang, W. W., Xie, X., Ye, L., Wang, L., Zhu, L., Ding, S., Su, Y., Zhou, J., Han, M., Zhuang, Y., Xu, T., Wu, X. H., Gu, N., **Zhong, Y.** PBmice: an integrated database system of *piggyBac* (PB) insertional mutations and their characterizations in mice. 2008. *Nucleic Acids Research*, 36: D729-D734.

MPSS: an integrated database system for surveying a set of proteins（摘要）
MPSS：用于多蛋白搜索的数据库整合系统

　　本文设计实现了名为MPSS（Multi-Protein Survey System）的多蛋白搜索数据库整合系统，为一次提取多个蛋白信息提供了一个平台。该系统整合了多个重要和被广泛使用的数据库，包括SwissProt，TrEMBL，PDB和InterPro，并附加了某些数据库如GO和KEGG等。使用者可以在MPSS的网络界面输入一组蛋白的ID号、登录名、SwissProt/TrEMBL的索引号或者GenBank的GI号，就可快速从公共数据库获得蛋白注释信息和蛋白分子特性。MPSS也可以显示所查询蛋白的复杂信息，包括3D结构，结构域，代谢通路，基因本体和可视化的基因本体树和KEGG代谢通路映射，提供了在研蛋白质的结构和分子功能的最新视图和可用知识。

　　Hao, P., He, W. Z., Huang, Y., Ma, L. X., Xu, Y., Xi, H., Wang, C., Liu, B. S., Wang, J. M., Li, Y. X., **Zhong, Y.** MPSS: an integrated database system for surveying a set of proteins. 2005. *Bioinformatics*, 21(9): 2142-2143.

II. 分子进化分析方法及应用

钟扬教授同时也致力于复杂进化模型和计算机模拟模型的建立。2000年，钟扬教授同美国密歇根州立大学的桑涛教授合作，提出了一种新的统计检测模型，以区分产生不一致基因树的复杂因素。以该模型为基础，发展出可检验杂交物种形成假说的自展技术，并应用计算机模拟确定了该模型对基因树各分支进化速率不变与可变情况下的适用范围。论文发表于分子进化领域的顶尖杂志 Systematic Biology（Testing hybridization hypotheses based on incongruent gene trees），受到了广泛关注。2001年，澳大利亚昆士兰大学 M. A. Ragan 教授在其发表于 Current Opinion in Genetics & Development 上的权威综述文章中将该工作列为"未来的方向"。该工作也是中国高校自然科学一等奖（2002年）的重要组成部分。

二十余年来，钟扬教授与中山大学施苏华教授等学者合作，完成了一系列植物分子系统发育分析工作，其中包括：a. 重建了红树科中海生的红树植物及其陆生近缘类群之间的系统发育关系，确定红树科中真红树为单系类群，而陆生类群则为多系起源，这项研究应用改进的分子进化分析方法，检测了红树植物及其近缘物种在分子进化速率上的异质性，是教育部自然科学一等奖（2008年）的重要组成部分，研究论文发表于目前生态学与生物多样性领域影响因子最高的学术期刊 Ecology Letters（影响因子15.3），是中国学者在该期刊发表的第一篇研究论文（Detecting evolutionary rate heterogeneity among mangroves and their close terrestrial relatives）；b. 重建了木兰科的分子系统发育关系，为该科起源与演化提供了新的证据；c. 探讨了使君子科、阿丁枫亚科、十齿花和四棱草等类群的系统发育关系；d. 开展了水稻雄性不育基因的进化分析等。这些工作发表于 Molecular Biology and Evolution 以及 Plant Cell 等权威杂志。

2006至2009年间，钟扬教授积极参与了上海交通大学附属瑞金医院陈竺院士领衔的日本血吸虫全基因组分析工作，主要负责动物分子进化树构建和适应性进化检测。在项目的推进过程中，钟扬教授课题组在算法上进行改进，克服了数据分析上的难题，使用了 300 000bp 以上长度序列进行了系统发育分析，对 5 000 个功能基因进行了适应性进化检测，获得了血吸虫进化及其与宿主间相互作用的分子证据，推动了重要成果在 *Nature* 杂志上发表（The *Schistosoma japonicum* genome reveals features of host-parasite interplay），钟扬教授作为论文作者 PI 之一。

2005 年，钟扬教授在与中国科技大学施蕴渝院士课题组的合作中，提出了基于结构信息的进化分析新方法。针对泛素超家族的起源问题，利用新方法确认 Urm1 蛋白就是泛素超家族的"分子化石"。论文于 2006 年以共同通讯作者发表于 *PNAS* (Solution structure of Urm1 and its implications for the origin of protein modifiers)，并被 *Nature*、*Nature Chemical Biology* 和 *EMBO Journal* 等顶尖杂志的论文所引用，特别是瑞士联邦高工（ETH）的 Matthias Peter 教授领导的研究小组基于该"分子化石"的结论，在 *Nature* 杂志撰文进一步明确了 Urm1 在真核生物转运 RNA 硫醇化中的作用。

2003 至 2004 年间，钟扬教授参与了由中国科学院上海生命科学研究院赵国屏院士领衔、国内外数十家单位加盟的 SARS 冠状病毒分子进化分析工作。对 61 个 SARS 病毒全基因组序列进行了系统发育分析。通过对流行病学所划分的 SARS 在中国流行期间不同阶段的同义置换与非同义置换比率进行分析，发现 SARS 冠状病毒在早期传播时存在不同选择式样。该论文发表于 *Science* (Molecular evolution of the SARS coronavirus during the course of the SARS epidemic in China)，钟扬教授为并列第一作者。

2009 年，钟扬教授在一项合作研究中，建立了 C_3 植物光合作用代谢的系统生物学分析模型，并进行了计算机模拟，发现了环境扰动下代谢通路间协调性增强的规律，这在植物系统生物学研究中属开创性工作。论文在 *PNAS* (Photosynthetic metabolism of C-3 plants shows highly cooperative regulation under changing environments: a systems biological analysis) 上发表后被 *Journal of Biological Chemistry*、*BMC Systems Biology* 和 *Plant Physiology* 等杂志多次引用。该工作是国家技术发明二等奖（2013 年）重要组成部分，为节水抗旱稻的创制提供了理论基础。

Testing hybridization hypotheses based on incongruent gene trees
检验基于不一致基因树的杂交假说

杂交在植物中是一个非常重要的进化机制，在动物中也已有了越来越多的记录。然而，重建网状进化的困难，是系统发育学中长期存在的问题。因此，杂交成种可能在导致基因树间的拓扑不一致上起着关键作用。反过来，这种不一致性提供了一个检测杂交成种的可能性。本文描述了杂交同其他生物学过程之间的一些差异，包括谱系分选、旁系同源，以及基因水平转移，它们是基因树间不一致性的原因。我们考虑了三个类群（A、B和C）的两个不一致树，B在基因树1中是A的姐妹群，但在基因树2中是C的姐妹群。利用一个基于分子钟的理论模型，我们证明了在杂交（类群B为杂交）或基因水平转移的情况中，类群A和C的每一个基因的分歧时间几乎相同，但在谱系分选或旁系同源的情况中则有显著差异。用自展检验（bootstrap test）测试了这些备选假设后，我们对这一模型进行了扩展并通过检验来解释多类群的不一致基因树。计算机模拟研究支持了当每个基因以固定速率进化时理论模型和自展检验的有效性。计算机模拟也验证了当同一类群的两个基因的速率的差异遵循比例时模型仍然有效。虽然这个模型不能检测不一致性的原因是杂交假说还是基因水平转移，但这两个过程可以通过比较多个不连锁的基因的系统发育关系来区分。

Sang, T., **Zhong, Y.** Testing hybridization hypotheses based on incongruent gene trees. 2000. *Systematic Biology*, 49(3): 422-434. by permission of Oxford University Press.

Testing hybridization hypotheses based on incongruent gene trees

Tao Sang[1] And Yang Zhong[2]

Abstract

Hybridization is an important evolutionary mechanism in plants and has been increasingly documented in animals. Difficulty in reconstruction of reticulate evolution, however, has been a long standing problem in phylogenetics. Consequently, hybrid speciation may play a major role in causing topological incongruence between gene trees. The incongruence, in turn, offers an opportunity to detect hybrid speciation. Here we characterized certain distinctions between hybridization and other biological processes, including lineage sorting, paralogy, and lateral gene transfer, that are responsible for topological incongruence between gene trees. Consider two incongruent gene trees with three taxa, A, B, and C, where B is a sister group of A on gene tree 1 but a sister group of C on gene tree 2. With a theoretical model based on the molecular clock, we demonstrate that time of divergence of each gene between taxa A and C is nearly equal in the

---

[1] Department of Botany and Plant Pathology, Michigan State University, East Lansing, Michigan 48824, USA; E-mail: sang@pilot.msu.edu.
[2] Institute of Biodiversity Science, School of Life Sciences, Fudan University, Shanghai 200433, China; E-mail: yangzhong@fudan.edu.

case of hybridization (B is a hybrid) or lateral gene transfer, but differs significantly in the case of lineage sorting or paralogy. After developing a bootstrap test to test these alternative hypotheses, we extended the model and test to account for incongruent gene trees with numerous taxa. Computer simulation studies supported the validity of the theoretical model and bootstrap test when each gene evolved at a constant rate. The computer simulation also suggested that the model remained valid as long as the rate heterogeneity was occurring proportionally in the same taxa for both genes. Although the model could not test hypotheses of hybridization versus lateral gene transfer as the cause of incongruence, these two processes may be distinguished by comparing phylogenies of multiple unlinked genes.

Keywords

Gene tree; hybridization; phylogeny; species tree; topological incongruence.

Hybridization, especially when coupled with polyploidization, is an important evolutionary mechanism in plants [Stebbins, 1950; Grant, 1981]. Masterson (1994) has suggested that ~70% of angiosperms are polyploids, implying the possibility of a tremendous amount of hybridization in the evolutionary history of flowering plants. In animals, hybrid species are increasingly being documented with the application of molecular markers [Bullini, 1994].

Although hybridization has received considerable attention from evolutionary biologists [e.g., Arnold, 1997; Rieseberg, 1997], accurate reconstruction of hybrid speciation remains challenging. The reticulate nature of hybrid speciation fails to meet the basic assumption of cladistics that speciation occurs in a bifurcated manner [Hennig, 1966]. Consequently, a phylogenetic tree containing unidentified hybrids remains an inaccurate reconstruction of the phylogeny. Furthermore, including hybrids in a phylogenetic analysis may result in an increased amount of homoplasy and possibly disrupt relationships of the other taxa [McDade, 1995].

Because of theoretical and technical difficulties involved in direct reconstruction of reticulate evolution, efforts have been made to explore alternative methods and new sources of data for improving our ability to reconstruct hybrid speciation. Three approaches have been used most frequently: (1) identifying hybrids before the phylogenetic analyses, (2) detecting hybrids according to their cladistic behavior, and (3) reconstructing hybrid speciation by comparing discordant positions of hybrids between phylogenetic trees obtained from independent data sets.

With the realization that inclusion of unidentified hybrids in a cladistic analysis may distort the relationships of other taxa, it is logical to follow the approach of identifying hybridization before starting the analysis, excluding the hybrids from the analysis, and adding the hybrids back onto the tree by connecting them with the putative parents. Morphological and molecular intermediacy or combination (or both) have served as criteria for identification of hybrids and their putative parents. In particular, continued development of new molecular markers, such as allozyme, restriction fragment length polymorphism, RAPD, ISSR, and DNA sequences, has contributed enormously to the accuracy of identification of hybrids and their parentage [Rieseberg & Ellstrand, 1993; Sang, et al., 1995; Campbell, et al., 1997; Wolfe, et al., 1998]. Difficulty arises, however, when the hybrid is too ancient to maintain recognizable morphological or molecular intermediacy/combination.

Instead of avoiding the potential problems caused by hybrids in a cladistic analysis, the second approach chooses to identify a hybrid based on its behavior on a cladogram [Funk, 1985]. Similar to the first approach, it relies on the assumption that a hybrid maintains morphological or molecular intermediacy/combination between the parents. The intermediacy or combination can be reflected in cladistic characters that are intermediate or polymorphic relative to those of the parents. Consequently, inclusion of the hybrid will increase the amount of homoplasy, largely because of parallelism between the hybrid and the parents [Funk, 1985]. It is expected, therefore, that homoplasy can be considerably diminished by removing the hybrid from the data matrix. A computer program, RETICLAD, has been designed to implement this approach [Rieseberg & Morefield, 1995]. This approach, however, may also be problematic when dealing with ancient hybrids in which morphological or molecular intermediacy/combination has been obscured. In addition, other factors, such as ancestral polymorphism or convergent evolution, can introduce homoplasy.

Extensive uses of chloroplast DNA (cpDNA) in plant phylogenetic studies have led to the third approach, which detects hybridization by comparing phylogenetic trees. In the majority of angiosperm species, cpDNA is maternally inherited [Mogensen, 1996]. In the case of biparental inheritance, the polymorphism can be fixed relatively rapidly because of the small effective population size of cpDNA, which is one-quarter of a nuclear gene [Moore, 1995]. Therefore, a cpDNA tree will most likely represent a uniparental phylogeny, in most cases, the maternal genealogy. When the cpDNA phylogeny is compared with another phylogenetic tree (morphology or molecular), conflicting positions of a taxon between these trees may be viewed as evidence for the hybrid origin of this taxon. A taxon possessing the cpDNA from a morphologically distinct taxon is known as cpDNA capture [Rieseberg & Soltis, 1991].

Besides cpDNA, nuclear ribosomal DNA (nrDNA) has been used most frequently in molecular phylogenetic studies of plants. When a species has substantially discordant positions between the cpDNA and nrDNA phylogenies, there is a possibility that the species may be a hybrid [Soltis & Kuzoff, 1995; Kellogg, *et al.*, 1996]. The hybrid could have inherited cpDNA form the maternal parent and fixed nrDNA sequences of the paternal parent by way of concerted evolution and thus would have different sister group relationships between the cpDNA and nrDNA trees [Wendel, *et al.*, 1995; Sang, *et al.*, 1997]. Comparing incongruence between gene trees thus opens the opportunity of reconstructing ancient hybridization, an event for which morphological intermediacy and molecular additivity in the hybrid subsequently have been obscured.

However, factors other than hybridization, such as random sorting of ancestral polymorphism (lineage sorting), gene duplication/deletion (paralogy), lateral gene transfer, or erroneous phylogenetic reconstruction, can also cause topological incongruence between gene trees. For example, paralogy in nrDNA repeats could potentially lead to inaccurate phylogenetic reconstructions in some plant groups, although nrDNA evolves together through concerted evolution [Buckler, et al., 1997]. Understanding and dealing with incongruence among gene trees are among the most acute theoretical issues in phylogenetics and have attracted considerable attention [e.g., de Queiroz, et al., 1995; Huelsenbeck, et al., 1996; Doyle, 1997; Maddison, 1997; Wendel & Doyle, 1998]. Determining whether topological incongruence is caused by hybridization or by other factors, therefore, represents a major challenge to the approach of detecting hybridization by gene tree comparison. Here we will focus our discussion on incongruence caused by different phylogenetic histories of data sets and leave out the issue of erroneous phylogenetic reconstruction.

Maddison (1997) developed a theoretical model to distinguish the biological processes that may potentially cause incongruence between gene trees, including lineage sorting, gene duplication/extinction, and lateral gene transfer/hybridization. He used a cladistic approach in which the number of events required to convert a gene tree to the species tree by assuming a certain process was counted; the process that required the fewest events was considered to be the cause of the incongruence between gene trees. However, at least two practical problems are associated with this model: the algorithmic difficulty of assessing all the possible topologies for a large number of species, and the determination of the appropriate weight of each event of different processes [Maddison, 1997].

In this paper, we attempt to depict patterns of topological incongruence between gene trees. Then, we focus on characterization of certain distinctions between hybridization and other processes as causes of topological incongruence. At first, to develop theoretical models, we contrast hybridization and lineage sorting, two processes often considered to be competing hypotheses for the incongruence of gene trees at lower taxonomical levels. Then we discuss how the models can be extended for testing other hypotheses, such as paralogy and lateral gene transfer, and ultimately how the hybridization hypothesis can be tested versus the rest of the hypotheses responsible for topological incongruence between gene trees.

Models for comparing incongruent gene trees

Three-Species Model

Let us first compare incongruent gene trees with only three ingroup species. Figures 1a and 1b show two gene trees generated from sequences of gene 1 and gene 2 of the ingroup species A, B, and C and the outgroup O. A_1, B_1, C_1, and O_1 are sequences of gene 1 from species A, B, C, and O, respectively; A_2, B_2, C_2, and O_2 are sequences of gene 2 from species A, B, C, and O, respectively. The two gene trees have incongruent topologies because of the discordant positions of the genes of species B; that is, B_1 forms a sister group with C_1 on gene tree 1, whereas B_2 is a sister group of A_2 on gene tree 2 (Figs. 1a, 1b).

Based on this simple example, a basic theoretical model can be constructed for testing the likelihood of hybridization versus lineage sorting or paralogy as the cause of the incongruence between the gene trees. On both gene trees, let T_0 be the time (million years before present) when genes of the in group species diverged from those of the outgroup species. T_0 is also the time when the ingroup species diverged from the outgroup species on the species tree, if the outgroup is selected in such a way that its relationship with the ingroup is most likely to represent a true species relationship. On gene tree 1, let T_i be the time when gene A_1 diverged from genes B_1 and C_1, and T_j be the time when B_1 diverged from C_1. On gene tree 2, let T_k

be the time when gene C_2 diverged from genes A_2 and B_2, and T_m be the time when A_2 diverged from B_2.

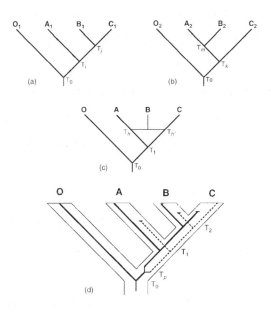

Fig. 1 Gene trees and species trees of three-species model, where A, B, and C are ingroup species and O is an outgroup species. T_i, T_j, T_k, and Tm represent divergence times between genes A_1 and (B_1, C_1), B_1 and C_1, C_2 and (A_2, B_2), and A_2 and B_2, respectively. T_h and $T_{h'}$ represent times when the lineages that hybridized to give rise to B diverged from A and C, respectively. T_0, T_1, and T_2 represent times of speciation. T_p represents time of occurrence of polymorphic alleles. (a) Tree of gene 1. A_1, B_1, C_1, and O_1 are sequences of gene 1 of species A, B, C, and O, respectively. (b) Tree of gene 2. A_2, B_2, C_2, and O_2 are sequences of gene 2 of species A, B, C, and O, respectively. (c) Species tree inferred based on the hypothesis that B is a hybrid species. (d) Species tree (out lined by thin solid lines) inferred on the basis that gene 2 has undergone lineage sorting. Phylogeny of gene 2 is illustrated by thick solid and broken lines representing two ancestral alleles.

If this incongruence is caused by hybridization, then the species tree, with B being the hybrid and A and C being the parental lineages, can be inferred (Fig. 1c). The hybrid species B fixed sequences of gene 1 that are similar to those of species C, and thus genes B_1 and C_1 form a sister group on gene tree 1. Meanwhile, the hybrid B inherited gene 2 from a parental lineage that is closely related to A, resulting in a sister group relationship between B_2 and A_2. Let T_1 be the time when species A and C diverged, and T_h and $T_{h'}$ be the times when the lineages that hybridized to give rise to B diverged from A and C, respectively. T_h will not equal $T_{h'}$ when the hybridization is ancient and the parental lineages that hybridized to produce B are extinct and diverged from A and C at different times.

Alternatively, lineage sorting of one gene may be the cause of the incongruence between the gene trees. Assume that gene tree 1 represents the species tree and gene 2 has undergone lineage sorting. The species tree and the phylogeny of gene 2 are illustrated in Figure 1d. Two polymorphic alleles of gene 2 arose in the common ancestor of species A, B, and C. Subsequently, one allele is maintained only in A and B, and the other is maintained only in C (Fig. 1d). Gene tree 2, reconstructed from these alleles, thus differs from the species tree. Let T_p be the time when the polymorphic alleles of gene 2 arose, T_1 be the time when species A diverged from species B and C, and T_2 be the time when species B and C diverged.

Under the hybridization hypothesis, $T_i=T_1$ and $T_j=T_{h'}$; $T_k=T_1$ and $T_m=T_h$ (Figs. 1a-c). Then, $T_i=T_k$. Under the lineage sorting hypothesis, $T_i=T_1$ and $T_j=T_2$; $T_k=T_p$, $T_m=T_1$ (Figs. 1a, 1b, 1d). Because $T_p>T_1$, then $T_k>T_i$. Therefore, we can test hypotheses of hybridization versus lineage sorting by testing $T_k=T_i$, or $T_k-T_i=0$. Defining $\Delta(x, y) = T_x-T_y$, we can test:

$$\Delta(k, i) = \begin{cases} < 0, & \text{if hybridization;} \\ a(a > 0), & \text{if lineage-sorting for gene 2.} \end{cases} \quad (1)$$

In the real case, we do not know which gene has undergone lineage sorting before the test. If $\Delta(k, i)<0$, lineage sorting may have occurred in gene 1.

To calculate $\Delta(k, i)$, the following expressions can be defined and inferred. Define $d(U_i, V_i)$ as the estimated number of nucleotide substitutions between sequences U_i and V_i of gene i of species U and V, respectively. Define r_i as the rate of nucleotide substitutions of gene i. Assuming the existence of a molecular clock, that is, a constant rate of nucleotide substitutions for both genes [Li, 1997], the substitution rates of gene 1 and gene 2 can be calculated as follows (Figs. 1a, 1b):

$$r_1 = \frac{d(O_1, A_1) + d(O_1, C_1)}{4T_0} = \frac{d(A_1, C_1)}{2T_i}, \quad (2)$$

$$r_2 = \frac{d(O_2, A_2) + d(O_2, C_2)}{4T_0} = \frac{d(A_2, C_2)}{2T_k}. \quad (3)$$

From Equations 2 and 3, we obtain

$$T_i = \frac{2d(A_1, C_1)T_0}{d(O_1, A_1) + d(O_1, C_1)}, \quad (4)$$

$$T_k = \frac{2d(A_2, C_2)T_0}{d(O_2, A_2) + d(O_2, C_2)}. \quad (5)$$

Therefore,

$$\Delta(k, i) = \left[\frac{2d(A_2, C_2)}{d(O_2, A_2) + d(O_2, C_2)} - \frac{2d(A_1, C_1)}{d(O_1, A_1) + d(O_1, C_1)} \right] T_0, \quad (6)$$

or

$$\Delta(k, i) = \Delta_0 T_0 \quad (7)$$

where

$$\Delta_0 = \frac{2d(A_2, C_2)}{d(O_2, A_2) + d(O_2, C_2)} - \frac{2d(A_1, C_1)}{d(O_1, A_1) + d(O_1, C_1)}. \quad (8)$$

Four-Species Model

The four-species model contains an additional ingroup species, D. The times of gene divergence are labeled on the two gene trees (Figs. 2a, 2b). The two genes of species C, C_1 and C_2, have discordant positions on the two gene trees, which leads to incongruence between the gene trees. Like the three-species model, the incongruence can be explained by either hybridization or lineage sorting. The species tree of hybridization (C is a hybrid between A and D) and the times of speciation and hybridization are shown in Fig. 2c. Under this hypothesis, $T_i=T_m=T_1$, $T_j=T_n=T_2$, $T_k=T_{h'}$, and $T_q=T_h$.

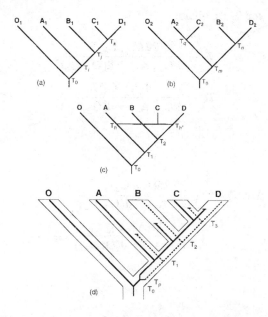

Fig. 2 Gene trees and species trees of four-species model, where A, B, C, and D are ingroup species and O is an outgroup species. (a) Tree of gene 1. (b) Tree of gene 2. (c) Species tree inferred based on the hypothesis that species C is a hybrid. (d) Species tree inferred on the basis that gene 2 has undergone lineage sorting.

Alternatively, the incongruence between the gene trees is caused by lineage sorting. Assuming that gene 1 represents the species tree and gene 2 has undergone lineage sorting, the species tree and the contained phylogeny of gene 2 are illustrated in Figure 2d. Under this hypothesis, $T_i=T_q=T_1$, $T_j=T_n=T_2$, $T_k=T_3$, and $T_m=T_p$. Because $T_p>T_1$, then $T_m>T_i$.

Similar to the three-species model, these two hypotheses can be tested by testing $\Delta(m, i)=0$, where

$$\Delta(m, i) \stackrel{\ddagger}{=} \frac{2d(A_2, D_2)}{d(O_2, A_2) + d(O_2, D_2)} - \frac{2d(A_1, D_1)}{d(O_1, A_1) + d(O_1, D_1)} \cdot T_0. \quad (9)$$

Five-Species Model

The observed gene trees and hypothetical species trees of the five-species model are shown in Figure 3. Genes of species D, D_1 and D_2, display discordant positions on the two gene trees. Under the hybridization hypothesis, in which D is a hybrid between B and C (Fig. 3c), $T_i=T_n=T_1$, $T_j=T_r=T_2$, $T_k=T_q=T_3$, $T_m=T_{h'}$,

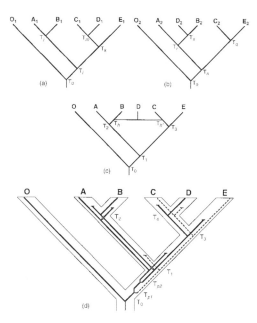

Fig. 3 Gene trees and species trees of five-species model, where A, B, C, D, and E are ingroup species and O is an outgroup species. (a) Tree of gene 1. (b) Tree of gene 2. (c) Species tree inferred based on the hypothesis that species D is a hybrid. (d) Species tree inferred on the basis that gene 2 has undergone lineage sorting.

and $T_s=T_h$. To explain the incongruence by lineage sorting of gene 2 (Fig. 3d), it is necessary to assume that polymorphic alleles of the gene arose twice, at times, T_{p1} and T_{p2}, in the common ancestor of the ingroup species. Under this hypothesis, $T_i=T_s=T_1$, $T_j=T_2$, $T_k=T_q=T_3$, $T_m=T_4$, $T_n=T_{p1}$, and $T_r=T_{p2}$. Because $T_{p1}>T_1$ and $T_{p2}>T_2$, then $T_n>T_i$ and $T_r>T_j$.

In this model, alternative hypotheses can be tested at two branching points of the gene trees: $\Delta(n, i)=0$ and $\Delta(r, j)=0$. The differences are calculated as follows:

$$\Delta(n, i) = \left[\frac{2d(B_2, C_2)}{d(O_2, B_2) + d(O_2, C_2)} - \frac{2d(B_1, C_1)}{d(O_1, B_1) + d(O_1, C_1)}\right] T_0, \quad (10)$$

$$\Delta(r, j) = \left[\frac{2d(A_2, B_2)}{d(O_2, A_2) + d(O_2, B_2)} - \frac{2d(A_1, B_1)}{d(O_1, A_1) + d(O_1, B_1)}\right] T_0. \quad (11)$$

If both differences are not significantly different from 0, the hybridization hypothesis is favored. The lineage-sorting hypothesis is supported if both differences are significantly larger than 0.

Numerous-Species Model

If incongruence between two gene trees is caused by one species, the incongruent gene trees with any given number of species can be simplified to fit one of the three-, four-, or five-species models. In other words, these simple models can be extended to accommodate topological incongruence between gene trees with any given number of species after a simplification process, that is, converting a monophyletic or paraphyletic group into an individual species. For example, species h has different positions between two gene trees of 11 ingroup species, a, b, c, d, e, f, g, h, i, j, and k (Fig. 4). First, positions of some of the sister groups were switched without altering the topology of the gene trees. Second, the monophyletic groups, (d, e) and (a, (b, c)), are designated as single species, A and B, respectively; the paraphyletic group, (f((j,k),g)), is designated as E. With these rearrangements and conversions, the new trees fit the five-species model (Figs. 3, 4).

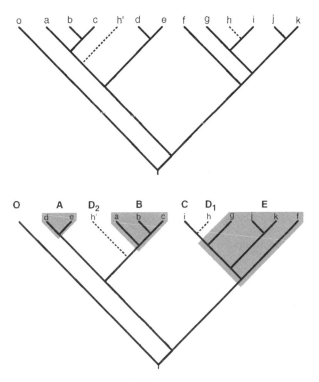

Fig. 4 **Simplification of incongruent gene trees with eleven ingroup species (a-k). (Top) Incongruence between two gene trees caused by conflicting positions of species h. Dashed branches represent different positions of h on two gene trees, whereas relationships of the remaining species are the same between the gene trees. (Bottom) Simplified gene trees that fit the five-species model.**

With regard to hypothesis testing, one or more species can be chosen from a reduced group to represent this group. If two or more species are chosen, the hypotheses can be tested multiple times. Conducting the test over all possible combinations of the species may increase the amount of confidence of the hypothesis testing. Sequences that do not evolve in a clocklike fashion, however, should not be chosen to represent the group.

Statistical test

A statistical test was designed to test hypotheses of hybridization versus lineage sorting, based on the theoretical models described above. For the three-species model, we need to test whether $\Delta(k, i)$ is significantly larger than 0 (Equation 1). According to Equation 7, testing $\Delta(k, i)=0$ is equal to testing $\Delta_0=0$.

The number of nucleotide substitutions, d, can be estimated directly from sequence divergence when the sequence divergence is low. Otherwise, correction models, such as the Jukes-Cantor (1969) model or the Kimura (1980) two-parameter model, should be used to correct for possible multiple hits. Each estimated number of nucleotide substitutions has a variance that can be calculated depending on which model is used. However, calculation of the variance of Δ_0 based on these individual variances and their covariances (e.g., by using Taylor progression) remains difficult. The variance of the ratio of nucleotide substitutions tends to be too large to permit any reasonable power of a significance test (B. Gaut, pers. comm.; Z. Yang, pers. comm.).

Therefore, we chose the bootstrap method to carry out the statistical test [Efron, 1979; Felsenstein, 1985; B. Gaut, pers. comm.; Z. Yang, pers. comm.]. Define $\Delta_0^{(j)}$ as a Δ_0 calculated from the jth bootstrapped data sets, where $j=1, 2,..., n$. To estimate the deviation of Δ_0 from the resampled information, the absolute values of $\Delta_0^{(j)}$ were used in the following calculation. The mean of $|\Delta_0^{(j)}|$ can be estimated as $M=\Sigma|\Delta_0^{(j)}|/n$. The variance of $|\Delta_0^{(j)}|$ is calculated as $V=(\Sigma|\Delta_0^{(j)}|-M)^2/n$. The statistics of the significance test can be calculated as $Z=[|\Delta_0|-M]/V^{1/2}$.

Table 1 Combinations of changes of parameters studied by computer simulation when $T_0=15$. In the case of the lack of a molecular clock, a correlation of rate heterogeneity is assumed between two genes; that is, the substitution rates of both genes in species A are either 1.5 times faster or 0.8 times slower than in the other species.

	With molecular clock						Without molecular clock $r_1=1, r_2=1$			
	$r_1=1, r_2=1$		$r_1=1, r_2=2$		$r_1=2, r_2=1$		$r_A=1.5$		$r_A=0.8$	
	T_i	T_k	T_i	T_k	T_i	T_k	T_i	T_k	T_i	T_k
Hybridization Lineage sorting	10	10	10	10	10	10	10	10	10	10
	10	11	10	11	10	11	10	11	10	11
	10	12	10	12	10	12	10	12	10	12

Computer simulation

Computer simulation increasingly has been applied to exploration of theoretical issues in phylogenetics [e.g., Hillis, *et al.*, 1994; Huelsenbeck, 1995]. A computer simulation is performed here to examine the validity of the theoretical models and the power of the statistical test under various evolutionary assumptions. Only the three-species model is examined with the computer simulation study. The implications of the simulation results should apply to the four- and five-species models, however, because of the similar nature of the test for all the models.

Two 1-kb DNA sequences, S_1 and S_2, were randomly generated and assigned to gene 1 and gene 2 of the most recent common ancestor of the outgroup and ingroup species (Figs. 1a, 1b), respectively. The G-C content of each sequence was set at 50%. Rates of nucleotide substitutions (r_i) were assumed to be 1 or 2 substitutions per 1 000 sites per million years. This corresponds to 1 (or 2)$\times 10^{-9}$ substitutions per site per year, which is biologically reasonable [Li, 1997]. When $r_i=1$, one substitution was generated randomly for a sequence during a one-million=year interval along a gene tree. All types of substitutions were treated at an equal probability. Through this process, gene sequences of the species A_1, B_1, C_1, O_1, A_2, B_2, C_2, and O_2 were generated.

The simulation experiment, with the same initial sequences of S_1 and S_2, was performed 20 times for testing hypotheses of hybridization versus lineage sorting and for testing the impact of several parameters on the model. The parameters, which were tested under the hypotheses of hybridization (T_k-T_i=0) or lineage sorting (T_k-T_i=1, 2), include relative substitution rates between genes 1 and 2, the time of divergence between ingroup and outgroup, and the molecular clock. With the molecular clock engaged (constant substitution rates for both genes), we tested hypotheses of hybridization and lineage sorting when the relative substitution rates of the genes were the same or when one gene evolved twice as fast as the other (Table 1). We also tested the hypotheses when substitution rates were not constant for either gene, but with the restriction that the rate heterogeneity was correlated between two genes; that is, the substitution rates of both genes increased or decreased proportionally in the same taxa (Table 1). These tests were done under the conditions of T_0=15 and T_j=T_m=5 (Fig. 1; Table 1). Assigning different values to T_j and T_m would not affect testing results (data not shown).

We explored the impact of increasing the overall sequence divergence on the hypothesis testing. When the overall divergence was doubled — T_0=30 and T_j=T_m=10 — we tested the following combinations under condition of r_1=r_2=1: T_i=20 and T_k=20; T_i=20 and T_k=21; T_i=20 and T_k=22; T_i=20 and T_k=23; and T_i=20 and T_k=24.

Using sequences generated from the computer simulation, each Δ_0 was calculated with Equation 8. Given the relatively low divergence between the sequences (up to 3%), the number of nucleotide substitutions was estimated directly from the sequence divergence. The difference between each Δ_0 and 0 was tested through 100 bootstrap replications.

Results and discussion

The computer simulation study demonstrated that the statistical test is able to support either hybridization or lineage-sorting hypotheses predicted by the theoretical model. If each gene evolves at a constant rate, then Δ_0 equals the ratio of $(T_k-T_i)/T_0$ (see proof in Appendix). Thus, when T_0 is fixed, the model predicts that the greater the difference between T_k and T_i, the greater the Δ_0 value, that is, the more likely that lineage sorting has occurred. When T_0 was set at 15 and the hybridization hypothesis is engaged ($T_i=T_k=10$), the Δ_0 values fluctuate around 0 (Fig. 5a), and > 80% of $\Delta 0$ values are not significantly larger than 0 ($P<0.05$) (Fig. 6a). Therefore, the hybridization hypothesis was confirmed by the test in >80% of simulation runs.

Under the lineage-sorting hypothesis and when $T_k-T_i=1$, almost all the Δ_0 values are >0 (Fig. 5a), and 45% ($P<0.01$) to 85% ($P<0.05$) of the values are significantly >0 (Fig. 6a). These results imply that when the ingroup has diverged from the outgroup for 15 million years and when the ancestral polymorphic alleles occurred one million years before the divergence of species A and C, the chance that the test is able to confirm the lineage sorting is ~50% or more. When $T_k-T_i=2$, all the Δ_0 values are >0 (Fig. 5a), and >90% of test results are significant at $P<0.01$ (Fig. 6a). This suggests that the chance of detecting lineage sorting increases rather rapidly as the time of origination of the ancestral polymorphism becomes more ancient relative to the time of the divergence of species A and C (Figs. 1d, 6a).

A slight fluctuation of Δ_0 during 20 runs of simulation for each combination is due to the small number of parallel substitutions and multiple hits. We tried to correct each of the sequence divergence by using the one parameter model [Jukes & Cantor, 1969], but the degree of fluctuation of Δ_0 values was not reduced significantly thereby (data not shown). This is probably a result of the calculation of the ratio of the sequence divergence, which canceled the effect of the correction. The fact that the simulation studies supported the theoretical models in the presence of multiple hits suggests that the models and statistical test can tolerate a certain number of multiple hits.

The computer simulation study demonstrated that doubling the overall divergence did not alter the test results markedly. When T_0 was increased to 30, the number of significant test results when T_k-T_i=2 was comparable with that for when T_k-T_i=1 and T_0=15 (Figs. 5a, 5d, 6a, 6d). Because Δ_0 remains the same when T_k-T_i and T_0 increase or decrease proportionally, the model should work at various values for sequence divergence, which correspond, in most cases, to various taxonomical values. However, if the divergence of the ingroup is fixed, that is, if T_k-T_i remains constant, increasing T_0 (choosing a more distantly related outgroup) will lead to a smaller Δ_0. The number of significant test values when T_0=30 and T_k-T_i=1 is intermediate between those of T_k-T_i=0 and T_k-T_i=1 when T_0=15 (Figs. 5a, 5d, 6a, 6d), which suggests that choosing an outgroup that is more closely related to the ingroup will permit a more sensitive test for lineage sorting.

Because rates of nucleotide substitution vary widely among genes [Li, 1997], one must determine whether the theoretical model can still hold when two genes evolve at different rates. Theoretically, Δ_0 should not be affected by rate differences between the two genes, as long as the substitution rate of each gene is constant (see proof in Appendix). The simulation results supported this prediction. In the case where the substitution rate of either gene 1 or gene 2 was doubled, the test results were similar in both cases and were also similar to that obtained by assuming the equal substitution rate of the two genes (Figs. 5a-c, 6a-c).

However, the lack of constant substitution rates of one or both genes among the studied species will violate the basic assumption of the model. Using simulation studies, we explored the validity of the model when the rate heterogeneity was correlated between the two genes; that is, rates increased or decreased proportionally in both genes of the same species. Satisfying this condition should extend application of the model, given that the rate heterogeneity caused by factors such as generation time effect can be correlated between genes or even across different genomes [Wu & Li, 1985; Gaut, et al., 1992]. For example, grasses evolve more rapidly than palms at synonymous sites in a mitochondrial, a nuclear, and a

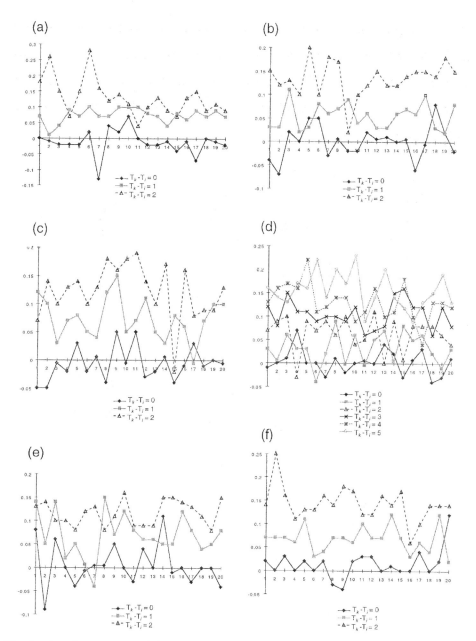

Fig. 5 Δ_0 **values generated from computer simulation.** Each Δ_0 value is represented by an abscissa; Δ_0 values resulting from 20 simulation runs for a combination of a hypothesis and various parameters are connected by a certain type of line for clearer visualization (Table 1). (a) $T_0=15$, $r_1=r_2=1$. (b) $T_0=15$, $r_1=1$, $r_2=2$. (c) $T_0=15$, $r_1=2$, $r_2=1$. (d) $T_0=30$, $r_1=r_2=1$. (e) $T_0=15$, $r_1=r_2=1$, $r_A=1.5$. (f) $T_0=15$, $r_1=r_2=1$, $r_A=0.8$.

chloroplast gene, and the rate increases in grasses are correlated among these genes [Eyre-Walker & Gaut, 1997]. The results of the simulation study (Figs. 5e-f, 6e-f), in which substitution rates of both genes of species A were proportionally higher or lower than those of species C and O, suggested that the model should be valid under such a condition.

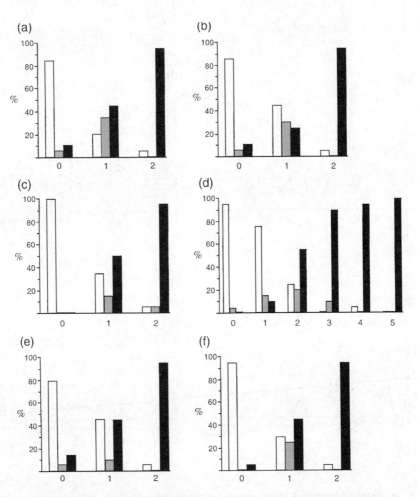

Fig. 6 Test results of computer simulation study. (a), (b), (c), (d), (e), and (f) show test results of Δ_0 presented in Figures 5a, 5b, 5c, 5d, 5e, and 5f, respectively. Histograms illustrate percentages of probabilities of the test results of 20 simulation runs for each combination of a hypothesis and various parameters; $P>0.05$ (open), $0.05>P>0.01$ (shaded), and $P<0.01$ (solid). Numbers 0, 1, 2, 3, 4, and 5 below the histograms represent corresponding values of (T_k-T_i) in Figure 5.

Because the validity of the theoretical models relies on the molecular clock for both genes, a relative rate test must be performed before the hybridization hypothesis can be tested. However, the degree of sensitivity to rate heterogeneity may differ between the relative rate test [Wu & Li, 1985] and the present bootstrap test. Problems may arise if the bootstrap test is more sensitive than the relative rate test to rate heterogeneity. In other words, the rate heterogeneity can be small enough to pass the relative rate test but large enough to affect the results of hypothesis testing with the bootstrap test. To address this question, computer simulation was designed to assess the relative sensitivity of the two tests to the degree of rate heterogeneity. Under the hybridization hypothesis (T_k-T_i=0), the substitution rate of gene A_2 was increased while those of the other genes were kept the same (Table 2). The probabilities obtained from both tests are compared in Table 2.

Table 2 Comparison of probabilities obtained from relative rate [Wu & Li, 1985] under the Jukes-Cantor (1969) model and bootstrap tests based on the three-species model. For both gene trees, T_0=15 and T_i=T_k=10; r_1=r_2=1 for all the sequences except r_{A_2}, which varies from 1.2, 1.3, or 1.4 to 1.5, meaning that the substitution rate of sequence A_2 is 1.2, 1.3, 1.4, or 1.5 times faster than those of the remaining sequences.

	r_{A_2}			
	1.5	1.4	1.3	1.2
Relative rate test	0.16	0.21	0.27	0.34
Bootstrap test	<0.01	<0.01	0.04	0.20

When substitution rates are variable among the species and the rate variation is not correlated between the two genes, the model is rather sensitive to rate heterogeneity. We explored the extent to which rate heterogeneity misleads the test results. Under the hybridization model (T_k-T_i=0), the hybridization hypothesis was rejected ($P<0.01$) when gene A_2 evolved >1.4 times more rapidly than the other genes (Table 2). When A_2 evolved 1.3 times more rapidly, the hybridization was still rejected ($P=0.04$). When the substitution rate of A_2 was 1.2 times greater than that of the other genes, the hybridization hypothesis could not be rejected.

Therefore, although our model and test can tolerate a certain degree of rate heterogeneity, they appear to be quite sensitive to rate heterogeneity, at least more sensitive than the relative rate test of Wu and Li (1985). Consequently, the results of the bootstrap test may not be free of bias attributable to the rate heterogeneity, even though the molecular clock is not rejected by the relative rate test. Future work should explore further the impact of the degree of rate heterogeneity and the way of measuring it on the efficiency of the bootstrap test. For example, a maximum likelihood method represents another approach to reconstructing a phylogeny and estimating branch lengths. If sequences of two genes are determined by a likelihood ratio test to have evolved among the studied species in a clocklike fashion [Huelsenbeck & Rannala, 1997; Baldwin & Sanderson, 1998], then the branch lengths estimated by maximum likelihood can be used in the bootstrap test.

So far we have discussed lineage sorting only as an alternative explanation to hybridization in causing incongruence between gene trees. Now we extend our discussion to account for the other two biological processes that are also responsible for the departure of gene trees from the species tree: gene duplication/extinction (paralogy), and lateral gene transfer [de Queiroz, et al., 1995; Doyle, 1997; Maddison, 1997; Wendel & Doyle, 1998].

Paralogy has an impact on the topology of gene trees similar to that of lineage sorting and can be tested by using the same model: $\Delta(k, i) > 0$. In these models, paralogy is analogous to lineage sorting when T_k is viewed as the time of duplication of two loci instead of two alleles of the same locus. The subsequent extinction of one allele from a species under the lineage-sorting hypothesis is analogous to extinction or incomplete sampling of one of the loci in the paralogy hypothesis. The same species tree can be inferred no matter which of the two processes caused the incongruence.

Despite the similar behavior of lineage sorting and paralogy in the models, they are different biological processes that can be distinguished in certain aspects [Maddison, 1997; Wendel & Doyle, 1998]. Basically, the occurrence of lineage sorting depends on a persistence of ancestral alleles through a common ancestor of the diverged species. Therefore, lineage sorting can often be found when an ancestral branch leading to the species is short (few generations) and wide (large effective population size) [Pamilo & Nei, 1988; Hudson, 1992; Maddison, 1997]. Therefore, lineage sorting is topology-dependent and is expected to occur often along the short branches of a gene tree. The occurrence of paralogy, on the other hand, depends largely on the dynamics of gene duplication and extinction during its evolution [Morton, et al., 1996]. Understanding molecular evolution of individual genes can help assess the likelihood of occurrence of paralogy at each locus.

Lateral gene transfer follows the same test result as hybridization, $\Delta(k, i)=0$, for the theoretical model. However, lateral gene transfer often happens to transposable elements [Clark, et al., 1994; Syvanen, 1994]. It is also very unlikely that many unlinked genes can be transferred from one species to the other in parallel. In contrast, hybridization, particularly by way of allopolyploidy, combines two entire genomes. Therefore, if $\Delta(k, i)=0$ is the test result for multiple unlinked genes, the incongruence is much more likely to result from hybridization than from lateral gene transfer.

Acknowledgments

We thank B. Gaut, F. Kong, and Z. Yang for valuable suggestions on statistical tests, and D. Ferguson, R. Olmstead, and two anonymous reviewers for valuable suggestions and comments on the manuscript.

References

1. Arnold, M. L. 1997. Natural hybridization and evolution. Oxford Univ. Press, New York.
2. Baldwin, B. G., M. J. Sanderson. 1998. Age and rate of diversification of the Hawaiian silversword alliance (Compositae) *Proc. Natl. Acad. Sci.* USA 95: 9402-9406.

3. Buckler, E. S., IV, A. Anthony, T. P. Holtsford. 1997. The evolution of ribosomal DNA: Divergent paralogues and phylogenetic implications. *Genetics*. 145: 821-832.

4. Bullini, L. 1994. Origin and evolution of animal hybrid species. *Trends Ecol. Evol.* 9: 422-426.

5. Campbell, C. S., M. F. Wojciechowski, B. G. Baldwin, L. A. Alice, M. J. Donoghue. 1997. Persistent nuclear ribosomal DNA sequence polymorphism in the *Amelanchier* agamic complex (Rosaceae). *Mol. Biol. Evol.* 14: 81-90.

6. Clark, J. B., W. P. Maddison, M. G. Kid-Well. 1994. Phylogenetic analysis supports horizontal transfer of *P* transposable elements. *Mol. Biol. Evol.* 11: 40-50.

7. De Queiroz, A., M. J. Donoghue, J. Kim. 1995. Separate versus combined analysis of phylogenetic evidence. *Annu. Rev. Ecol. Syst.* 26: 657-681.

8. Doyle, J. J. 1997. Trees within trees: Genes and species, molecules and morphology. *Syst. Biol.* 46: 537-553.

9. Efron, B. 1979. Bootstrap methods: Another look at the jacknife. *Ann. Stat.* 7: 1-26.

10. Eyre-Walker, A., B. S. Gaut. 1997. Correlated rates of synonymous site evolution across plant genomes. *Mol. Biol. Evol.* 14: 455-460.

11. Felsenstein, J. 1985. Confidence limits on phylogenetics: An approach using the bootstrap. *Evolution*. 39: 783-791.

12. Funk, V. A. 1985. Phylogenetic pattern and hybridization. *Ann. Mo. Bot. Gard.* 72: 681-715.

13. Gaut, B. S., S. V. Muse, W. D. Clark, M. T. Clegg. 1992. Relative rates of nucleotide substitution at the *rbc*L locus of monocotyledonous plants. *J. Mol. Evol.* 35: 292-303.

14. Grant, V. 1981. Plant speciation, 2nd edition. Columbia Univ. Press, New York.

15. Hennig, W. 1966. Phylogenetic systematics. Univ. Illinios Press, Urbana-

Champaign.

16. Hillis, D. M., J. P. Huelsenbeck, C. W. Cunningham. 1994. Application and accuracy of molecular phylogenetics. *Science*. 264: 671-677.

17. Hudson, R. R. 1992. Gene trees, species trees, and the segregation of ancestral alleles. *Genetics*. 131: 509-512.

18. Huelsenbeck, J. P. 1995. Performance of phylogenetic methods in simulation. *Syst*. Biol. 44: 17-48.

19. Huelsenbeck, J. P., J. J. Bull, C. W. Cunningham. 1996. Combining data in phylogenetic analysis. *Trends Ecol. Evol.* 11: 152-158.

20. Huelsenbeck, J. P., B. Rannala. 1997. Phylogenetic methods come of age: Testing hypotheses in an evolutionary context. *Science*. 276: 227-232.

21. Jukes, T. H., C. R. Cantor. 1969. Evolution of protein molecules. 21-132 *in* Mammalian protein metabolism (H. N. Munro, ed.). Academic Press, New York.

22. Kellogg, E. A., R. Appels, R. J. Mason-Gamer. 1996. When genes tell different stories: The diploid genera of *Triticeae* (Gramineae). *Syst. Bot.* 21: 321-347.

23. Kimura, M. 1980. A simple method for estimating evolutionary rates of base substitutions through comparative studies of nucleotide sequences. *J. Mol. Evol.* 16: 111-120.

24. Li, W.-H. 1997. Molecular evolution. Sinauer, Sunderland, Massachusetts.

25. Masterson, W. P. 1997. Gene trees in species trees. *Syst. Biol.* 46: 523-536.

26. Masterson, J. 1994. Stomatal size in fossil plants: Evidence for polyploidy in majority of angiosperms. *Science*. 264: 421-424.

27. McDade, L. A. 1995. Hybridization and phylogenetics. 305-331 *in* Experimental and molecular approaches to plant biosystematics (P. C. Hoch and A. G. Stephenson, eds.), Missouri Botanical Garden, St. Louis.

28. Mogensen, H. L. 1996. The hows and whys of cytoplasmic inheritance in seed plants. *Am. J. Bot.* 83: 383-404.

29. Moore, W. S. 1995. Inferring phylogenies from mtDNA variation: Mitochondrial-gene trees versus nuclear-gene trees. *Evolution*. 49: 718-726.

30. Morton, B. R., B. Gaut, M. Clegg. 1996. Evolution ofalcohol dehydrogenase genes in the Palm and Grass families. *Proc. Natl. Acad. Sci.* USA 93:11735-11739.

31. Pamilo, P., M. Nei. 1988. Relationship between gene trees and species trees. *Mol. Biol. Evol.* 5: 568-583.

32. Rieseberg, L. H. 1997. Hybrid origins of plant species. *Annu. Rev. Ecol. Syst*. 28: 359-389.

33. Rieseberg, L. H., N. C. Ellstrand. 1993. What can molecular and morphological markers tell us about plant hybridization? *Crit. Rev. Plant Sci.* 12: 213- 241.

34. Rieseberg, L. H., J. D. Morefield. 1995. Character expression, phylogenetic reconstruction, and the detection of reticulate evolution. 333-353 *in* Experimental and molecular approaches to plant biosystematics (P. C. Hoch and A. G. Stephenson, eds.), Missouri Botanical Garden, St. Louis.

35. Rieseberg, L. H., D. E. Soltis. 1991. Phylogenetic consequences of cytoplasmic gene flow in plants. *Evol. Trends Plants*. 5: 65-84.

36. Sang, T., D. J. Crawford, T. F. Stuessy. 1995. Documentation of reticulate evolution in peonies (*Paeonia*) using sequences of internal transcribed spacer of nuclear ribosomal DNA: Implications for biogeography and concerted evolution. *Proc. Natl. Acad. Sci.* USA 92: 6813-6817.

37. Sang, T., D. J. Crawford, T. F. Stues SY. 1997. Chloroplast phylogeny, reticulate evolution, and biogeography of *Paeonia* (Paeoniaceae). *Am. J. Bot*. 84: 1120-1136.

38. Soltis, D. E., R. K. Kuzoff. 1995. Discordance between nuclear and chloroplast phylogenies in the *Heuchera* group (Saxifragaceae). *Evolution*. 49: 727-742.

39. Stebbins, G. L., J. R. 1950. Variation and evolution in plants. Columbia Univ. Press, New York.

40. Syvanen, M. 1994. Horizontal gene transfer: Evidence and possible consequences. *Annu. Rev. Genet.* 28: 237-261.

41. Wendel, J. F., J. J. Doyle. 1998. Phylogenetic incongruence: Window into genome history and molecular evolution. 265-296 *in* Molecular systematics of plants, II: DNA sequencing (D. Soltis, P. Soltis, J. Doyle, eds.) Kluwer Academic Publishers, Boston.

42. Wendel, J. F., A. Schnabel, T. Seelanan. 1995. Bidirectional interlocus concerted evolution following allopolyploid speciation in cotton (*Gossypium*). *Proc. Natl. Acad. Sci.* USA 92: 280-284.

43. Wolfe, A. D., Q. Y. Xiang, S. R. Kephart. 1998. Diploid hybrid speciation in *Penstemon* (Scro-phulariaceae). *Proc. Natl. Acad. Sci.* USA 95: 5112-5115.

44. Wu, C.-I., AND W.-H. Li. 1985. Evidence for higher rates of nucleotide substitution in rodents than in man. *Proc. Natl. Acad. Sci.* USA 82: 1741-1745.

Detecting evolutionary rate heterogeneity among mangroves and their close terrestrial relatives

检测红树及其陆生近缘种间的进化速率异质性

红树林是潮间带生态系统的主要组成，其从形态学与生理学上都有别于陆生的近缘种。基于系统发育分析与相对速率检验，我们研究了典型红树的科（如红树科，Rhizophoraceae）的分子进化模式，以及该科内不同物种的叶绿体 *matK* 和 *rbcL* 基因的进化速率异质性。我们的研究分析了红树科中这两个基因的进化速率异质性：红树的木榄属（*Bruguiera*）在 *matK* 基因序列上的同义替换与非同义替换位点上比陆生组的竹节树属（*Carallia*）的替换率相对要慢，并且同义与非同义替换矩阵具有相关性。然而，*rbcL* 基因在红树和近缘的陆生组的非同义替换位点上表现出了高度的速率异质性，并且与其同义替换位点的速率不相偶联。进化选择可能大大影响着进化速率的差异，进一步的研究可帮助我们更好地理解对红树物种的进化速率异质性与分子适应性产生影响的各种压力。

Zhong, Y., Zhao, Q., Shi, S. H., Huang, Y. L., Hasegawa, M. Detecting evolutionary rate heterogeneity among mangroves and their close terrestrial relatives. 2002. *Ecology Letters*, 5(3): 427-432. by permission of Wiley.

Detecting evolutionary rate heterogeneity among mangroves and their close terrestrial relatives

Yang Zhong,[1,*] Qiong Zhao,[2] Suhua Shi,[3] Yelin Huang[3] and Masami Hasegawa[4]

Abstract

Mangroves form the dominant intertidal ecosystems and differ morphologically and physiologically from their close terrestrial relatives. We investigate the molecular evolutionary pattern of the typical mangrove family, i.e., Rhizophoraceae, and rate heterogeneity for the plastid *mat*K and *rbc*L genes in different species of the family, as revealed by phylogenetic analyses and relative-rate tests. Our study documents evolutionary rate heterogeneity in the Rhizophoraceae for the two genes: the mangrove genus *Bruguiera* has relatively slow substitution rates compared to the

---

* Correspondence: Institute of Biodiversity Science, Fudan University, Shanghai 200433, People's Republic of China. E-mail: yangzhong@fudan.edu.cn.
1 Ministry of Education Key Laboratory for Biodiversity Science and Ecological Engineering, and Institute of Biodiversity Science, Fudan University, Shanghai 200433, People's Republic of China.
2 Department of Genetics, University of Wisconsin- Madison, WI 53706, USA.
3 Key Laboratory of Gene Engineering of Ministry of Education, School of Life Sciences, Zhongshan University, Guangzhou 510275, People's Republic of China.
4 The Institute of Statistical Mathematics, 4-6-7 Minami-Azabu, Minato-ku, Tokyo 106, Japan.

terrestrial genus *Carallia* at both synonymous and non-synonymous sites in the *mat*K sequences, and the synonymous and non-synonymous substitution matrices are correlated. However, the *rbc*L non-synonymous sites exhibit a high degree of rate heterogeneity among mangroves and related terrestrial groups, and uncoupling of rates with the synonymous sites. Selection is probably an important influence on the rate variation, suggesting further investigation for better understanding of various forces contributing to the rate heterogeneity and molecular adaptation in mangroves.

Keywords

Mangrove, *mat*K, molecular phylogeny, non-synonymous substitution, *rbc*L, relative-rate test, selection, synonymous substitution.

Introduction

Phylogenetic analyses have been a standard tool in systematic and evolutionary biology, with goals ranging from reconstructing evolutionary histories of organisms or genes [Li, 1997] to measuring biodiversity [Purvis & Hector, 2000], and have also become a comparative approach to ecological studies, e.g. exploring the phylogenetic structure of ecological communities [Webb, 2000] and detecting adaptation to stress in various environments [Yang & Bielawski, 2000; Schulte, 2001]. In particular, molecular phylogenetics has begun to attract more attention of ecologists to some difficult problems regarding adaptive evolution of plants and animals in different ecosystems, such as aquatic and terrestrial ecosystems.

Mangroves form the dominant intertidal ecosystems throughout the tropical regions of the world and enter the subtropics, with a total area of about 182 000 km^2 worldwide [Spalding *et al.*, 1997]. Current knowledge of the evolution of mangroves is still limited in terms of rangewide distributions and is mostly restricted to morphological and physiological analyses, which have indicated a high degree of homogeneity [Dodd *et al.*, 1998]. For example, mangroves differ from their terrestrial related plants in that the former are characterized by their peculiar adaptive viviparous fruits and their pneumatophores and knee roots, etc. [Geh & Keng, 1974]. However, less attention has been paid to molecular evolution of mangroves and their terrestrial related groups [Zhong et al., 2000].

Recently, molecular phylogenetic analyses of the mangrove family Rhizophoraceae, which comprise 15 genera and approximately 130 species including eumangrove ("true" mangrove) and terrestrial plants, have been conducted using several kinds of DNA sequences [Huang et al., 1999; Setoguchi et al., 1999; Schwarzbach & Ricklefs, 2000; Zhong et al., 2000; Shi et al., 2002]. Here, we present detailed analyses of relative evolutionary rates of two chloroplast genes, matK (maturase K) and rbcL (large subunit of the enzyme ribulose-1,5-bisphosphate carboxylase), among the mangroves and their close terrestrial relatives in Rhizophoraceae. The results have implications both for evolutionary biology and ecology of mangroves and related groups, especially for better understanding of molecular adaptation in mangroves.

Materials and methods

The sequences of chloroplast DNA (cpDNA) matK and rbcL genes used in this study are listed in Table 1. According to the viewpoint of the APG [Angiosperm Phylogeny Group, 1998], the family Rhizophoraceae belongs to the order Malpighiales, so that *Byrsonima crassifolia* (Malpighiaceae) can be selected as an outgroup for our phylogenetic analyses.

Each sequence data set was aligned using the program CLUSTAL-X [Thompson et al., 1997]. The phylogenetic analyses of the nucleotide sequences of the rbcL and matK genes were performed using the NucML program in the MOLPHY Package version 2.3 [Adachi & Hasegawa, 1996] with the HKY 85 model [Hasegawa et al., 1985]. Bootstrap proportions (BPs) were estimated by the RELL (Resampling of Estimated Log-Likelihoods) method [Kishino et al., 1990; Hasegawa & Kishino, 1994] with 10 000 bootstrap resamplings.

Table 1 Sequences of *mat*K and *rbc*L genes used in this study

Genus	Abbreviation	Species	Habitat and distribution[2]	GenBank Accession No.	
				*mat*K	*rbc*L
Blepharistemma	BL	*B. membranifolia*	T; AS		AF006761
Bruguiera	BR	*B. cylindrica*	M; AS, AU		AF127694
		B. exaristata	M; AU		AF127695
		B. gymnorhiza	M; AS, AF, AU	AF105088	AF127693
		B. parviflora	M; AS, AU		AF127692
		B. sexangula	M; AS, AU	AF105091	AF127691
Carallia	CR	*C. brachiata*	T; AS	AF105086	AF006757
		C. graciniifolia	T; AS	AF126370	AF127373
		C. pectinifolia	T; AS	AF105087	AF127372
Cassipourea	CS	*C. ceylanica*	T; AF		AF127674
		C. guianensis	T; AM		AF127673
		C. elliptica	T; AM		AF127672
		C. rotundifolia	T; AF		AF006762
Ceriops	CE	*C. australis*	M; AS, AF, AU		AF127683
		C. tagal	M; AS, AF, AU	AF105089	AF127684
Crossostylis	CO	*C. biflora*	T; SPI		AF127679
		C. grandiflora	T; SPI		AF006760
Dactylopetalum	DA	*D. verticillatum*	T; AF		AF127676
Gynotroches	GY	*G. axillaris*	T; AS, SPI		AF127678
Kandelia	KA	*K. candel*	M; AS	AF105090	AF127682
Macarisia	MA	*M. emarginata*	T; AF		AF129130
Pellacalyx	PE	*P. axillaris*	T; AS		AF127681
		P. saccardians	T; AS		AF127680
		P. yunnanensis	T; AS	AF126371	

续表

Genus	Abbreviation	Species	Habitat and distribution[2]	GenBank Accession No.	
				*mat*K	*rbc*L
Rhizophora	RH	*R. apiculata*	M; AS, AF, AU		AF127685
		R. mangle	M; AM, AF		AF127689
		R. mucronata	M; AS, AF, AU		AF127687
		R. racemosa	M; AM		AF127690
		R. stylosa	M; AS	AF105092	AF127686
Sterigmapetalum	ST	*S. guianense*	T; AM		AF127671
Byrsonima[1]	BY	*B. crassifolia*	T; AM	Unpublished[3] L01892	

[1] Outgroup.

[2] M, mangrove; T, terrestrial; AS, Asia; AF, Africa; AM, America; AU, Australia; SPI, South Pacific Islands.

[3] Provided by Charles C. Davis at Harvard University.

The rates of synonymous and non-synonymous substitutions were estimated by using the method of Li (1993) and Pamilo & Bianchi (1993). The relative-rate tests of between the lineages at generic level in the phylogenetic trees were performed with the method proposed by Li & Bousquet 1992) and using the RRTree program [Robinson *et al.*, 1998; Robinson-Rechavi & Huchon, 2000]. The closest sister group of each genus pair to be compared in a phylogenetic tree was selected as the reference group for a relative-rate test. The binary transferred matrices were constructed with the results of the relative-rate tests, and Mantel tests [Sokal & Rohlf, 1995; Liedloff, 1999] were conducted to detect the associations among these matrices. Significance values of the Mantel tests were based on 100 random iterations.

Results and discussion

The phylogenetic trees of the Rhizophoraceae based on the sequences of *mat*K and *rbc*L genes are shown in Figs 1 and 2, respectively. The *mat*K tree strongly supports the idea that the mangroves form a monophyletic group with a relatively high BP value (91%). The BP value for the monophyly of mangroves is 81% in the highest likelihood tree among 945 trees, and the BP value for the monophyly based on the total *rbc*L and *mat*K sequences is 97%. However, the terrestrial plants form different phylogenetic patterns: monophyly in the *mat*K tree (Fig. 1) but paraphyly in the *rbc*L tree (Fig. 2).

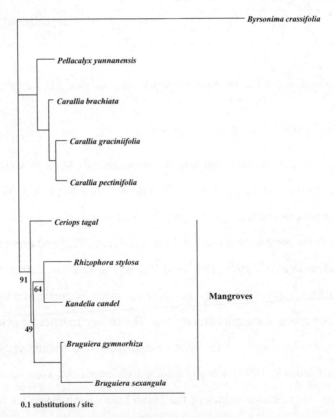

Fig. 1 A phylogenetic tree of Rhizophoraceae based on the *mat*K sequences constructed by the maximum likelihood method with MOLPHY. Numbers indicate the bootstrap proportions (percentage over 10 000 replicates).

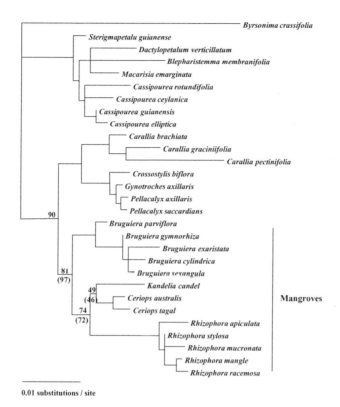

Fig. 2 A phylogenetic tree of Rhizophoraceae based on the *rbc*L sequences constructed by the maximum likelihood method with MOLPHY. Numbers without the parentheses indicate the bootstrap proportions (percentage over 10 000 replicates). Numbers in the parentheses indicate the bootstrap proportions for the total of *rbc*L and *mat*K sequences.

The results obtained from the relative-rate tests at synonymous and non-synonymous sites in the sequences of *mat*K and *rbc*L genes are given in Table 2 and Table 3, respectively. Our exploratory analyses detect a number of significant rate differences at synonymous and non-synonymous sites in the two plastid sequences. First, two significant contrasts are that the mangrove *Bruguiera* has relatively slower substitution rates than the terrestrial *Carallia* at both synonymous and non-synonymous sites in the *mat*K sequences (Table 2). The Mantel tests show that the synonymous and non-synonymous relative-rate matrices are correlated at the *mat*K ($P=0.01$). This observation suggests that selective constraint at non-synonymous sites is fairly constant among evolutionary lineages of the *mat*K locus.

Table 2 Results of relative-rate tests for *mat*K sequences[1]

	PE	CR	RH	KA	CE	BR
PE	–	−0.62	0.03	0.09	−0.56	1.08
CR	0.48	–	−0.04	0.56	−0.63	−4.41**
RH	−0.62	−1.75	–	0.36	−0.45	1.11
KA	0.92	0.82	1.60	–	−0.99	0.77
CE	0.19	−0.57	1.60	−0.59	–	1.16
BR	0.12	2.04*	1.23	−0.31	−0.72	–

K_{ij}/SE is given as $(K_i-K_j)/SE$ so that a $K_{ij}/SE > 0$ means that the *i*th column is estimated to have faster substitution rates than the row. Each lower-triangular matrix represents the results at synonymous sites, and each upper-triangular matrix represents the results at non-synonymous sites.

[1] Abbreviations of genera see Table 1.

*$P<0.05$, **$P<0.01$.

Table 3 Results of relative-rate tests for *rbc*L sequences1

	BL	ST	DA	CS	MA	CO	GY	PE	CR	RH	KA	CE	BR
BL	-	0	−0.49	−1.29	−1.42	−1.76	−2.01*	−1.61	1.15	0.30	−0.71	−0.51	−0.82
ST	1.36	-	1.97*	2.57*	0.51	0.22	−0.23	0.21	1.63	1.95*	1.24	1.32	1.19
DA	0.20	−0.78	-	−0.78	−2.00*	−1.37	−1.81	−1.41	−0.39	0.35	−0.29	−0.33	−0.33
CS	1.55	0.42	1.15	-	−1.87	−1.12	−1.26	−0.78	0.76	−0.16	0.44	0.44	0.36
MA	1.22	−0.54	0.20	−1.03	-	−0.13	−0.27	0.24	1.30	2.25*	1.08	1.38	1.19
CO	−0.03	−0.62	−0.18	−0.83	−0.33	-	−0.44	0.45	2.44*	2.39*	1.29	1.61	1.31
GY	0.13	−0.57	−0.02	−0.77	−1.16	0.48	-	1.44	2.06*	2.70**	2.98**	1.77	1.42
PE	0.03	−0.66	−0.10	−0.86	−0.26	0.19	−0.35	-	1.37	2.26*	1.22	1.26	1.02
CR	0.00	−1.36	−0.70	−1.82	0.00	−1.41	−1.50	−1.32	-	−0.40	−2.11*	−0.77	−0.96
RH	0.03	$−0.47^2$	$−0.10^2$	$−0.42^2$	0.19^2	0.87^2	$−1.18^2$	$−1.19^2$	1.08^2	-	−0.92	−1.06	−0.94
KA	−0.55	−1.11	−0.55	−1.31	−0.87	−0.68	−1.01	−0.59	0.92	$−0.13^2$	-	−0.04	−0.06
CE	−0.30	−0.91	−0.30	−1.13	−0.62	−0.35	−0.39	−0.26	1.36	0.75^2	0.50	-	−0.03
BR	$−0.13^2$	$−0.41^2$	0.04^2	$−0.59^2$	$−0.35^2$	$−1.10^2$	$−1.10^2$	$−1.11^2$	0.68^2	0.19^2	0.30^2	$−0.62^2$	-

[1] Abbreviations of genera see Table 1.

K_{ij}/SE is given as $(K_i - K_j)/SE$ so that a $K_{ij}/SE>0$ means the *i*th column is estimated to have faster substitution rates than the row. Each lower-triangular matrix represents the results at synonymous sites, and each upper-triangular matrix represents the results at non-synonymous sites.

[2] Synonymous transversion substitution when K_s is saturated.

*$P<0.05$, **$P<0.01$.

Second, there are 13 significant contrasts at non-synonymous sites in the *rbc*L sequences (Table 3). Among them, six indicate that the mangrove genera have relatively faster non-synonymous substitution rates than the related terrestrial groups. However, the terrestrial *Carallia* still shows a relatively faster non-synonymous rate than the mangrove *Kandelia*. In our previous paper [Zhong et al., 2000], we noted that there were significant total substitution rate differences between *Carallia brachiata* and other *Carallia* species in the *mat*K sequences. Obviously, these *Carallia* species need to be further studied as important reference taxa relative to mangroves.

Moreover, the *rbc*L non-synonymous sites also exhibit rate heterogeneity among the terrestrial groups, regardless of their geographical distributions (Table 3). The Mantel tests show that the *rbc*L rates at synonymous and non-synonymous sites are uncorrelated ($P=1.00$). This can be also seen to some extent by visual examination of Table 3, e.g. there are independently increased non-synonymous rates against synonymous rates in the two genera *Carallia* and *Rhizophora*.

Our analyses of the two chloroplast genes not only clearly indicate that the rates of synonymous and non-synonymous substitution vary among evolutionary lineages of mangroves and their terrestrial related groups, but also increase the understanding of possible forces contributing to the patterns. It is well known that synonymous substitutions primarily reflect the process of neutral evolution but non-synonymous substitutions reflect either positive selection or purifying selection over evolutionary time [Gaut et al., 1992; Gaut et al., 1997; Li, 1997; Yang & Bielawski, 2000]. Recently, Gaut et al. (1997) reported the uncoupling of *rbc*L non-synonymous and synonymous rates in the grass family, *i.e.*, Poaceae, suggesting that locus-specific selection coefficients can change significantly among lineages. Our study reveals molecular evolutionary patterns of mangroves and their terrestrial relatives in which non-synonymous and synonymous substitution rates are uncoupled, and therefore selection is probably an important influence on the rate variation.

The speciation rate hypothesis has previously also been invoked to explain rate heterogeneity at *rbc*L [Gaut *et al*., 1997]. Mayr (1954) postulated that rates of genetic change should influence rates of speciation, and Bousquet *et al*. (1992) and Barraclough *et al*. (1996) revealed a correlation between speciation rates and nucleotide substitution rates. Gaut *et al*. (1997) analysed the *rbc*L sequences in the grass family and found that the basal grass lineage (subfamily Anomochlooideae) had slower nucleotide substitution rates than other grass subfamilies, that is, their prediction based on the speciation rate hypothesis was consistent with the results of synonymous relative-rate tests. However, our relative-rate tests reveal that no significant contrast at the synonymous sites in the *rbc*L sequences is detected (Table 3), yet there is no basal lineage of mangroves in the *rbc*L tree (Fig. 2).

Excluding speciation rate hypothesis, two possible explanations for the rate heterogeneity are about different ecological and biogeographical forces to the mangroves and related terrestrial groups. On the one hand, all mangrove species, regardless of geographical distribution, form a monophyletic group on both the *mat*K and *rbc*L trees. The rate variations do not occur within the mangrove group but between the mangrove and related terrestrial groups, indicating that the marine environment might be a major selection force to mangroves. On the other hand, the related terrestrial groups show different phylogenetic patterns. In particular, there are also relatively complex patterns of rate variation at *rbc*L among the groups, e.g., significant differences between *Carallia* (CR) distributed in Asia and *Crossostylis* (CO) distributed in the South Pacific Islands, and between *Dactylopetalum* (DA) and *Macarisia* (MA), both distributed in Africa. These cases may reflect the effects of selection imposed by different climates and habitats, adding a biogeographical dimension to the problem.

In general, the environmental mechanisms responsible for molecular evolutionary patterns of mangroves and related groups require examination. Positive selection, such as we describe here, is believed to reflect powerful selective forces acting on protein-coding sequences, and may stem from such factors as changed ecological conditions [Ford, 2001] or an arms race scenario [Dawkins & Krebs, 1979]. We have uncovered the traces of such a scenario here, and note that both ecological and molecular studies will be needed to elucidate the causes of the phenomenon in mangroves.

Acknowledgements

We would like to thank Dr. Ying Cao and Ms. Li Wang for technical assistance. Financial support was provided by the National Natural Science Foundation of China (39825104, 30170071) and the Chinese Ministry of Education Grant for Ph. D. programmes.

References

1. Adachi, J. & Hasegawa, M. (1996). *MOLPHY*: Programs for Molecular Phylogenetics, Version 2.3. Computer Science Monographs 28. Institute of Statistical Mathematics, Tokyo.
2. Angiosperm Phylogeny Group. (1998). An ordinal classification for the families of flowering plants. *Ann. Missouri Bot. Gard.*, 85: 531-553.
3. Barraclough, T. G., Harvey, P. H. & Nee, S. (1996). Rate of *rbc*L gene sequence evolution and species diversification in flowering plants. *Proc. R. Soc. Lond. B.*, 263: 589-591.
4. Bousquet, J., Strauss, S. H., Doerksen, A. H. & Price, R. A. (1992). Extensive variation in evolutionary rate of *rbc*L gene sequences among seed plants. *Proc. Nat. Acad. Sci. USA*, 89: 7844-7848.
5. Dawkins, R. & Krebs, J. R. (1979). Arms races between and within species. *Proc. R. Soc. Lond. B.*, 205: 489-511.
6. Dodd, R. S., Rafii, Z. A., Fromard, F. & Blasco, F. (1998). Evolutionary diversity among Atlantic coast mangroves. *Acta Oecologica.*, 19: 323-330.

7. Ford, M. J. (2001). Molecular evolution of transferrin: evidence for positive selection in salmonids. *Mol. Biol. Evol.*, 18: 639-647.
8. Gaut, B. S., Clark, L. G., Wendel, J. F. & Muse, S. V. (1997). Comparisons of the molecular evolutionary process at *rbc*L and *ndh*F in the grass family (Poaceae). *Mol. Biol. Evol.*, 14: 769-777.
9. Gaut, B. S., Morton, B. R., McCaig, B. M. & Clegg, M. T. (1992). Relative rates of nucleotide substitution at the *rbc*L locus of monocotyledonous plants. *J. Mol. Evol.*, 35: 292-303.
10. Geh, S. & Keng, H. (1974). Morphological studies on some inland Rhizophoraceae. *Gard. Bull. Singapore*, 27: 183-221.
11. Hasegawa, M. & Kishino, H. (1994). Accuracies of the simple methods for estimating the bootstrap probability of a maximum likelihood tree. *Mol. Biol. Evol.*, 11: 142-145.
12. Hasegawa, M., Kishino, H. & Yano, T. (1985). Dating of the human-ape splitting by a molecular clock of mitochondrial DNA. *J. Mol. Evol.*, 22: 160-174.
13. Huang, Y., Qiu, X., Shi, S., Tan, F. & Chang, H. T. (1999). The molecular phylogeny in Rhizophoraceae in China. *Acta Sci. Nat. Univ. Sunyatseni*, 38: 39-45 (in Chinese with English summary).
14. Kishino, H., Miyata, T. & Hasegawa, M. (1990). Maximum likelihood inference of protein phylogeny, and the origin of chloroplasts. *J. Mol. Evol.*, 31: 151-160.
15. Li, W. H. (1993). Unbiased estimation of the rates of synonymous and nonsynonymous substitution. *J. Mol. Evol.*, 36: 96-99.
16. Li, W. H. (1997). *Molecular Evolution*. Sinauer Associates, Sunderland, MA.
17. Li, P. & Bousquet, J. (1992). Relative-rate test for nucleotide substitutions between two lineages. *Mol. Biol. Evol.*, 9: 1185-1188.
18. Liedloff, A. (1999). MANTEL: *Mantel Nonparametric Test Calculator*, Version 2.0. Queensland University of Technology, Brisbane, Australia.

19. Mayr, E. (1954). Change of genetic environment and evolution. In: *Evolution as a Process* (eds Huxley, J., Hardy, A. C. & Ford, E. B.). George Allen & Unwin, London, 157-180.

20. Pamilo, P. & Bianchi, N. O. (1993). Evolution of the *Zfx* and *Zfy* genes: rates and interdependence between the genes. *Mol. Biol. Evol.*, 10: 271-281.

21. Purvis, A. & Hector, A. (2000). Getting the measure of biodiversity. *Nature*, 405: 212-219.

22. Robinson, M., Gouy, M., Gautier, C. & Mouchiroud, D. (1998). Sensitivity of the relative-rate test to taxonomic sampling. *Mol. Biol. Evol.*, 15: 1091-1098.

23. Robinson-Rechavi, M. & Huchon, D. (2000). RRTree: relative-rate tests between groups of sequences on a phylogenetic tree. *Bioinformatics*, 16: 296-297.

24. Schulte, P. M. (2001). Environmental adaptations as windows on molecular evolution. *Comp. Biochem. Physiol. B.*, 128: 597-611.

25. Schwarzbach, A. E. & Ricklefs, R. E. (2000). Systematic affinities of Rhizophoraceae and Anisophylleaceae, and intergeneric relationships within Rhizophoraceae, based on chloroplast DNA, nuclear ribosomal DNA, and morphology. *Am. J. Bot.*, 87: 547-564.

26. Setoguchi, H., Kosuge, K. & Tobe, H. (1999). Molecular phylogeny of Rhizophoraceae based on *rbc*L gene sequences. *J. Plant Res.*, 112: 443-455.

27. Shi, S., Zhong, Y., Huang, Y., Qiu, X. & Chang, H. T. (2002). Phylogenetic relationships of the Rhizophoraceae in China based on sequences of the chloroplast gene *mat*K and ITS regions of nuclear ribosomal DNA and combined data set. *Biochem. Syst. Ecol.*, 30: 309-319.

28. Sokal, R. R. & Rohlf, F. J. (1995). *Biometry*. W.H. Freeman, New York.

29. Spalding, M., Blasco, F. & Field, C. (1997). *World Mangrove Atlas*. Samara Publishing Co., Cardigan, UK.

30. Thompson, J. D., Gibson, T. J., Plewniak, F., Jeanmougin, F. & Higgins,

D. G. (1997). The Clustal-X windows interface: flexible strategies for multiple sequence alignment aided by quality analysis tools. *Nucl. Acids Res.*, 25: 4876-4882.

31. Webb, C. O. (2000). Exploring the phylogenetic structure of ecological communities: an example for rain forest trees. *Am. Nat.*, 156: 145-155.

32. Yang, Z. & Bielawski, J. P. (2000). Statistical methods for detecting molecular adaptation. *Trends Ecol. Evol.*, 15: 496-503.

33. Zhong, Y., Shi, S., Tang, X., Huang, Y., Tan, F. & Zhang, X. (2000). Testing relative evolutionary rates and estimating divergence times among six genera of Rhizophoraceae using cpDNA and nrDNA sequences. *Chinese Sci. Bull.*, 45: 1011-1015.

The *Schistosoma japonicum* genome reveals features of host-parasite interplay（摘要）

日本血吸虫基因组揭示宿主-寄生虫相互作用的特性

  日本血吸虫（*Schistosoma japonicum*）是一种可引起人类血吸虫病的寄生虫，它也是该病在中国和菲律宾的重要病原体。本文公布了日本血吸虫的全基因组序列草图，这是扁形动物门，甚至是冠轮动物超门中第一个被报道的基因组序列。该基因组为这一复杂的多细胞病原体的分子结构和与宿主的相互作用提供了全局视角，揭示了日本血吸虫能够利用宿主的营养，神经内分泌激素和生长发育信号通路。日本血吸虫有复杂的神经系统和发达的感受系统，所以能接受不同环境下（如淡水中或中间宿主和哺乳动物宿主的组织里）产生的生理反应所对应配体的刺激。而它体内数量众多的蛋白酶，包括血吸虫尾蚴的弹性蛋白酶，可能在经皮侵染哺乳动物和降解血红蛋白中发挥作用。该基因组信息将为促进血吸虫病新疗法的开发提供一个宝贵平台。

  The Schistosoma Japonicum Genome Sequencing and Functional Analysis Consortium. The *Schistosoma japonicum* genome reveals features of host-parasite interplay. 2009. *Nature*, 460(7253): 345-352.（作为PI之一）

Solution structure of Urm1 and its implications for the origin of protein modifiers（摘要）

Urm1蛋白的溶液结构及其在修饰蛋白进化起源中的意义

　　修饰蛋白在多种生物过程中普遍存在，并通过与目标蛋白共价结合来调控其活性和功能。虽然我们已经在真核生物中发现了泛素和一些类泛素修饰蛋白，但在原核细胞中还没有发现修饰蛋白；因此，它们的进化起源仍是个难题。为了推断修饰蛋白和载硫蛋白之间的进化关系，我们用核磁共振方法解析了从酿酒酵母中提取的泛素相关修饰蛋白Urm1的溶液结构。对泛素超家族（尤其是Urm1家族）的结构比较和系统发育分析结果表明，Urml是一种独一无二的"分子化石"，保有整个超家族共同祖先所具备的最保守的结构和序列特征。Urml和钼喋呤合成酶小亚基MoaD之间在3D结构、疏水性和表面静电等特征上的相似性都说明这两种蛋白与配体作用的方式类似，而Urml-Uba4和MoaD-MoeB之间的相似性也在三磷酸腺苷依赖性蛋白结合（真核生物中）和三磷酸腺苷依赖性辅因子硫化之间建立了进化关联。

　　Xu, J. J., Zhang, J. H., Wang, L., Zhou, J., Huang, H. D., Wu, J. H., **Zhong, Y.**, Shi, Y. Y. Solution structure of Urm1 and its implications for the origin of protein modifiers. 2006. *Proceedings of the National Academy of Sciences of the USA*, 103(31): 11625-11630.

Molecular evolution of the SARS coronavirus during the course of the SARS epidemic in China（摘要）
中国SARS流行期间SARS冠状病毒的分子进化

该文分析了来源于SARS（重症急性呼吸综合征）流行的早、中和晚期的61个SARS冠状病毒基因组序列以及狸猫的两个病毒序列。揭示了病毒流行期间各阶段的基因型特性，并发现最早期的基因型和动物的类SARS冠状病毒相似。在病毒流行的早期和晚期均发现，主要的片段缺失存在于基因组的Orf8区。病毒基因组的中性突变率稳定，但编码序列中的氨基酸置换率随流行进程有所减缓。刺突蛋白对正选择压力表现出了最强烈的初始应答，随后发生净化选择并最终稳定。

The Chinese SARS Molecular Epidemiology Consortium. Molecular evolution of the SARS coronavirus during the course of the SARS epidemic in China. 2004. *Science*, 303(5664): 1666-1669.

Photosynthetic metabolism of C-3 plants shows highly cooperative regulation under changing environments: A systems biological analysis（摘要）
系统生物学分析发现C3植物光合代谢在变化环境下的高度协调

我们利用了一种新方法，对干旱胁迫和高CO_2浓度条件下C_3植物叶绿体中的光合代谢的稳定性进行了分析，这种方法称为M_DFBA，即最小化代谢调节动态流平衡分析。C_3植物叶绿体中的光合代谢能使用一种高度协调机制，在瞬变扰动的情况下将代谢物浓度的波动最小化。我们的研究表明，这种高度协调机制确保了生物系统的稳定，而相较正常情况，在扰动情况下的协作更为密切，这可以将代谢物浓度的波动降至最小，而后者是维持系统功能的关键。我们的研究方法有助于了解此类现象和动态过程中的复杂代谢网络的稳定机制。

Luo, R. Y., Wei, H. B., Ye, L., Wang, K. K., Chen, F., Luo, L. J., Liu, L., Li, Y. Y., Crabbe, M. J. C., Jin, L., Li, Y. X., **Zhong, Y.** Photosynthetic metabolism of C-3 plants shows highly cooperative regulation under changing environments: A systems biological analysis. *2009. Proceedings of the National Academy of Sciences of the USA*, 106(3): 847-852.

III. 生物多样性与植物基因组分析

2000年，钟扬教授来到复旦大学后从事了大量的生物多样性的研究工作，特别是从2001年起，每年赴西藏进行野外科学考察，针对青藏高原特殊的环境以及特有植物，开展了基于基因组信息的生物多样性与分子进化研究，同时为西部地区的高校培养了一批少数民族高级人才。在近二十年的探索中，钟扬教授做了大量基础性的种质资源调查工作，也在相关领域发表了一系列重要论文。

拟南芥是研究分子遗传、生态进化的模式植物，2013年，钟扬教授课题组历经十余年艰苦探索，首次发现了位于青藏高原的高海拔（海拔4 000米以上）拟南芥群体，并在全基因组测序基础上检测了功能基因的适应性进化，结果表明西藏拟南芥为目前世界上发现的所有野生拟南芥的基部（原始）类群，说明其在青藏高原隆升之后经历了长期的隔离和适应性进化。由于拟南芥这一模式植物的极端重要性，青藏高原高海拔拟南芥类群的发现已引起国内外植物学界高度关注，而后续对该类群的深入研究将极大地促进对植物适应逆境的生理和发育机制的认识。基于该拟南芥生态型的初步研究工作于2017年底发表在 Science Bulletin（《科学通报英文版》）(Discovery of a high-altitude ecotype and ancient lineage of Arabidopsis thaliana from Tibet) 上。

诠释全基因组水平的遗传变异对于了解生物如何适应外界环境具有十分重要的科学意义。近年来，在国家自然科学基金"微进化重大研究计划"重大研究计划支持下立项的"青藏高原极端环境下植物基因组变异及适应性进化机制研究"项目资助下，钟扬教授对西藏特有植物西藏沙棘和山岭麻黄等的微进化特征进行了分析，并通过对青藏高原鱼腥藻的全基因组测序和转录组分析，发现了蓝藻高原适应的分子遗传学证据 (The genome and transcriptome of Trichormus sp. NMC-1: insights into adaptation to extreme environments on the Qinghai-Tibet Plateau)，这是在国际上除动物之外首次利用了全基因组研究生物适应青藏高原极端环境的遗传机制，为青藏高原野生植物或微生物的起源与演化提供一个可资借鉴的案例。

钟扬教授及其课题组，通过对青藏高原十余年的探索和植物收集，初步了解了我国重要生物资源特别是西藏生物资源的分布特点。其中，西藏沙棘是研究青藏高原植物的遗传结构、生物适应性进化机制和气候动态变化非常好的植物材料，是钟扬教授指导其藏族研究生开展的研究之一，也是钟扬教授课题组探讨青藏高原生物多样性形成机制的代表性论文 [Testing the effect of the Himalayan mountains as a physical barrier to gene flow in *Hippophae tibetana* Schlect. (Elaeagnaceae)]。此外，对青藏高原的重要植物资源，如药用植物沙棘、红景天和独一味 [Genetic diversity and population structure of *Lamiophlomis rotata* (Lamiaceae), an endemic species of Qinghai-Tibet Plateau] 等以及垫状植物点地梅 [Fine- and landscape-scale spatial genetic structure of cushion rockjasmine, *Androsace tapete* (Primulaceae), across southern Qinghai-Tibetan Plateau] 等的遗传多样性和化学多样性研究取得了重要进展，为青藏高原特有植物的保护与利用提供了科学依据。作为藏药中用量最大的植物种类，独一味的遗传多样性分析结果还引起了有关植物种植供应商的极大兴趣，为这种宝贵药用资源的稳产和高产提供了技术保障。

此外，在中国的经济植物资源上，钟扬教授课题组早期也进行了大量的植物遗传多样性研究，如作为重要的室内花卉观赏植物的报春花属植物。通过对生长于欧洲及中国的报春花属的若干植物是否含有致敏化合物及其遗传多样性的研究，为今后报春花的引种驯化和人工栽培育种材料的选择等提供了科学依据 (Genetic diversity in *Primula obconica* from Central and Southwest China as revealed by ISSR markers)。

Discovery of a high-altitude ecotype and ancient lineage of *Arabidopsis thaliana* from Tibet

在西藏发现一种拟南芥的高海拔生态型和古老分支库系统

拟南芥是植物研究中的重要模式生物，在全球广泛分布。目前，已有上千个拟南芥群体的基因组完成测序，但研究大多集中在欧洲的群体。本研究通过十年的寻找，发现了分布于青藏高原海拔4 200米以上的野生拟南芥群体，经过形态学和分子鉴定，确定为拟南芥新生态型，定名为Tibet-0。本研究随后对Tibet-0进行了全基因组重测序，将结果与47个不同地区拟南芥生态型进行比较和系统发育分析。系统发育树和溯祖分析都表明，Tibet-0具备原始的遗传特征，是研究涉及的48个生态型中最原始的一支。研究还估计了Tibet-0和其他生态型的分歧时间，发现正位于冰期和青藏高原隆起的地质事件时期，推测Tibet-0的独特基因型可能是冰期气候和青藏高原隆起共同作用的结果。

Zeng, L. Y., Gu, Z. Y., Xu, M., Zhao, N., Zhu, W. D., Yonezawa, T., Liu, T. M., Lhag, Q., Tashi, T., Xu, L. L., Zhang, Y., Xu, R. Y., Sun, N. Y., Huang, Y. Y., Lei, J. K., Zhang, L., Xie, F., Zhang, F., Gu, H. Y., Geng, Y. P., Hasegawa, M., Yang, Z. H., Crabbe, M. J. C., Chen, F., **Zhong, Y.** Discovery of a high-altitude ecotype and ancient lineage of *Arabidopsis thaliana* from Tibet. 2017. *Science Bulletin*, 62(24): 1628-1630. with permission from Elsevier.

Discovery of a high-altitude ecotype and ancient lineage of *Arabidopsis thaliana* from Tibet

Liyan Zeng[1,2,†], Zhuoya Gu[1,†], Min Xu[3,4,†], Ning Zhao[3,†], Weidong Zhu[3,5], Takahiro Yonezawa[3,6], Tianmeng Liu[3], Lha Qiong[3], Tashi Tersing[3,7], Lingli Xu[1], Yang Zhang[8], Rongyan Xu[1], Ningyu Sun[1], Yanyan Huang[1], Jiankun Lei[9], Liang Zhang[9], Feng Xie[10], Fang Zhang[11], Hongya Gu[12], Yupeng Geng[13], Masami Hasegawa[1,6], Ziheng Yang[14], M. James C. Crabbe[15,16], Fan Chen[11,*], Yang Zhong[1,3,17,*]

* Corresponding authors.
1 Ministry of Education Key Laboratory for Biodiversity Science and Ecological Engineering, School of Life Sciences, Fudan University, Shanghai 200433, China.
2 Shanghai Public Health Clinical Center, Fudan University, Shanghai 201508, China.
3 Institute of Biodiversity Science and Geobiology, College of Sciences, Tibet University, Lhasa 850000, China.
4 Institute of Forest Inventory, Planning and Research of Tibet Autonomous Region, Lhasa 850010, China.
5 Three Gorges Center for Food and Drug Control, Yichang 443005, China.
6 Institute of Mathematical Statistics, Midori-cho 10-3, Tachikawa, Tokyo 190-8562, Japan.
7 Tibet Museum of Natural Science, Lhasa 850000, China.
8 Department of Bioengineering, University of Illinois at Urbana-Champaign, Champaign, IL 61801, USA.
9 School of Computer Science, Fudan University, Shanghai 200433, China.
10 School of Urban Rail Transportation, Soochow University, Suzhou 215131, China.
11 Institute of Genetics and Developmental Biology, Chinese Academy of Sciences, Beijing 100101, China.
12 School of Life Sciences, Peking University, Beijing 100871, China. （转下页）

Arabidopsis thaliana(A. thaliana) has long been a model species for dicotyledon study, and was the first flowering plant to get its genome completed sequenced[1]. Although most wild *A. thaliana* are collected in Europe, several studies have found a rapid *A. thaliana* west-east expansion from Central Asia[2]. The Qinghai-Tibet Plateau (QTP) is close to Central Asia and known for its high altitude, unique environments and biodiversity[3]. However, no wild-type *A. thaliana* had been either discovered or sequenced from QTP. Studies on the *A. thaliana* populations collected under 2 000m asl have shown that the adaptive variations associated with climate and altitudinal gradients[4]. Hence a high-altitude *A. thaliana* provides a precious natural material to investigate the evolution and adaptation process.

Here, we present the genome of a new ecotype of *A. thaliana* collected in the Gongga County, Tibet (4 200 m asl) (Fig. 1a), to demonstrate its evolutionary history and adaptation to highaltitude regions. The Tibetan samples were identified as *A. thaliana* by comparing the nuclear internal transcribed spacer (ITS), four chloroplast genes (*mat*K, *rbc*L, *rpo*B, and *rps*16), and three chloroplast intergenic spacers (IGS, *trn*L-*trn*F, and *trn*T-*trn*L) with *A. thaliana* (Col-0) and *A. lyrata* (Supplementary Fig. 1). This is the first report that an *A. thaliana* population has been collected in the QTP over 4 000 m asl and identified by molecular analysis. Moreover, the new Tibetan

---

（接上页）

13 School of Life Sciences, Yunnan University, Kunming 650091, China.

14 Department of Genetics, Evolution and Environment, University College London, Darwin Building, Gower Street, London WC1E 6BT, United Kingdom.

15 Department of Zoology, University of Oxford, Tinbergen Building, South Parks Road, Oxford OX1 3PS, United Kingdom.

16 Institute of Biomedical and Environmental Science & Technology, Department of Life Sciences, University of Bedfordshire, Park Square, Luton LU1 3JU, United Kingdom.

17 Shanghai Center for Bioinformation Technology, Shanghai 201203, China.

†   These authors contributed equally to this work.

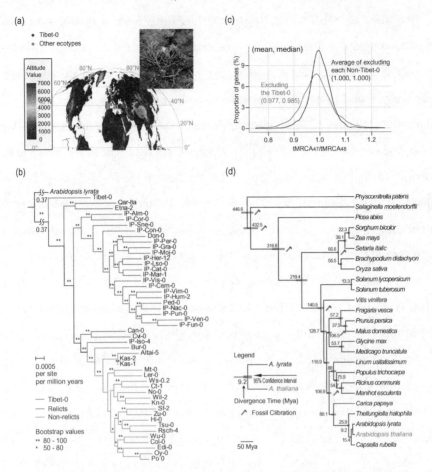

Fig. 1 Collection, phylogentic analysis and adaptation analysis of the Tibet-0 and other 47 *A. thaliana* ecotypes. *a,* Origins of Tibet-0 and other 47 *A. thaliana* ecotypes that we have analyzed in this paper (Supplementary Table 4). Elevation data were downloaded from WorldClim (http://www.worldclim.org/). Colors indicate altitudes, going from low to high elevation: Deep blue, blue, white, yellow, orange, and brown. *b,* The maximum likelihood phylogenetic trees based on 5611 orthologues of Tibet-0, 47 *A. thaliana* ecotypes and *A. lyrata* used as outgroup. Bootstrap values based on 100 replications are listed as percentages at each node. The Tibet-0 was marked in red. *c,* The time of most recent common ancestor (tMRCA) based on 2788 single-copy orthologues. The red line represents the tMRCA47/tMRCA48 distribution where the tMRCA47 values exclude Tibet-0. The black line represents the mean of tMRCA47/tMRCA48 distribution where the tMRCA47 values exclude each Non-Tibet-0. *d,* Phylogenetic affinities inferred from the maximum likelihood analysis of nucleotide sequence of 334 single copy orthologues in 25 plants. The divergence time of *A. thaliana* and *A. lyrata* was about 9.2Mya (million years ago). 7 fossil calibrations used in the study were marked as the hammer symbol. Branch lengths are proportional to the number of expected nucleotide substitutions. The number on the branch is the divergence time and unit is Mya.

ecotype (herein referred to as "Tibet-0") is diploid (2n=10) according to karyotype analysis of its pollen mother cells during meiosis (Supplementary Fig. 2 online), suggesting that the ploidy of the Tibet-0 is stable and capable of further sequence analysis.

We then conducted genome-wide resequencing of Tibet-0 with a mean coverage of 40x of the reference genomes Col-0 and TAIR10, by using Illumina Hiseq2000 (Supplementary Tables 5, 6, online). We compared Tibet-0 with 47 other *A. thaliana* ecotypes that have been genome-wide sequenced, and found that Tibet-0 was of high divergence, including a higher proportion of SNPs (Supplementary Tables 7-9, online). Evolutionary relationships between Tibet-0 and other ecotypes were evaluated by the following two independent approaches based on 5611 single-copy orthologues in 47 *A. thaliana* ecotypes including 26 relicts and 21 non-relicts defined by the 1001 Genomes Consortium[5]. The first approach is the phylogenetic method. The genealogy among the individuals was inferred based on the concatenated genomic data, and Tibet-0 was placed at the root of the *A. thaliana* populations with high support value (Fig. 1b)[5,6]. It makes Tibet-0 the most ancestral lineage.

However, since this phylogenetic approach assumes that all gene loci have the same genealogy, coalescent method was also applied as a cross check[7]. In this method, 2 788 single-copy orthologues were independently analyzed, and the distributions of the tMRCAs (the time to the most recent common ancestor) for these genes were estimated. If Tibet-0 is the most basal lineage among the *A. thaliana* populations and Tibet-0 specific alleles has generally older histories than others, tMRCAs excluding Tibet-0 will be smaller than tMRCAs of all *A. thaliana* populations. Otherwise, if there is no such genetic structure and Tibet-0 specific alleles are included within the genetic diversity of other *A. thaliana* populations, tMRCAs excluding Tibet-0 will be equal to the tMRCAs of all *A. thaliana* populations. To examine the differences among the distributions, the tMRCAs were first estimated based on 48 *A. thaliana* (tMRCA48). Subsequently, each ecotype was excluded once, and the tMRCAs of 47 remaining *A. thaliana* were estimated (tMRCA47: there are 48 combinations of tMRCA47). Finally, the relative tMRCAs (tMRCA47/tMRCA48) were estimated. Fig. 1C illustrates the distributions of the relative tMRCAs. When Tibet-0 was excluded, the distribution of the relative tMRCAs ($tMRCA47_{excluding\ Tibet-0}/tMRCA48$) significantly shifted (t test, $P = 9.77E-32$), while the average of $tMRCA47_{excluding\ one\ ecotype\ other\ than\ Tibet-0}/tMRCA48$ showed no significant change (Fig. 1c). These findings confirm that Tibet-0 has the most ancestral positions among *A. thaliana* populations.

To understand the correlation between the evolution of *A. thaliana* and major geological events, especially Tibetan uplifts, the divergence time between Tibet-0 and other ecotypes were estimated. Since there is no suitable fossil calibrations within *A. thaliana*, the divergence time between *A. lyrata* and *A. thaliana* was estimated based on the genomic data in the framework of whole land plant evolution with reliable fossil records, and it was estimated to be about 9 million years ago (Fig.1d, Supplementary Fig. 3). Then, the time of the common ancestor of *A. thaliana* was estimated by multiplying the divergence time between *A. lyrata* and *A. thaliana* and the ratio of the divergence time between *A. lyrata* and *A. thaliana*. The divergence time between Tibet-0 and other ecotypes was found to be 126-149 Ka (kili annum: thousand years ago). Interestingly, the Gonghe movement, which was the last phase of Tibetan uplift, isolated the Qinghai Lake and raised the QTP to its present height began at about 150 Ka[8]. Besides, the divergence time of Tibet-0 and other ecotypes is in the middle Pleistocene late Middle Pleistocene from 781 to 126 Ka[9].

A. thaliana has been widely used in studies of plant biology. By collecting and sequencing *A. thaliana* collected from the QTP over 4 200 m asl, we have found that the Tibet-0 is a new and divergent ecotype that isolated from other *A. thaliana* ecotypes since the last uplift of the QTP. After 126-149 thousands years evolution in the extreme plateau environment, Tibet-0 possesses a distinctive genome with a high proportion of SNPs compared to other ecotypes. According to the strongly negatively skewed Tajima's D of 5 611 single-copy orthologues, a recent selective sweep or population expansion might have occurred in the *A. thaliana*, which is consistent with previous studies (Supplementary Fig. 4, online)[10]. Considering the ancestral position of Tibetan populations as well as the subsequent selective sweep or population expansion, possibly in the Last Glacial Period, suggested by the negative Tajima's D, we suppose that some mutations might have emerged in the ancient *A. thaliana* population located around the QTP, and then spread to most other populations. Following step is investigating phenotypic traits of Tibet-0 to study the adaptive evolution of *A. thaliana* to high altitudes. As a new model plant, the Tibet-0 from QTP would provide an invaluable material for further study.

Author contributions

F. C. and Y. Zhong conceived the project. L. Zeng, Z. G., T. Y., F. C. and Y. Zhong contributed to the design of the project and extensive discussions. M. X., N. Z., W. Z., L. Q. and T. T. collected samples from Tibet. L. Zeng and H. G. helped with sample identification. L. X., R. X., F. X., J. L., L. Z., Z. G., N. Z., Y. H., T. Y., M. H., F. Z., F. C., Y. G., L. Zhang, Y. Zhang, Z. Y., M. J. C. C. and Y. Zhong performed the common garden experiments, sequence analyses and evolutionary analyses. L. Zeng, Z. G., Y. Zhang, M. J. C. C., N. S., F. C. and Y. Zhong wrote the manuscript. Other authors revised the manuscript.

Conflict of interest

The authors declare that they have no conflict of interest.

Availability of data and materials

The genomic DNA of Tibet-0 has been deposited in the Sequence Read Archive (SRA, http://www.ncbi.nlm.nih.gov/sra/) under accession number SRP052218.

Acknowledgments

This study was supported by the National Natural Science Foundation of China (91131901), the specimen platform of China (teaching specimens sub-platform) and PSCIRT project.

Appendix A. Supplementary data

Supplementary data associated with this article can be found, in the online version, at https://doi.org/10.1016/j.scib.2017.10.007.

References

1. Arabidopsis Genome I. Analysis of the genome sequence of the flowering plant *Arabidopsis thaliana*. *Nature*. (2000) 408: 796-815.
2. Yin P., Kang J., He F., Qu L. J., Gu H. The origin of populations of *Arabidopsis thaliana* in China, based on the chloroplast DNA sequences. *BMC Plant Biol*. (2010);10: 22.
3. Liu S. W. N., Duan K., Xiao C., Ding Y. Recent progress of glaciological studies in China. *J Geog Sci*. (2004) 14: 401-410.
4. Suter L, Ruegg M, Zemp N, Hennig L, Widmer A. Gene regulatory variation mediates flowering responses to vernalization along an altitudinal gradient in Arabidopsis. *Plant Physiol*. (2014) 166: 1928-1942.
5. Consortium G. 1135 genomes reveal the global pattern of polymorphism in *Arabidopsis thaliana*. *Cell*. (2016) 166: 481-491.
6. Gan X. *et al*. Multiple reference genomes and transcriptomes for *Arabidopsis thaliana*. *Nature*. (2011) 477: 419-423.
7. Nakagome S., Mano S., Hasegawa M. Comment on "Nuclear genomic sequences reveal that polar bears are an old and distinct bear lineage". *Science*. (2013) 339: 1522.
8. Li J. Late Cenozoic intensive uplift of Qinghai-Xizang Plateau and its impacts

on environments in surrounding area. *Quat Sci.* (2001) 21: 381-391.

9. Cohen K. M. G. P. Global chronostratigraphical correlation table for the last 2.7 million years. *Subcommission on Quaternary. Stratigraphy.* (2011) 31: 243-247.

10. Shimizu K. K. *et al.* Darwinian selection on a selfing locus. Science . (2004) 306: 2081-2084.

The genome and transcriptome of *Trichormus* sp. NMC-1: insights into adaptation to extreme environments on the Qinghai-Tibet Plateau

鱼腥藻属株系 NMC-1 的基因组和转录组：探究在青藏高原极端环境下的适应性

青藏高原是全世界极端环境下物种最丰富的区域，为研究生物适应性进化提供了一个理想的天然实验室。本研究中我们得到了青藏高原三离藻属（*Trichormus*）某株系 NMC-1 的基因组序列草图，并在低温条件下进行了全转录组的测序，以此研究三离藻 NMC-1 株系能适应特殊环境的遗传机制。纳木错鱼腥藻的基因组为 5.9Mb，G+C 含量为 39.2%，包含 5 362 个基因。基于全基因组的系统树表明该株系属于三离藻属（*Trichormus*）和鱼腥藻属（*Anabaena*）的分支，也支持形态学的鉴定结果。通过将纳木错鱼腥藻与近缘的六种蓝藻基因组进行比较，发现该藻种具有更高比例的未知功能基因。并找到了 2 204 个直系同源基因，利用自然选择模型找到了 70 个受到强烈正选择的基因。综合该藻种特有基因、扩张与收缩基因、正选择基因和低温下表达差异基因的结果都说明信号转导途径、细胞膜合成、次生代谢和能量相关基因对鱼腥藻适应性进化有重要作用。这些途径的基因（比如 CheY-like 基因、胞外多糖、MAAs 合成相关基因）明显参与了鱼腥藻对青藏高原低温和强紫外线环境的适应。总之，我们的研究表明纳木错鱼腥藻株系通过进化出一整套复杂的策略来适应青藏高原低温强紫外线的极端环境。

Qiao, Q., Huang, Y. Y., Qi, J., Qu, M. Z., Jiang, C., Lin, P. C., Li, R. H., Song, L. R., Yonezawa, T., Hasegawa, M., Crabbe, M. J. C., Chen, F., Zhang, T. C., **Zhong, Y.** The genome and transcriptome of *Trichormus* sp. NMC-1: insights into adaptation to extreme environments on the Qinghai-Tibet Plateau. 2016. *Scientific Reports*, 6: 29404.

The genome and transcriptome of *Trichormus* sp. NMC-1: insights into adaptation to extreme environments on the Qinghai-Tibet Plateau

Qin Qiao[1,2,*], Yanyan Huang[1,*], Ji Qi[1], Mingzhi Qu[1], Chen Jiang[1], Pengcheng Lin[3], Renhui Li[4], Lirong Song[4], Takahiro Yonezawa[1], Masami Hasegawa[1], M. James C. Crabbe[5,6], Fan Chen[7], Ticao Zhang[8] & Yang Zhong[9,1]

---

* These authors contributed equally to this work. Correspondence and requests for materials should be addressed to T.Z. (email: zhangticao@mail.kib.ac.cn) or Y.Z. (email: yangzhong@fudan.edu.cn).

1 Ministry of Education Key Laboratory for Biodiversity Science and Ecological Engineering, School of Life Sciences, Fudan University, Shanghai, 200433, China.
2 School of Agriculture, Yunnan University, Kunming, 650091, China.
3 College of Chemistry and Life Sciences, Qinghai University for Nationalities, Xining, 810007, China.
4 Institute of Hydrobiology, Chinese Academy of Sciences, Wuhan, 430072, China.
5 Department of Zoology, University of Oxford, Tinbergen Building, South Parks Road, Oxford, OX1 3PS, UK.
6 Institute of Biomedical and Environmental Science & Technology, Faculty of Creative Arts, Technologies and Science, University of Bedfordshire, Park Square, Luton, LU1 3JU, UK.
7 Institute of Genetics and Developmental Biology, Chinese Academy of Sciences, Beijing, 100101, China.
8 Key Laboratory for Plant Diversity and Biogeography of East Asia, Kunming Institute of Botany, Chinese Academy of Science, Kunming, 650204, China.
9 Institute of Biodiversity Science and Geobiology, Tibet University, Lhasa, 850012, China.

Abstract

The Qinghai-Tibet Plateau (QTP) has the highest biodiversity for an extreme environment worldwide, and provides an ideal natural laboratory to study adaptive evolution. In this study, we generated a draft genome sequence of cyanobacteria *Trichormus* sp. NMC-1 in the QTP and performed whole transcriptome sequencing under low temperature to investigate the genetic mechanism by which *T.* sp. NMC-1 adapted to the specific environment. Its genome sequence was 5.9 Mb with a G+C content of 39.2% and encompassed a total of 5 362 CDS. A phylogenomic tree indicated that this strain belongs to the *Trichormus* and *Anabaena* cluster. Genome comparison between *T.* sp. NMC-1 and six relatives showed that functionally unknown genes occupied a much higher proportion (28.12%) of the *T.* sp. NMC-1 genome. In addition, functions of specific, significant positively selected, expanded orthogroups, and differentially expressed genes involved in signal transduction, cell wall/membrane biogenesis, secondary metabolite biosynthesis, and energy production and conversion were analyzed to elucidate specific adaptation traits. Further analyses showed that the CheY-like genes, extracellular polysaccharide and mycosporine-like amino acids might play major roles in adaptation to harsh environments. Our findings indicate that sophisticated genetic mechanisms are involved in cyanobacterial adaptation to the extreme environment of the QTP.

The Qinghai-Tibet Plateau (QTP) is not only the highest and largest young plateau in the world, but also has the most variety of extreme environments, including rapid fluctuations in temperature, low oxygen concentration, low pressure, strong ultraviolet (UV) radiation, and severe winds. The QTP is also one of the global biodiversity hotspots with many unique environments, including snowy mountains, saline lakes, and arid deserts[1]. These environments provide an ideal natural laboratory for studies on adaptive evolution. Organisms that live in the QTP must have undergone a series of significant adaptive evolutionary genetic changes to produce a wide range of ecologically adaptive characters. Previous studies on adaptive evolution at the whole genome level on this region have focused mainly on Tibetans adapted to hypoxia (see review by Cheviron & Brumfield[2]). Recently, several genome-wide studies regarding the QTP adaptations have been conducted on non-model animals, such as yaks[3], ground tits[4], and Tibetan boars[5]. However, how other organisms (not animals) adapt to the QTP environments at the genomic level is still unclear.

Table 1　Statistics of assemble data in genome sequencing.

Statistics of contigs data		Statistics of scaffolds data	
No. of all contigs	123	No. of all scaffolds	58
Bases in all contigs	5 897 265 bp	Bases in all scaffolds	5 938 148 bp
No. of large contigs(>1000 bp)	92	No. of large scaffords (>1000 bp)	38
Bases in large contigs	5 876 594 bp	Bases in large scaffolds	5 922 589 bp
Largest length of contigs	317 061 bp	Largest length of scaffords	1 580 590 bp
N50 length of contigs	156 762 bp	N50 length of scaffolds	566 878 bp
N90 length of contigs	50 741 bp	N90 length of scaffolds	169 940 bp
		N rate	0.688%
		G+C content	39.18%
		No. of CDSs	5 362

Cyanobacteria are the earliest photosynthetic organisms; they have successfully colonized many varieties of habitats and have considerable global ecological importance[6]. Cyanobacteria can tolerate a broad range of stresses experienced in various environmental conditions, including variable osmolarity, persistent low temperatures, and high irradiance[7-9]. The genetic mechanisms of cyanobacterial responses to stress have been studied, especially with regard to their two-component regulatory systems, histidine and serine-threonine protein kinases, and DNA binding transcription factors[7,8]. Our previous field work revealed that cyanobacteria are also abundant in such extreme environments, including lakes on the QTP. The high diversity of cyanobacteria that live on the QTP indicates that they cope with harsh conditions. Among these cyanobacteria, *Trichormus* is a genus of filamentous cyanobacterium with nitrogen-fixing abilities in heterocysts. Based on akinete development, *Trichormus* was recently split from *Anabaena*[10], which is a genus that has traditionally been used to study the genetics and physiology of cellular differentiation, pattern formation, and nitrogen fixation[11]. However, the phylogenetic relationship of these two genera was not firmly established.

To better understand how cyanobacteria evolved specific adaptations to unfavorable abiotic stress factors on the QTP, we sequenced the genome and performed deep transcriptome analysis of *T.* sp. NMC-1. This strain was isolated from Namucuo Lake, which is the largest (1 920 km^2) and highest (a.s.l. 4 741 m) saltwater lake in the world. Genomic comparison between *T.* sp. NMC-1 and related species was also conducted to reveal the adaptive evolutionary pattern.

Results

Genome assembly and annotation. We sequenced the draft genome of *T.* sp. NMC-1 using the Illumina Genome Analyzer II platform and generated a total of 4 118 651×2 high qualities paired-end (PE) reads and 2 849 093×2 high quality mate-pair (MP) reads after the raw data were cleaned. *T.* sp. NMC-1 genome sequencing data have been deposited at NCBI BioProject under accession PRJNA324543. All sequence types provided 183-fold coverage of the genome (108- and 75-fold coverage of PE and MP data, respectively; Supplemental Table S1). After cleaned contamination, the assembly consisted of 58 scaffolds with an N50 length of 567 kb and a total genome length of 5.9 Mb (Table 1). Among these scaffolds, the 10 longest, which ranged from 1.58 to 0.17 Mb (Supplemental Table S2), covered approximately 90.5% of the assembled genome. GC content (39.2%) distributions were similar to those of other related species (Table 2). Within the genome, a total of

5 362 CDS were identified (Table 2), and 1 346 (28.12%) of these protein-encoding genes have unknown functions based on Clusters of Orthologous Groups (COG) of proteins functional categories (Fig. 1A). We then surveyed 102 housekeeping genes that were previously identified as nearly universal in bacteria[12], and found all of these genes were present in the *T.* sp. NMC-1 draft genome. In addition, a survey of 682 core orthologous protein families from 13 cyanobacterial genomes[13] indicated that all of these genes but *dnaA* were present in the draft genome, reported to be absent from *Synechocystis* sp. PCC6803[14] and *T. azollae* 0708[15]. Therefore, the sequencing and assembly results were sufficiently accurate for further comparative and evolutionary genomics analysis.

Phylogenetic analysis based on whole genome sequences. Based on morphological analysis under both light and fluorescence microscopy (Supplemental Fig. S1), and 16S rRNA BLAST search in NCBI, the cyanobacterial strain from the Namucuo Lake exhibited typical morphological features of, and high sequence similarity to, *Trichormus* and *Anabaena*, which belong to the family Nostocaceae. However, because of the low rate of 16S rRNA evolution and to avoid the effect of horizontal gene transfer in cyanobacterial genomes, the CVTree without sequence alignment approach was applied to construct a phylogenetic tree. Previous research suggested that this method resolves the relationships among closely related strains better than 16S rRNA[16,17]. The phylogenetic tree shows that Nostocales was divided into two clusters; one cluster mainly included species from *Nostoc* and the species *A. variabilis* ATCC29413 (Fig. 2). The strain we studied was located in the other cluster and was grouped with *A.* sp. PCC7108, *A. cylindrical* PCC7122, and *T. azollae* 0708 (Fig. 2).

Table 2 Genome structure of *T.* sp. NMC-1 and six close relatives.

Genome Features	Length (Mb)	G+C content(%)	Total ORF	Homologs	rRNA	tRNA
T. sp. NMC-1	5.94	39.18	5362	4590	12	45
T. azollae 0708	5.49	38.3	5380	3093	12	44
A. sp. PCC 7108	5.89	38.78	5169	4571	12	43
A. cylindrica PCC 7122	7.06	38.79	6182	5187	12	61
N. sp. 7120	7.21	41.2	6213	4852	12	48
N. punctiforme 73102	9.06	41.3	7164	4935	12	88
T. variabilis ATCC 29413	7.11	41.4	5813	4762	12	47

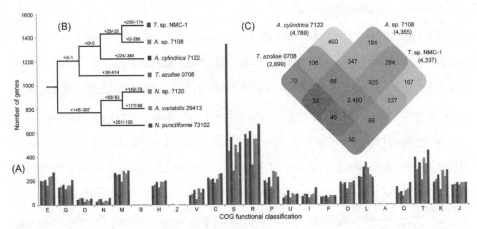

Fig. 1 Comparative genomic analysis between *T.* sp. NMC-*1* and close relatives. (A) comparison of COG functional classification among the seven relatives. (B) The significantly (P<0.05) expanded and contracted COG clusters in *T.* sp. NMC-1 compared with the six close relatives. (C) Comparison of orthogroups among the four closest relatives.

Evolution of COG clusters. A total of 3 295 genes were assigned to 1 540 COG clusters in the *T.* sp. NMC-1 genome. Compared with the six close relatives (9.79%-13.03%), an unexpectedly high proportion (28.12%) of proteins with unknown functions was detected in the *T.* sp. NMC-1 genome (Fig. 1A). Additionally, COG clusters underwent obvious expansion or contraction in the *T.* sp. NMC-1 genome. A total of 209 COG clusters substantially expanded and

174 contracted in the *T.* sp. NMC-1 genome compared with the six other genomes (Fig. 1B). In particular, 18 COG clusters were the most expanded and 10 were the most contracted (P < 0.0001, Fig. 3, Supplemental Table S4). The most significant expanded COG clusters were related to signal transduction, secondary metabolite biosynthesis, cell wall synthesis, posttranslational modification, and defense mechanisms; these clusters included COG0784 (CheY-like receiver), COG2203 (GAF domain), COG2202 (PAS/PAC domain), COG2199 (GGDEF domain, diguanylatecyclase), COG2931 (RTX toxins and related Ca^{2+}-binding proteins), COG0500 (SAM-dependent methyltransferases), COG0845 (AcrA Membrane-fusion protein), COG0526 (thiol-disulfide isomerase or thioredoxin), COG2214 (DnaJ-class molecular chaperone), COG1002 (type II restriction enzyme), and COG0732 (restriction endonuclease S subunits) (Fig. 3, Supplemental Table S4).

Identified orthogroups and genes under positive selection. A total of 5 452 orthogroups (homologous gene clusters, similar to gene families) shared by *T.* sp. NMC-1 and six other related species were detected. Figure 1C shows the statistical results of the four most closely related species; there are 4 170 orthogroups (including 4 358 genes) in *T.* sp. NMC-1 shared with six other species, whereas 167 orthogroups (including 232 genes) are specific to *T.* sp. NMC-1. Combined with the genes (772) of *T.* sp. NMC-1 not clustered in orthogroups, there are 1 004 genes specific to *T.* sp. NMC-1. Of these specific genes, 236 genes have known COG functions and are involved in adaptation, such as cell wall/membrane biogenesis, and defense mechanisms (Supplemental Table S3).

In 5 452 gene clusters, 2 204 single-copy number (one-to-one) orthologs were shared by all seven species. For these single-copy number orthologs, the branch-site model of the PAML 4 package[18] was used to detect genes with signals of positive selection. Finally, 491 possible genes under positive selection were identified in the *T.* sp. NMC-1 genome ($\omega>1$); of these genes, 70 showed highly significant evidence of positive selection (P<0.01) (Supplementary Table S5). These 70 positively selected genes were also enriched in functions related to adaptation, such as amino acid/nucleotide/carbohydrate/coenzyme transport and metabolism (26 genes), cell wall/ membrane biogenesis (10 genes), signal transduction (five genes), and posttranslational modification, protein turnover, and chaperones (five genes), by functional annotation (Fig. 4).

Transcriptome sequencing and analyses. Global gene expression profiles of *T.* sp. NMC-1 under cold conditions were examined using transcriptome sequencing (Table 3). Finally, we generated 31.1-40.35 million clean reads and 3.12-4.04 Gb of RNA-seq data in treated and control strains after quality filtering (Table 3). The clean data were submitted to the NCBI Sequence Reads Archive (SRA) database (no. SRR3597124). All of the Pearson correlations between biological replicates were greater than 0.95, which indicates high reliability of the experiment and rationality of sample selection (Supplemental Fig. S2). The transcriptome data of control and treatment were mapped to our *T.* sp. NMC-1 genome assembly and yielded 5 362 predicted protein-coding genes and 1 023 novel transcripts. Compared with the control strain, the cold-treated strain had 312 genes with significantly altered expression after 3 d (FDR <= 0.001). COG and KEGG enrichment analyses were carried out for the up- and down-regulated genes, respectively (Supplemental Table S6). According to the COG categories, signifi- cantly up-regulated genes included those involved in processes such as membrane biogenesis, translation, ribosomal structure and biogenesis, secondary metabolites biosynthesis, amino acid transport and metabolism, and defense mechanisms (Fig. 4; Supplemental Table S6). In contrast, down-regulated genes were primarily involved in processes such as signal transduction mechanisms, membrane biogenesis, and energy production and conversion; however, some down-regulated genes had unknown functions (Fig. 4; Supplemental Table S6).

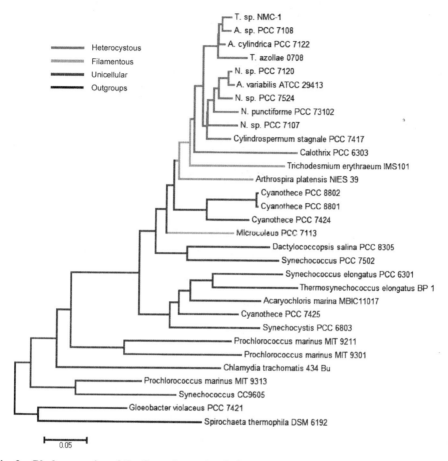

Fig. 2　Phylogenomics of the Cyanobacteria phylum as determined using CVTree software.

Discussion

In this study, an alignment-free method, CVTree, was used to construct a phylogeny based on 31 cyanobacteria whole genomes (Fig. 2). The topology of our phylogenetic tree is consistent with morphological classification (unicellular, filamentous and heterocystous cyanobacteria) as well as previous studies that analyzed 16S rRNA and dozens of conserved proteins[10,13,19]. *Trichormus* sp. NMC-1 was most closely related to and had a similar genome size and amount of gene content as *A.* sp. PCC 7108. In our phylogenetic tree, it is difficult to distinguish *Trichormus*, *Anabaena*, and *Nostoc* strains, which is consistent with a previous phylogenetic study[10]. The results indicate that these three genera are phylogenetically heterogeneous and genetically inconsistent with the morphological taxonomy. It is notable that there were similar G+C contents within each of two distinct clusters (*Trichormus* & *Anabaena* vs. *Nostoc* & *Anabaena*) (Table 2). This similarity in the same clusters indicates that G+C content could be used as a potential character for taxonomic identification in Nostocaceae.

The *T.* sp. NMC-1 genome possessed a very high proportion of genes with unknown functions compared with the other six related species based on COG category comparison. This result indicates that the *T.* sp. NMC-1 genome might have rapidly evolved after diverging from a common ancestor of *T.* sp. NMC-1 and *A.* sp. PCC7108 to adapt to the extreme conditions of the QTP. Except for the genes with unknown functions, the obviously expanded genes in *T.* sp. NMC-1 were involved in signal transduction pathways (e.g., CheY-like receiver and related genes), secondary metabolites biosynthesis (e.g., SAM-dependent methyltransferases), cell wall/membrane biogenesis (e.g., membrane-fusion proteins), and energy production and conversion (e.g., thiol-disulfide isomerase and thioredoxins) (Supplemental Table S6). Similarly, some genes involved in the above mentioned pathways also significantly contracted. It has been reported that significant changes of gene number in one gene family was related to a major mechanism underlying the adaptive divergence of closely related species[20,21]. Therefore, dramatic fluctuation of these categories of gene families might reflect adaptation of the *T.* sp. NMC-1 to the extreme conditions of the QTP.

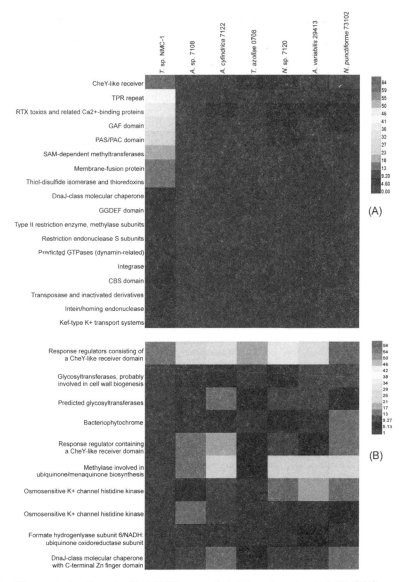

Fig. 3 The most significantly (P<0.0001) expanded (A) and contracted (B) COG clusters in *T.* sp. NMC-1 genome compared to six close relatives.

Orthologs are homologous genes that have evolved from one ancestral gene by speciation, and orthologs that show positive selection have usually undergone adaptive divergence[22]. Our results revealed 70 genes that underwent significant positive selection in the *T.* sp. NMC-1 genome based on a branch-site model. Most of these genes were related to specific adaptation traits, such as cold resistance (10 genes related to cell wall/membrane biogenesis), signal transduction mechanisms (CheA signal transduction histidine kinase), energy metabolism (fructose-1,6-bisphosphatase, alpha-mannosidase, and Fe-S-cluster-containinghydrogenase), and UV radiation resistance (caffeoyl-CoA O-methyltransferase). Interestingly, these enriched gene functions are similar to those of specific genes, and significantly expanded and contracted orthologs. Therefore, all of these consistent results indicate that *T.* sp. NMC-1 evolved complex strategies for adapting to the extreme environments in the QTP. In the following paragraphs, we will discuss the relationship between functions of these genes and the adaptation of *T.* sp. NMC-1 on the QTP.

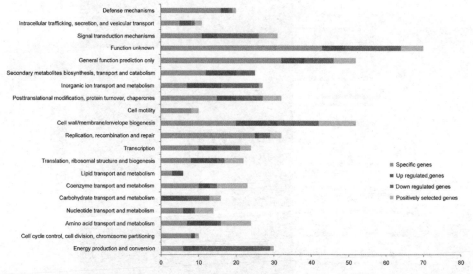

Fig. 4 COG enrichment analysis of species-specific genes, differentially expressed genes and positively selected genes in *T.* sp. NMC-1.

Cyanobacteria that live in the QTP must sense and respond to various external stimulus signals from the harsh environment. Resistance to all of these stimuli should first start at signal transduction. The histidine kinase two-component systems are conserved as potential candidates of sensors and transducers of environmental

Table 3 Number and length of reads and number of expressed genes detected by RNA sequencing in control and cold treated samples of *T.* sp. NMC-1.

Sample name	Total clean reads	Clean bases (G)	Q20 (%)	Q30 (%)	Num. of expressed genes	Num. of highly expressed genes (RPKM>60)
Control_1	34444484	3.54	97.62	92.01	5091	2323
Control_2	35690982	3.56	97.72	92.22	5118	2499
Cold_1	31104996	3.12	97.52	91.57	5115	2558
Cold_2	40350844	4.04	97.51	91.59	5151	2698
Cold_3	32592572	3.26	97.58	91.81	5142	2708

signals in cyanobacteria[7,9]. The histidine kinase CheA and its substrate, the response regulator CheY, are partners in the two-component signaling pathway. It is noteworthy that there were five genes involved in signaling that showed significant positive selection in the *T.* sp. NMC-1 genome; of these, two genes contained the CheY-like receiver domain. The positive selection could induce gene functional changes under natural environmental stress. These data are consistent with previous studies, which suggested that CheA and CheY proteins, both contain Che receiver domain, are involved in response and adaptation to external stimuli[23]. In addition to positive selection, transcriptional changes of CheY-like receiver genes also highlighted the importance of CheY-like receiver genes contributing to cold stimuli adaptation. As the transcriptome sequencing results show that five CheY-like receiver and related regulator genes had significantly up-regulated and down-regulated expression during cold treatment (Supplementary Table S6). Furthermore, the COG0784 (CheY-like receiver) was the most expanded COG cluster in *T.* sp. NMC-1, and COG0745 (response regulators that include a CheY-like receiver domain) was the most significantly contracted COG cluster compared with its close relatives. The dramatic fluctuation of CheY gene families in *T.* sp. NMC-1 suggests adaptive divergence of closely related species. Based on the above analysis, positively selected, and differentially expressed CheY-like receiver genes with drastic, frequent turnover in

T. sp. NMC-1 further corroborates that these genes are involved in response to cold stress, and correlated with adaptation to the harsh conditions of the QTP.

Organisms that live on the QTP must face a number of growth-related challenges from the rapid temperature changes, including decreased rates of enzyme activity, reduced fluidity of lipid membranes, and enhanced stability of nucleic acids[24,25]. T. sp. NMC-1 cells grow as aggregates and are often surrounded by a mucilaginous sheath, which is composed of components such as extracellular polysaccharide (EPS), cell surface-associated proteins, and pigments. Based on the transcriptome of T. sp. NMC-1 exposed to low temperatures, the category of cell wall/membrane biogenesis had the most up-regulated and major down-regulated genes, including those related to glycosyltransferases, the δ-70-transcription factor, and a membrane-bound lytic mureintransglycosylase (Fig. 4), which are involved in EPS biosynthesis[26]. In addition, out of the 70 genes that showed positive selection in the T. sp. NMC-1 genome, nine were involved in cell wall/membrane biogenesis and included five glycosyltransferase genes (COG0438) and one δ-70 transcriptional factor. These results indicated that EPS is important for cold adaptation of T. sp. NMC-1, unfavorable environmental conditions such as temperature, UV radiation, or osmotic pressure[8,27]. Similarly, EPS and the cell wall metabolism protein family were previously shown to be expanded for cold adaptation in the genome of *Coccomyxa subellipsoidea*, which is a polar unicellular micro alga[28]. Apart from EPS, we also identified four up-regulated genes involved in lipid transport and metabolism under cold treatment, two of which are fatty acid desaturases, namely *desA* and *desB* (>2.5-fold, Supplementary Table S6). Moreover, one RNA-binding protein (RbpB) was also induced, which has been reported to maintain *desA* and *desB* mRNA levels[29]. These expression changes could increase the proportion of unsaturated fatty acids with decreasing temperature. Therefore, these results indicate that genes related to aspects of the cell wall/membrane (e.g., EPS and membrane lipid biogenesis) in T. sp. NMC-1 play major roles in response to cold conditions.

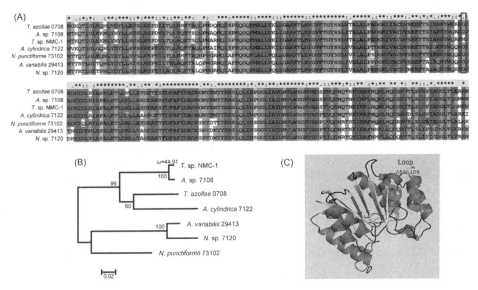

Fig. 5 Positive selection analysis of O-methyltransferase in *T.* sp. NMC-1. (A) Multiple sequence alignment of O-methyltransferase. (B) Positive selection test of seven relatives using the branch-site model in the PAML 4 package. (C) Predicted three-dimensional structure of O-methyltransferase in *T.* sp. NMC-1. The positive-selection site (Asn-109) is labeled.

The QTP has the strongest UV-B radiation during the summer in the world[30]. The highly energetic UV radiation is harmful to all organisms, because it damages DNA and proteins. Marine organisms, including some cyanobacteria, have evolved to prevent UV-induced damage by synthesizing UV-absorbing/screening compounds such as mycosporine-like amino acids (MAAs)[31-33]. MAAs belong to a family of more than 20 compounds that absorb UV radiation. Some species of cyanobacteria have the ability to biosynthesize MAAs, whereas others lack this ability[34,35]. Of six closely related species, only *T. variabilis* ATCC 29413 was able to synthesize MAAs[34]. A short four-enzyme pathway of MAA biosynthesis was identified and included dehydroquinate synthase (DHQS), O-methyltransferase (O-MT), ATP-grasp, and nonribosomal peptide synthetase (NRPS) homologs[36]. BLAST searches of both DNA and protein sequences in *T.* sp. NMC-1 revealed that three genes (*dhqs*, *o-mt*, and *nrps*) and several other genes that contain highly similar ATP-grasp domains were identified. Furthermore, complex MAA compositions that included shinorine, palythine-serine, asterina330, and palythenic acid were identified in *T.* sp. NMC-1using HPLC-ESI-MS/MS methods (Supplemental Table S8, Fig. S3). It is notable that the gene encoding O-MT showed significant positive selection in *T.* sp. NMC-1 compared with related species. Additionally, an amino acid residue at position 109 in the loop of O-MT three-dimensional structure, asparagine (N), was detected under significant positive selection in *T.* sp. NMC-1; the typical amino acid residue at this site is glycine (G) in other related species (Fig. 5). Loops often play an important role in a protein's three-dimensional structure and act as the active site of an enzyme or binding site of a receptor. Therefore, we speculate that *o-mt* in *T.* sp. NMC-1 underwent positive selection during adaptation to the QTP environment, and this site may have a specific function. In fact, in addition to their role as sunscreen compounds, MAAs are also involved in antioxidant, osmotic stress, and desiccation resistance[37-39]. Therefore, MAAs might play multiple roles in the adaptation of *T.* sp. NMC-1 to the various extreme environments of Namucuo Lake, which has high radiation exposure and salinity.

Conclusion

Interpretation of genetic variation at the whole genome level can contribute to understanding how organisms adapt to changing environments. Organisms that live in the QTP must have undergone a series of significant adaptive evolutionary changes to produce a wide range of ecologically adaptive characters. In this study, we sequenced the draft genome and whole transcriptome of *T.* sp. NMC-1 strain in Tibet and conducted evolutionary analyses based on comparative genomics. Our findings show that positively selected and enhanced genes were involved in signal transduction mechanisms, cell wall/membrane biogenesis, secondary metabolite biosynthesis, defense mechanisms, and energy production and conversion, all of which relate to the specific adaptation traits found in this challenging environment. In particular, we found that the CheY-like genes, extracellular polysaccharide and mycosporine-like amino acids may play major roles in responding to external harsh environments. Our findings indicate that sophisticated genetic mechanisms are involved in *T.* sp. NMC-1 adaptation to the extreme environment of the QTP.

Material and methods

Isolation, culture and identification of strains. The original collection of *Trichormus* sp. strain (NMC-1) was conducted on June 15, 2011 in the Namucuo Lake (N30° 46.45', E90° 52.01') in the QTP of the South West of China. The lake water was stored at 4 °C at night and then transferred to the School of Life Sciences, Fudan University in Shanghai by air the next day. The *T.* sp. NMC-1 was isolated using previously described micropipette washing methods[40]. Then each single trichome looking like *Trichormus* under the microscope was transferred to 250 ml sterilized glass Erlenmeyers containing 50 ml of MA medium[41] and maintained at 28°C/23 °C with a 16/8 h (light/dark) diurnal cycle (light intensity 2 200lux). High quality genomic DNA was extracted from the sample using the Genomic DNA Kit (Tiangen Biotech Co., China) following the manufacturer's instructions. The *T.* sp. NMC-1 purification procedure was validated by PCR using cyanobacterial 16S rRNA gene specific primers[42].

DNA library construction, genome sequencing, assembly and annotation. The 300 bp paired-end (PE) and 3500 bp mate-pair (MP) DNA libraries were sequenced on the Illumina Genome Analyzer II system. Library preparation, sequencing and base calling were performed according to the manufacturer's user guide (Illumina, Inc). The raw sequence reads with adapter contamination, PCR duplicates, and low-quality sequences (Q<20) were cleaned from the initial sequencing output using custom scripts.

The genome sequence of the *T.* sp. NMC-1 was assembled using SOAP *de novo*[43], which employs the *de Bruijn* graph algorithm in order to reduce computational complexity. We first assembled the reads from the short insert size of 300 bp into contigs using Kmer (31-mers) overlap information, and then used the mate-pair libraries, step by step from the shortest to the longest insert size, to join the contigs into scaffolds. We also cleaned contamination scaffolds from other bacteria based on a combination of protein annotation, percent GC nucleotide composition and assembly coverage depth.

ORFs and amino acid sequences were predicted from all scaffolds using the gene finding program GeneMark[44] and Glimmer3[45]. Functional annotation of CDSs was performed through Blastp searches against GenBank's non-redundant (nr) protein database and UniProtKB/Swiss-Prot protein database[46]. COG (Clusters of Orthologous Groups of proteins)[47] functional categories were assigned to CDS according to DOE-JGI Standard operating procedures[48]. These data sources were combined to assert a product description for each predicted protein.

Phylogenetic tree based on whole genome sequences. The phylogenetic tree was produced based on whole genome sequences with an alignment-free and parameter-free phylogenetic tool, CVTree ver. 2.0[49]. This method circumvents the ambiguity of choosing the genes for phylogenetic reconstruction and avoids the necessity of aligning sequences of essentially different length and gene content[17]. In total, 31 sequenced cyanobacterial genomes were chosen to produce the phylogenetic tree, using *Gloeobacter violaceus* PCC 7421 and *Spirochaeta thermophila* DSM 6192 as outgroups.

Identification of COG clusters and orthologs between *T.* sp. NMC-1 and close relatives. Based on the phylogenetic tree, we selected genomes of six close relatives (*T. azollae* 0708, *T. variabilis* ATCC 29413, *A.* sp. PCC 7108, *A. cylindrical* PCC 7122, *Nostoc.* sp. 7120, and *N. punctiforme* PCC 73102) and *T.* sp. NMC-1 to identify COG clusters and orthologs. To identify COG clusters that had undergone expansion or contraction along each branch of the phylogenetic tree, the software package Café[50] was applied, which is based on a likelihood model, to infer the change in gene family size. The COG clusters of the *T.* sp. NMC-1 were compared with the other six genomes, and significant levels of expansion and contraction were determined at 0.05 with lambda value equal 12 based on node numbers in the phylogenetic tree.

Furthermore, to define a set of conserved genes for cross-taxa comparison, we used OrthoMCL software[51] to identify orthologous gene clusters (orthogroups) among the seven genomes. OrthoMCL was run with an e-value cut-off of 1e-5 and an inflation parameter of 1.5. Genes that were not included in any orthogroups, or only present in one species of orthogroups, were defined as species specific genes. The set of genes recovered from this procedure are listed in Supplementary Table S3. A Venn diagram of shared or specific gene families in *T*. sp. NMC-1 and the closest relatives (*T. azollae* 0708, *A.* sp. 7108, and *A. cylindrical* 7122; for convenience, we omitted the resource center name) according to our phylogenetic results (Fig. 1C) was conducted using R 2.2.1.

Positive selection analysis. Positive selection can be inferred from a higher proportion of nonsynonymous (Ka) over synonymous substitution (Ks) per site (Ka/Ks>1). In this analysis, only single-copy genes that were shared by all seven genomes were considered. To calculate the nonsynonymous and synonymous substitution rates for each one-to-one ortholog, alignment at amino acid level for each orthogroup was generated in MUSCLE[52] using default settings. Then the resulting protein alignments were reverse-translated to codon-based nucleotide alignments with PAL2NAL[53]. For each alignment, a gene tree was constructed by RAxML software[54] using GTR+GAMMA model, the maximum likelihood criteria. Using each gene tree topology, we applied the improved branch-site model[55] implemented in codeml from PAML 4 package[18] to estimate the Ka/Ks substitution rates (ω value) for each orthogroup respectively. A foreground branch was specified as the clade of *T*. sp. NMC-1. A significant likelihood ratio test (LRT) was conducted to determine whether positive selection is operating in the foreground branch. In this study, the highly significant positively selected genes were inferred only if the P-value was less than 0.01. We also detected positively selected sites if their posterior probability was greater than 95% based on empirical Bayes analysis[56].

Transcriptome sequencing and analyses. *T.* sp. NMC-1 cells were grown to the mid-logarithmic phase before cold stress treatment. Then the strains were cultured at 10°C (treatment) for six hours each day within three days, and these experiments were performed as three biological replicates. Low temperature (10°C) was selected to reflect the *in situ* low temperature (according to our investigation) in the Namucuo Lake. Two samples were cultured at 28°C as the control group. After treatment, total RNA was extracted from *T.* sp. NMC-1 samples separately. The quality of the RNA samples was examined using the Agilent 2100 Bioanalyzer. Library construction and Illumina sequencing was performed at Novogene Bioinformatics Technology Co., Ltd (Beijing, China). An RNA-seq analysis was performed according to the protocol recommended by the manufacturer (Illumina Inc.). The reads from different conditions were mapped to the whole-genome assembly using Bowtie 2-2.0.6[57]. HTSeq 0.6.1[58] was used to count the read numbers mapped to each gene (Table 3). And then RPKM (reads per kilobase per million reads) of each gene was calculated based on the length of the gene and reads count mapped to this gene.

Differential expression analysis of two conditions (control and cold treatment with three biological replicates) was performed using the DESeq R package (1.10.1). Genes with an adjusted P-value<0.05 found by DESeq were assigned as differentially expressed. COG classification and Pfam domain assignment were conducted on the different expression genes (DEG). The KOBAS software[59] was used to test the statistical enrichment of differential expression genes in KEGG pathways.

Acknowledgements

We thank Xiaoni Gan for suggestions to improve the manuscript. This work is supported by grants from National Natural Science Foundation of China (91131901, 31300201, 31590820, 31590823), National High Technology Research and Development Program of China (2014AA020528), the specimen platform of China and the PSCIRT project.

Author Contributions

Y. Z., M. H., M. J. C. C. and T. Z. conceived and designed the project. Q. Q., T. Z., M. Q., Y. H., R. L. and L. S. prepared samples. Q. Q., T. Z., P. L., Y. H., F. C. and C. J. contributed to DNA sequencing and HPLC-ESI-MS/MS experiment. T. Z., Q. Q., J. Q., Y. H. and T. Y. performed data analyses. T. Z., Q. Q., M. J. C. C., T. Y. and Y. Z. wrote the paper. All authors read and approved the final manuscript.

Additional Information

Supplementary information accompanies this paper at http://www.nature.com/srep

Competing financial interests: The authors declare no competing financial interests.

How to cite this article: Qiao, Q. *et al.* The genome and transcriptome of *Trichormus* sp. NMC-1: insights into adaptation to extreme environments on the Qinghai-Tibet Plateau. *Sci. Rep.* **6**, 29404; doi: 10.1038/srep29404 (2016).

This work is licensed under a Creative Commons Attribution 4.0 International License. The images or other third party material in this article are included in the article's Creative Commons license, unless indicated otherwise in the credit line; if the material is not included under the Creative Commons license, users will need to obtain permission from the license holder to reproduce the material. To view a copy of this license, visit http://creativecommons.org/licenses/by/4.0/

References

1. Myers, N., Mittermeier, R. A., Mittermeier, C. G., Da Fonseca, G. A. & Kent, J. Biodiversity hotspots for conservation priorities. *Nature*. 403: 853-858 (2000).

2. Cheviron, Z. A. & Brumfield, R. T. Genomic insights into adaptation to high-altitude environments. *Heredity*. 108: 354-361 (2012).

3. Qiu, Q. *et al.* The yak genome and adaptation to life at high altitude. *Nature Genetics*. 44: 946-949 (2012).

4. Qu, Y. *et al.* Ground tit genome reveals avian adaptation to living at high altitudes in the Tibetan plateau. *Nature Communications*. 4 (2013).

5. Li, M. *et al.* Genomic analyses identify distinct patterns of selection in domesticated pigs and Tibetan wild boars. *Nature Genetics*. 45: 1431-1438 (2013).

6. Zehr, J. P. *et al.* Unicellular cyanobacteria fix N2 in the subtropical North Pacific Ocean. *Nature*. 412: 635-638 (2001).

7. Los, D. A. *et al.* Stress sensors and signal transducers in cyanobacteria. *Sensors*. 10: 2386-2415 (2010).

8. Vincent, W. & Quesada, A. Ecology of Cyanobacteria II (ed Brian A, Whitton) Ch. 13: 371-385 (Springer Netherlands, 2012).

9. Zorina, A. *et al.* Regulation systems for stress responses in cyanobacteria. *Russian Journal of Plant Physiology*. 58: 749-767 (2011).

10. Rajaniemi, P. *et al.* Phylogenetic and morphological evaluation of the genera *Anabaena*, *Aphanizomenon*, *Trichormus* and *Nostoc* (Nostocales, Cyanobacteria). *International Journal of Systematic and Evolutionary Microbiology*. 55: 11-26 (2005).

11. Kaneko, T. *et al.* Complete genomic sequence of the filamentous nitrogen-fixing cyanobacterium *Anabaena* sp. strain PCC 7120. *DNA Research*. 8: 205-213 (2001).

12. Puigbò, P., Wolf, Y. I. & Koonin, E. V. Search for a Life in "Tree of Life" in the

thicket of the phylogenetic forest. *Journal of Biology.* 8: 1-17 (2009).

13. Shi, T. & Falkowski, P. G. Genome evolution in cyanobacteria: the stable core and the variable shell. *Proceedings of the National Academy of Sciences, USA* 105: 2510-2515 (2008).

14. Richter, S., Hagemann, M. & Messer, W. Transcriptional Analysis and Mutation of a *dnaA*-Like Gene in *Synechocystis* sp. Strain PCC 6803. *Journal of Bacteriology.* 180: 4946-4949 (1998).

15. Ran, L. *et al.* Genome erosion in a nitrogen-fixing vertically transmitted endosymbiotic multicellular cyanobacterium. *PLoS One.* 5: e11486 (2010).

16. Qi, J., Luo, H. & Hao, B. CVTree: a phylogenetic tree reconstruction tool based on whole genomes. *Nucleic Acids Research.* 32: W45-W47 (2004).

17. Qi, J., Wang, B. & Hao, B.-I. Whole proteome prokaryote phylogeny without sequence alignment: a K-string composition approach. *Journal of Molecular Evolution.* 58: 1-11 (2004).

18. Yang, Z. PAML 4: phylogenetic analysis by maximum likelihood. *Molecular Biology and Evolution.* 24: 1586-1591 (2007).

19. Shih, P. M. *et al.* Improving the coverage of the cyanobacterial phylum using diversity-driven genome sequencing. *Proceedings of the National Academy of Sciences, USA.* 110: 1053-1058 (2013).

20. Dassanayake, M. *et al.* The genome of the extremophile crucifer *Thellungiella parvula. Nature Genetics.* 43: 913-918 (2011).

21. Sudmant, P. H. *et al.* Diversity of human copy number variation and multicopy genes. *Science.* 330: 641-646 (2010).

22. Fitch, W. M. Distinguishing homologous from analogous proteins. *Systematic Biology.* 19: 99-113 (1970).

23. Hess, J. F., Oosawa, K., Kaplan, N. & Simon, M. I. Phosphorylation of three proteins in the signaling pathway of bacterial chemotaxis. *Cell.* 53: 79-87 (1988).

24. Cavicchioli, R. Cold-adapted archaea. *Nature Reviews Microbiology.* 4: 331-

343 (2006).

25. Siddiqui, K. S. & Cavicchioli, R. Cold-adapted enzymes. *Annual Review of Biochemistry*. 75: 403-433 (2006).

26. Methé, B. A. *et al*. The psychrophilic lifestyle as revealed by the genome sequence of *Colwellia psychrerythraea* 34H through genomic and proteomic analyses. *Proceedings of the National Academy of Sciences, USA*. 102: 10913-10918 (2005).

27. Moyer, C. L. & Morita, R. Y. Psychrophiles and psychrotrophs. *eLS*. (2007).

28. Blanc, G. *et al*. The genome of the polar eukaryotic microalga Coccomyxa subellipsoidea reveals traits of cold adaptation. *Genome Biology*. 13: R39 (2012).

29. Tang, Q., Tan, X. & Xu, X. Effects of a type-II RNA-binding protein on fatty acid composition in *Synechocystis* sp. PCC 6803. *Chinese Science Bulletin*. 55: 2416-2421 (2010).

30. Norsang, G. *et al*. Ground-based measurements and modeling of solar UV-B radiation in Lhasa, Tibet. *Atmospheric Environment*. 43: 1498-1502 (2009).

31. Carreto, J. I. & Carignan, M. O. Mycosporine-like amino acids: relevant secondary metabolites. Chemical and ecological aspects. *Marine drugs*. 9: 387-446 (2011).

32. Shick, J. M. & Dunlap, W. C. Mycosporine-like amino acids and related gadusols: biosynthesis, accumulation, and UV-protective functions in aquatic organisms. *Annual review of Physiology*. 64: 223-262 (2002).

33. Singh, S. P., Häder, D.-P. & Sinha, R. P. Cyanobacteria and ultraviolet radiation (UVR) stress: mitigation strategies. *Ageing Research Reviews*. 9: 79-90 (2010).

34. Singh, S. P., Klisch, M., Sinha, R. P. & Häder, D.-P. Genome mining of mycosporine-like amino acid (MAA) synthesizing and nonsynthesizing cyanobacteria: A bioinformatics study. *Genomics*. 95: 120-128 (2010).

35. Singh, S. P., Klisch, M., Sinha, R. P. & Häder, D. Ä. Effects of Abiotic

Stressors on Synthesis of the Mycosporine-like Amino Acid Shinorine in the Cyanobacterium *Anabaena variabilis* PCC 7937. *Photochemistry and Photobiology*. 84: 1500-1505 (2008).

36. Balskus, E. P. & Walsh, C. T. The genetic and molecular basis for sunscreen biosynthesis in cyanobacteria. *Science*. 329: 1653-1656 (2010).

37. Kogej, T., Gostinčar, C., Volkmann, M., Gorbushina, A. A. & Gunde-Cimerman, N. Mycosporines in extremophilic fungi — novel complementary osmolytes? *Environmental Chemistry*. 3: 105-110 (2006).

38. Suh, H. J., Lee, H. W. & Jung, J. Mycosporine Glycine Protects Biological Systems Against Photodynamic Damage by Quenching Singlet Oxygen with a High Efficiency-∂. *Photochemistry and Photobiology*. 78: 109-113 (2003).

39. Wright, D. J. *et al*. UV irradiation and desiccation modulate the three-dimensional extracellular matrix of *Nostoc commune* (Cyanobacteria). *Journal of Biological Chemistry*. 280: 40271-40281 (2005).

40. Pereira, P. *et al*. Paralytic shellfish toxins in the freshwater cyanobacterium *Aphanizomenon flos-aquae*, isolated from Montargil reservoir, Portugal. *Toxicon*. 38: 1689-1702 (2000).

41. Ichimura, T. Isolation and culture methods of algae. *Methods in phycological studies*. (Ed. by Nishizawa, K. & Chihara, M.), 294-305 (1979).

42. Nübel, U., Garcia-Pichel, F. & Muyzer, G. PCR primers to amplify 16S rRNA genes from cyanobacteria. *Applied and Environmental Microbiology*. 63: 3327-3332 (1997).

43. R, L. *et al*. De novo assembly of human genomes with massively parallel short read sequencing. *Genome Research*. 20: 265-272 (2010).

44. Borodovsky, M. & McIninch, J. GeneMark: parallel gene recognition for both DNA strands. *Computers & Chemistry*. 17: 123-133 (1993).

45. Delcher, A. L., Harmon, D., Kasif, S., White, O. & Salzberg, S. L. Improved microbial gene identification with GLIMMER. *Nucleic Acids Research*. 27:

4636-4641 (1999).

46. Consortium, U. Update on activities at the Universal Protein Resource (UniProt) in 2013. *Nucleic Acids Research*. 41: D43-D47 (2013).

47. Tatusov, R. L., Galperin, M. Y., Natale, D. A. & Koonin, E. V. The COG database: a tool for genome-scale analysis of protein functions and evolution. *Nucleic Acids Research*. 28: 33-36 (2000).

48. Mavromatis, K. *et al.* The DOE-JGI Standard operating procedure for the annotations of microbial genomes. *Standards in Genomic Sciences*. 1: 63 (2009).

49. Xu, Z. & Hao, B. CVTree update: a newly designed phylogenetic study platform using composition vectors and whole genomes. *Nucleic Acids Research*. 37: W174-W178 (2009).

50. De Bie, T., Cristianini, N., Demuth, J. P. & Hahn, M. W. CAFE: a computational tool for the study of gene family evolution. *Bioinformatics*. 22: 1269-1271 (2006).

51. Li, L., Stoeckert, C. J. & Roos, D. S. OrthoMCL: identification of ortholog groups for eukaryotic genomes. *Genome Research*. 13: 2178-2189 (2003).

52. Edgar, R. C. MUSCLE: multiple sequence alignment with high accuracy and high throughput. *Nucleic Acids Research*. 32: 1792-1797 (2004).

53. Suyama, M., Torrents, D. & Bork, P. PAL2NAL: robust conversion of protein sequence alignments into the corresponding codon alignments. *Nucleic Acids Research*. 34: W609-W612 (2006).

54. Stamatakis, A., Ludwig, T. & Meier, H. RAxML-III: a fast program for maximum likelihood-based inference of large phylogenetic trees. *Bioinformatics*. 21: 456-463 (2005).

55. Zhang, J., Nielsen, R. & Yang, Z. Evaluation of an improved branch-site likelihood method for detecting positive selection at the molecular level. *Molecular Biology and Evolution*. 22: 2472-2479 (2005).

56. Yang, Z., Wong, W. S. & Nielsen, R. Bayes empirical Bayes inference of amino

acid sites under positive selection. *Molecular Biology and Evolution*. 22: 1107-1118 (2005).

57. Langmead, B. & Salzberg, S. L. Fast gapped-read alignment with Bowtie 2. *Nature Methods*. 9: 357-359 (2012).

58. Anders, S., Pyl, P. T. & Huber, W. HTSeq-A Python framework to work with high-throughput sequencing data. *bioRxiv* (2014).

59. Xie, C. *et al.* KOBAS 2.0: a web server for annotation and identification of enriched pathways and diseases. *Nucleic Acids Research*. 39: W316-W322 (2011).

Testing the effect of the Himalayan mountains as a physical barrier to gene flow in *Hippophae tibetana* Schlect. (Elaeagnaceae) （摘要）

检测喜马拉雅山对西藏沙棘基因流的阻碍效应

　　西藏沙棘是海拔分布最高达5 250m左右的青藏高原特有的雌雄异株、风媒、小型灌木植物。而喜马拉雅山脉是青藏高原的一个巨大的地理屏障，可把喜马拉雅地区分为南侧的尼泊尔和北侧的西藏自治区两大块。然而目前没有发现有关喜马拉雅山脉对高山植物的基因流和植物物种分化影响的相关报道。本研究利用8个核基因微卫星标记和cpDNA *trn*T-*trn*F基因序列检测了喜马拉雅山脉对其两侧西藏沙棘基因流造成的地理阻隔作用。同时，我们还对珠峰北侧的西藏沙棘居群内的精细尺度下空间遗传结构进行了研究。来自于7个居群（4个来自于尼泊尔境内，3个来自于西藏境内）的总计241个体进行了取样并进行了微卫星分析，其中包括取自于100m×100m空间自相关大样方内121个体；同时为了检测西藏沙棘的种子流，还对来自于6个居群的100个体*trn*T-*trn*F基因序列进行了测序。核基因和叶绿体基因分析结果都表明两个地区的西藏沙棘之间存在显著遗传分化。两组数据都支持喜马拉雅山脉对其南北两侧西藏沙棘基因流有显著阻隔作用。在精细尺度下，空间自相关分析表明在距离小于45米范围内西藏沙棘居群内存在显著空间遗传结构,这可能主要是由于其营养繁殖、生境破碎化和有限的基因流等因素造成的。

　　Lhag, C., Zhang, W. J., Wang, H., Zeng, L. Y., Birks, H. J. B., **Zhong, Y.** Testing the effect of the Himalayan mountains as a physical barrier to gene flow in *Hippophae tibetana* Schlect. (Elaeagnaceae). 2017. *PLoS ONE*, 12(5): e0172948.

Genetic diversity and population structure of *Lamiophlomis rotata* (Lamiaceae), an endemic species of Qinghai-Tibet Plateau（摘要）
青藏高原特有植物独一味的遗传多样性和种群结构

独一味是一种青藏高原特有的多年生草本植物，也是一种重要的药用植物。本文用ISSR和RAPD分子标记技术检测了来自青藏高原的8个自然居群（4个来自西藏，2个来自云南，2个来自青海）共188个个体，结果表明其遗传变异相对比较高（P=94.85%, I=0.440-0.220, HT=0.289-0.028）。来自三个不同区域的遗传结构也呈现出显著的地理相关性，这种进化分歧可能与青藏高原隆起和第四纪气候振荡相关。这些发现暗示了物种保护应该尽可能原地保护尽量多的居群。考虑到遗传变异和地缘分布，青海和云南的居群应该被优先保护。最优策略则是发展独一味的栽培和组织培养以确保可持续使用。

Liu, J. M., Wang, L., Geng, Y. P., Wang, Q. B., Luo, L. J., **Zhong, Y.** Genetic diversity and population structure of *Lamiophlomis rotata* (Lamiaceae), an endemic species of Qinghai-Tibet Plateau. 2006. *Genetica*, 128(1-3): 385-394.

Fine- and landscape-scale spatial genetic structure of cushion rockjasmine, *Androsace tapete* (Primulaceae), across southern Qinghai-Tibetan Plateau（摘要）

青藏高原垫状点地梅在精细尺度和景观尺度的空间遗传结构。

垫状点地梅是世界上海拔分布最高的被子植物之一。在青藏高原，垫状点地梅的分布可以高达5 300米，是高山荒漠和高山流石滩生境的先锋物种。在本研究中，我们使用ISSR标记来研究垫状点地梅在精细尺度和景观尺度的空间遗传结构。我们要验证的假设是从西到东，横贯青藏高原的雅鲁藏布江低海拔河谷对于垫状点地梅的空间结构有显著影响。从五个种群（两个位于雅鲁藏布江以北，三个位于雅鲁藏布江以南）总共采集了235个个体，其中158个个体位于一个30米×90米的样地，有精确的空间定位。在精细尺度上，空间自相关分析表明，在10米以内的短距离内，存在出显著的空间遗传自相关。这可能是因为种子或花粉介导的基因扩散的局限性。在景观尺度，AMOVA表明大部分遗传变异（85%）位于种群内个体之间。雅鲁藏布江在塑造空间遗传格局的过程中仅发挥了微弱的作用。而且PCA和STRUCTURE分析表明，在雅鲁藏布江两侧的种群之间具有显著的遗传关联。历史性的基因交换和缓慢的遗传漂变可能是这些地形上隔离的种群间缺乏显著遗传分化的原因。

Geng, Y. P., Tang, S. Q., Tashi, T., Song, Z. P., Zhang, G. R., Zeng, L. Y., Zhao, J. Y., Wang, L., Shi, J., Chen, J. K., **Zhong, Y.** Fine- and landscape-scale spatial genetic structure of cushion rockjasmine, *Androsace tapete* (Primulaceae), across southern Qinghai-Tibetan Plateau. 2009. *Genetica*, 135(3): 419-427.

Genetic diversity in *Primula obconica* from Central and South-west China as revealed by ISSR markers（摘要）

以ISSR标记揭示中国中部和西南部植物报春花的遗传多样性

　　本研究利用ISSR分子标记对我国中部湖北和西南的四川省4个野生种群和1个栽培种群的60个样本进行遗传多样性分析，获得了249个多态性条带和60个ISSR基因型，发现4个野生种群的遗传多样性远高于栽培种，而湖北三个自然种群的遗传多样性也高于四川的自然种群。UPGMA聚类分析发现种群间没有明显的遗传分化。该研究从种群遗传多样性角度为报春花的栽培和育种提供理论依据。

Nan, P., Shi, S. H., Peng, S. L., Tian, C. J., Zhong, Y. Genetic diversity in *Primula obconica* (Primulaceae) from Central and South-west China as revealed by ISSR markers. 2003. *Annals of Botany*, 91(3): 329-333.

附：钟扬教授学术论文总目

（按时间顺序排列）

1. 钟扬. 电子计算机在植物学中的应用.《武汉植物学研究》1986年第4卷第3期，311-320.

2. 钟扬，陈卓良. 湖北省油橄榄适生气候的主成分分析.《湖北农业科学》1986年第7期，21-23.

3. 钟扬，何芳良. 植物群落演替过程的预测模型.《预测》1986年第6期，13-15, 18.

4. 钟扬，张晓艳. 荷花品种的模糊聚类分析.《华中农业大学学报》1986年第5卷第4期，403-409.

5. **钟扬，张晓艳. 荷花品种的数量分类研究.《武汉植物学研究》1987年第5卷第1期，49-58.**（生物数据模型与信息系统代表性论文）

6. Zhong, Y., Chen, Z. L., Ke, S. Q., Moore, A. A fuzzy mathematical model for cultivable regions for olive. 1988. *Journal of Wuhan Botanical Research*, 6(3): 275-284.

7. 何芳良，钟扬. 生态系统演替过程的数学模型.《武汉植物学研究》1988年第6卷第2期，167-172.

8. 黄德世，钟扬，陈家宽. 图论在中国慈姑属数量分类研究中的应用.《武汉植物学研究》1988年第6卷第4期，405-406.

9. 陶光复，钟扬. 湖北樟属数量化学分类研究.《植物分类学报》1988年第26卷第6期，406-417.

10. 张晓艳，黄国振，钟扬. 荷花品种综合评选的数字模型.《北京林业大学学报》1988年第10卷第2期，12-16.

11. 钟扬. 国内植物数量生态学研究概况.《武汉植物学研究》1988年第6卷第1期，87-94.

12. 钟扬，柯善强，张晓艳. 植物生殖的定量细胞学研究进展.《大自然探索》1988年第7卷第3期（总第25期），67-71.

13. Clifford, H. T., 钟扬. 浸水条件下禾本科种子发芽力的分类学意义.《种子》1989年第5期（总第43期），71-73.（译文）

14. 洪树荣，钟扬，傅俊，柯善强. 正交拉丁方实验在猕猴桃组织培养中的应用.《武汉植物学研究》1990年第8卷第2期，171-177.

15. 李伟，陈家宽，钟扬，黄德世. 世界慈姑属植物的数量分类研究.《武汉大学学报（自然科学版）》1990年第3期，102-108.

16. 钟扬，黄德世. 计算机辅助分支分析：方法和程序.《武汉植物学研究》1990年第8卷第2期，199-200.

17. 钟扬，张晓艳，黄德世. 睡莲目的数量分支分类学研究.《生物数学学报》1990年第5卷第2期，156-161.

18. 何景彪，孙祥钟，钟扬，黄德世. 海菜花属的分支学研究.《武汉植物学研究》1991年第9卷第2期，121-129.

19. 何景彪，孙祥钟，钟扬，黄德世. 海菜花 Ottelia acuminata (Gagnep.) Dandy 的种下分类研究.《武汉大学学报（自然科学版）》1991年第3期，114-120.

20. 李伟，钟扬. 我国内陆水生植被研究概况.《武汉植物学研究》1991年第9卷第3期，281-288.

21. 钟扬，陈家宽. 矮慈姑居群的数量分类研究.《广西植物》1991年第11卷第4期，304-307.

22. 何景彪，孙祥钟，王徽勤，钟扬，黄德世. 中国海菜花属植物的性状分析.《武汉植物学研究》1992年第10卷第2期，101-108.

23. 李伟，周进，王徽勤，钟扬. 斧头湖挺水植被的群落学研究——Ⅰ. 菰群落（Com. Zizanina latifolia）的结构.《武汉植物学研究》1992年第10卷第2期，109-116.

24. 李伟，周进，王徽勤，钟扬. 斧头湖挺水植被的群落学研究——Ⅱ. 莲群落（Com.

Nelumbo nucifera）的结构.《武汉植物学研究》1992年第10卷第3期，273-279.

25. 吴立廉，刘世叶，张敬芬，钟扬，洪树荣，严光琼. 天然保鲜剂对猕猴桃贮藏效果的影响.《武汉植物学研究》1992年第10卷第4期，387-389.

26. 钟扬，陈家宽，邹洪才，李伟. 中国慈姑属系统发育的研究.《武汉植物学研究》1992年第10卷第3期，243-248.

27. 周进，钟扬. 湖南、江西普通野生稻居群变异的数量分类研究.《武汉植物学研究》1992年第10卷第3期，235-242.

28. 李伟，周进，王徽勤，钟扬. 斧头湖挺水植被的群落学研究——Ⅲ. 菰群落（Com. *Zizanina latifolia*）的生物量动态及生产量测定.《武汉植物学研究》1993年第11卷第3期，233-238.

29. 钟扬，李伟. 斧头湖水生植物考察.《植物杂志》1994年第1期，22-23.

30. 李伟，钟扬. 湖北斧头湖湖滨湿地植物的联结与相关分析.《武汉植物学研究》1995年第13卷第1期，65-69.

31. Jung, S. W., Perkins, S., Zhong, Y., Pramanik, S., Beaman, J. H. A new data model for biological classification. 1995. *Computer Applications in the Biosciences: CABIOS*, 11(3): 237-246.

32. 李伟，钟扬. 湖北斧头湖湖滨湿地水田碎米荠群落的定量分析.《水生生物学报》1995年第19卷第3期，250-256.

33. 钟扬. 植物分类信息系统概述.《植物学通报》1995年第12期（增刊），1-6.

34. **Zhong, Y., Jung, S. W., Pramanik, S., Beaman, J. H. Data model and comparison and query methods for interacting classifications in a taxonomic database. 1996. *Taxon*, 45(2): 223-241.**（生物数据模型与信息系统代表性论文）

35. Zhong, Y., Meacham, C. A., Pramanik, S. A general method for tree-comparison based on subtree similarity and its use in a taxonomic database. 1997. *Biosystems*, 42(1): 1-8.

36. 洪亚平，钟扬，陈之端. 交互分类信息系统和电子植物志的设计与实现——Ⅱ. 分类排列和"系统学处理"软件系统.《武汉植物学研究》1998年第16卷第4期，379-382.

37. 钟扬，余清清，彭勇，肖培根. 传统药物及天然产物信息系统间的交互运行技术.《中国中医药信息杂志》1998年第5卷第4期，59-60.

38. Zhong, Y., Luo, Y. N., Pramanik, S., Beaman, J. H. **HICLAS: a taxonomic database system for displaying and comparing biological classification and phylogenetic trees. 1999. *Bioinformatics*, 15(2): 149-156.**（生物数据模型与信息系统代表性论文）

39. 唐先华,黄德世,张晓艳,钟扬,杨继.相对速率检验方法与实用程序.《武汉植物学研究》1999年第17卷第4期, 353-356.

40. 余清清, 钟扬.一个中药信息系统原型的初步设计与实现.《武汉植物学研究》1999年第17卷第1期, 91-93.

41. 钟扬,唐先华,施苏华,黄椰林,谈凤笑.外类群对构建基因树的影响.《中山大学学报(自然科学版)》1999年第38卷第1期, 124-127.

42. Cui, X. H., Zhong, Y., Chen, J. K. Influence of a catastrophic flood on densities and biomasses of three plant species in Poyang Lake, China. 2000. *Journal of Freshwater Ecology*, 15(4): 537-541.

43. He, Z. C., Zhang, X. Y., Zhong, Y., Ye, L. Phylogenetic relationships of *Actinidia* and related genera based on micromorphological characters of foliar trichomes. 2000. *Genetic Resources and Crop Evolution*, 47(6): 627-639.

44. **Sang, T., Zhong, Y. Testing hybridization hypotheses based on incongruent gene trees. 2000. *Systematic Biology*, 49(3): 422-434.**（分子进化分析方法及应用代表性论文）

45. Shi, S. H., Jin, H., Zhong, Y., He, X., Huang, Y., Tan, F., Boufford, D. E. Phylogenetic relationships of the Magnoliaceae inferred from cpDNA *mat*K sequences. 2000. *Theoretical and Applied Genetics*, 101(5-6): 925-930.

46. Wang, J. B., Wang, C., Shi, S. H., Zhong, Y. ITS regions in diploids of *Aegilops* (Poaceae) and their phylogenetic implications. 2000. *Hereditas*, 132(3): 209-213.

47. Wang, J. B., Wang, C., Shi, S. H., Zhong, Y. Evolution of parental ITS regions of nuclear rDNA in allopolyploid *Aegilops* (Poaceae) species. 2000. *Hereditas*, 133(1): 1-7.

48. Zhong, Y., Shi, S. H., Tang, X. H., Huang, Y. L., Tan, F. X., Zhang, X. Y. Testing relative evolutionary rates and estimating divergence times among six genera of Rhizophoraceae using cpDNA and nrDNA sequences. 2000. *Chinese Science Bulletin*, 45(11): 1011-1015.

49. Zhong, Y., Zhang, L., Su, D. M. Collaborations tailored for bioinformatics projects. 2000.

Science, 290(5499): 2074. (Letter)

50. 崔心红，钟扬，李伟，陈家宽. 特大洪水对鄱阳湖水生植物三个优势种的影响.《水生生物学报》2000年第24卷第4期，322-325.

51. 何子灿，钟扬，刘洪涛，唐先华，叶力，黄德世，徐立铭. 中国猕猴桃属植物叶表皮毛微形态特征及数量分类分析.《植物分类学报》2000年第38卷第2期，121-136.

52. 王超，施苏华，王建波，钟扬. 山羊草属二倍体物种核rDNA ITS区序列及其系统发育关系分析.《植物学报》2000年第42卷第5期，507-511.

53. 王超，王建波，施苏华，钟扬. 山羊草属异源多倍体物种核rDNA ITS区的进化.《植物分类学报》2000年第38卷第3期，211-217.

54. 吴世安，吕海亮，杨继，饶广远，尤瑞麟，葛颂，钟扬. 叶绿体DNA片段的RFLP分析在黄精族系统学研究中的应用.《植物分类学报》2000年第38卷第2期，97-110.

55. 张弛，胡鸿钧，李中奎，李夜光，钟扬. 衣藻属的系统发育分析——基于形态形状和nrDNA ITS序列.《武汉植物学研究》2000年第18卷第3期，189-194.

56. 钟扬，张亮，任文伟，陈家宽. 生物多样性信息学：一个正在兴起的新方向及其关键技术.《生物多样性》2000年第8卷第4期，397-404.

57. 钟扬，施苏华，唐先华，黄椰林，谈凤笑，张晓艳. 红树科6属cpDNA和nrDNA序列相对速率检验及分歧时间估计.《科学通报》2000年第45卷第1期，40-44.

58. Shi, S. H., Huang, Y., Zhong, Y., Du, Y., Zhang, Q., Chang, H., Boufford, D. E. Phylogeny of the Altingiaceae based on cpDNA *mat*K, PY-IGS and nrDNA ITS sequences. 2001. *Plant Systematics and Evolution*, 230(1-2): 13-24.

59. Du, Y. Q., Shi, S. H., Zhong, Y., Gong, X., Liu, X. Phylogenetic position of *Schnabelia* (Lamiaceae): a preliminary analysis of chloroplast *mat*K sequences. 2001. *Journal of Genetics and Molecular Biology*, 12(3): 138-144.

60. 黄椰林，施苏华，钟扬，谈凤笑. 一种从特殊植物材料中制备PCR模板的新方法.《科学通报》2001年第46卷第24期，2055-2057.

61. 唐绍清，钟扬，施苏华，张宏达. *Camellia nitidissima*与*C. petelotii*之间的关系研究——来自nrDNA ITS的证据.《武汉植物学研究》2001年第19卷第6期，449-452.

62. 吴纪华，梁彦龄，钟扬，傅萃长，陈家宽. 线虫系统发育研究进展.《科学通报》2001

年第46卷第13期，1068-1073.

63. Huang, Y. L., Shi, S. H., Zhong, Y., Tan, F. X. A new method for preparation of template DNA for PCR from special plant materials. 2002. *Chinese Science Bulletin*, 47(9): 725-727.

64. Jian, S. G., Shi, S. H., Zhong, Y., Tang, T., Zhang, Z. H. Genetic diversity among South China *Heritiera littoralis* detected by inter-simple sequence repeats (ISSR) analysis. 2002. *Journal of Genetics and Molecular Biology*, 13(4): 272-276.

65. Nan, P., Peng, S. L., Ren, H., Shi, S. H., Tian, C. J., Zhong, Y. Genetic diversity of *Primula ovalifolia* from Central and Southwest China based on ISSR markers. 2002. *Journal of Genetics and Molecular Biology*, 13(2): 119-123.

66. Nan, P., Peng, S. L., Zhang, Y. H., Zhong, Y. Composition of volatile oil of *Primula obconica* in Central China. 2002. *Natural Product Letters*, 16(4): 249-253.

67. Shen, W., Zhao, G. J., Zhu, B., Xu, A. L., Zhong, Y. Phylogenetic analysis of homologous amino acid sequences of five conserved genes in bacteria. 2002. *Journal of Genetics and Molecular Biology*, 13(2): 108-118.

68. Shi, S. H., Zhong, Y., Huang, Y. L., Du, Y. Q., Qiu, X. Z., Chang, H. T. Phylogenetic relationships of the Rhizophoraceae in China based on sequences of the chloroplast gene *mat*K and the internal transcribed spacer regions of nuclear ribosomal DNA and combined data set. 2002. *Biochemical Systematics and Ecology*, 30(4): 309-319.

69. Tan, F. X., Shi, S. H., Zhong, Y., Gong, X., Wang, Y. G. Phylogenetic relationships of Combretoideae (Combretaceae) inferred from plastid, nuclear gene and spacer sequences. 2002. *Journal of Plant Research*, 115(1122): 475-481.

70. Tang, S. Q., Zhong, Y. A phylogenetic analysis of nrDNA ITS sequences from Ser. *Chrysantha* (Sect. *Chrysantha*, *Camellia*, Theaceae). 2002. *Journal of Genetics and Molecular Biology*, 13(2): 105-107.

71. Tian, C. J., He, X. Y., Zhong, Y., Chen, J. K. Effects of VA mycorrhizae and *Frankia* dual inoculation on growth and nitrogen fixation of *Hippophae Tibetana*. 2002. *Forest Ecology and Management*, 176(1-3): 307-312.

72. Wu, J. H., Liang, Y. L., Zhong, Y., Fu, C. Z., Chen, J. K. Advances in phylogenetic studies of

Nematoda. 2002. *Chinese Science Bulletin*, 47(1): 10-15.

73. Yang, J., Huang, J. X., Gu, H. Y., Zhong, Y., Yang, Z. H. Duplication and adaptive evolution of the chalcone synthase genes of *Dendranthema* (Asteraceae). 2002. *Molecular Biology and Evolution*, 19(10): 1752-1759.

74. Zhang, X. Y., Tang, X. H., Shi, S. H., Zhong, Y., Chen, J. K. Comparison of phylogenetic relationships of the water lily genera *Nymphaea*, *Nuphar* and *Euryale* inferred from sequences of nrDNA ITS regions, nuclear 18S rRNA and cpDNA gene *rbc*L. 2002. *Journal of Genetics and Molecular Biology*, 13(3): 169-176.

75. Zhao, B., Zhong, Y., Zhang, X. Y., Ren, W. W., Chen, J. K. 2002. Building a search engine for interoperability among taxonomic, phylogenetic, and molecular databases through the web. *Journal of Genetics and Molecular Biology*, 13(3): 209-214.

76. **Zhong, Y., Zhao, Q., Shi, S. H., Huang, Y. L., Hasegawa, M. Detecting evolutionary rate heterogeneity among mangroves and their close terrestrial relatives. 2002. *Ecology Letters*, 5(3): 427-432.**（分子进化分析方法及应用代表性论文）

77. 李婷，张文驹，卢宝荣，钟扬，陈家宽．小麦A，B，D基因组可能供体种的高分子量麦谷蛋白亚基基因进化式样的分析．《复旦学报（自然科学版）》2002年第41卷第5期，596-598．

78. 彭少麟，南蓬，钟扬．高等植物中的萜类化合物及其在生态系统中的作用．《生态学杂志》2002年第21卷第3期，33-38．

79. 王莉，南蓬，张晓艳，钟扬．生物复杂性研究动态．《生物多样性》2002年第10卷第2期，238-242．

80. 袁长春，施苏华，钟扬，龚洵．用核糖体DNA的ITS序列探讨滇桐属的系统学位置．《中山大学学报（自然科学版）》2002年第41卷第6期，73-77．

81. 张美云，钱吉，钟扬，郑师章．野生大豆若干耐盐生理指标的研究．《复旦学报（自然科学版）》2002年第41卷第6期，669-673．

82. Lei, Y. D., Nan, P., Tashi, T., Bai, Z. K., Tian, C. J., Zhong, Y. Chemical composition of the essential oils of two *Rhodiola* species from Tibet. 2003. *Zeitschrift fur Naturforschung C*, 58(3-4): 161-164.

83. Nan, P., Peng, S. L., Shi, S. H., Ren, H., Yang, J., Zhong, Y. Interpopulation congruence in Chinese *Primula ovalifolia* revealed by chemical and molecular markers using essential oils and ISSRs. 2003. *Zeitschrift fur Naturforschung C*, 58(1-2): 57-61.

84. **Nan, P., Shi, S. H., Peng, S. L., Tian, C. J., Zhong, Y. Genetic diversity in *Primula obconica* (Primulaceae) from Central and South-west China as revealed by ISSR markers. 2003. *Annals of Botany*, 91(3): 329-333.**（生物多样性与植物基因组分析代表性论文）

85. Pang, J. F., Wang, Y. Z., Zhong, Y., Hoelzel, A. R., Papenfuss, T. J., Zeng, X. M., Ananjeva, N. B., Zhang, Y. P. A phylogeny of Chinese species in the genus *Phrynocephalus* (Agamidae) inferred from mitochondrial DNA sequences. 2003. *Molecular Phylogenetics and Evolution*, 27(3): 398-409.

86. Peng, Y. L., Chen, Z. D., Gong, X., Zhong, Y., Shi, S. H. Phylogenetic position of *Dipentodon sinicus*: evidence from DNA sequences of chloroplast *rbc*L, nuclear ribosomal 18S, and mitochondria *mat*R genes. 2003. *Botanical Bulletin of Academia Sinica*, 44(3): 217-222.

87. Qin, L., Xiong, B., Luo, C., Guo, Z. M., Hao, P., Su, J., Nan, P., Feng, Y., Shi, Y. X., Yu, X. J., Luo, X. M., Chen, K. X., Shen, X., Shen, J. H., Zou, J. P., Zhao, G. P., Shi, T. L., He, W. Z., Zhong, Y., Jiang, H. L., Li, Y. X. Identification of probable genomic packaging signal sequence from SARS-CoV genome by bioinformatics analysis. 2003. *Acta Pharmacologica Sinica*, 24(6): 489-496.

88. Ren, W. W., Zhong, Y., Meligrana, J., Anderson, B., Watt, W. E., Chen, J. K., Leung, H. L. Urbanization, land use, and water quality in Shanghai: 1947-1996. 2003. *Enviroment International*, 29(5): 649-659.

89. Shi, S. H., Du, Y. Q., Boufford, D. E., Gong, X., Huang, Y. L., He, H. H., Zhong, Y. Phylogenetic position of *Schnabelia*, a genus endemic to China: evidence from sequences of cpDNA *mat*K gene and nrDNA ITS regions. 2003. *Chinese Science Bulletin*, 48(15): 1576-1580.

90. Tang, T., Zhong, Y., Jian, S. G., Shi, S. H. Genetic diversity of *Hibiscus tiliaceus* (Malvaceae)

in China assessed using AFLP markers. 2003. *Annals of Botany*, 92(3): 409-414.

91. Tian, C. J., He, X. Y., Zhong, Y., Chen, J. K. Effect of inoculation with ecto- and arbuscular mycorrhizae and *Rhizobium* on the growth and nitrogen fixation by black locust, *Robinia pseudoacacia*. 2003. *New Forests*, 25(2): 125-131.

92. Yu, X. J., Luo, C., Lin, J. C., Hao, P., He, Y. Y., Guo, Z. M., Qin, L., Su, J., Liu, B. S., Huang, Y., Nan, P., Li, C. S., Xiong, B., Luo, X. M., Zhao, G. P., Pei, G., Chen, K. X., Shen, X., Shen, J. H., Zou, J. P., He, W. Z., Shi, T. L., Zhong, Y., Jiang, H. L., Li, Y. X. Putative hAPN receptor binding sites in SARS-CoV spike protein. 2003. *Acta Pharmacologica Sinica*, 24(6): 481-488.

93. Yuan, C. C., Nan, P., Shi, S. H., Zhong, Y. Chemical composition of the essential oils of two Chinese endemic *Meconopsis* species. 2003. *Zeitschrift fur Naturforschung C*, 58(5-6): 313-315.

94. Zhang, X. Y., Zhong, Y., Chen, J. K. Fanwort in eastern China: an invasive aquatic plant and potential ecological consequences. 2003. *AMBIO*, 32(2): 158-159.

95. Zhong, Y., Zhang, X. Y., Ma, J., Zhang, L. Rapid development of bioinformatics education in China. 2003. *Journal of Biological Education*, 37(2): 75-78.

96. 唐先华，张晓艳，施苏华，钟扬，赖旭龙．睡莲类植物ITS nrDNA序列的分子系统发育分析．《地球科学——中国地质大学学报》2003年第28卷第1期，97-101．

97. 施苏华，杜雅青，David Bufford，龚洵，黄椰林，何航航，钟扬．用cpDNA *mat*K基因和nrDNA ITS区序列确定我国特有植物四棱草属的系统位置．《科学通报》2003年第48卷第11期，1176-1180．

98. 田春杰，陈家宽，钟扬．微生物系统发育多样性及其保护生物学意义．《应用生态学报》2003年第14卷第4期，609-612．

99. 张晓艳，钟扬，陈家宽．中国东部的水盾草：一种入侵的水生植物及其潜在的生态后果．《AMBIO-人类环境杂志》2003年第32卷第2期，158-159．

100. Cao, S. L., Qin, L., He, W. Z., Zhong, Y., Zhu, Y. Y., Li, Y. X. Semantic search among heterogeneous biological databases based on gene ontology. 2004. *Acta Biochimica et Biophysica Sinica*, 36(5): 365-370.

101. Chen, Z. Y., Li, B., Zhong, Y., Chen, J. K. Local competitive effects of introduced *Spartina alterniflora* on *Scirpus mariqueter* at Dongtan of Chongming Island, the Yangtze River estuary and their potential ecological consequences. 2004. *Hydrobiologia*, 528(1-3): 99-106.

102. **The Chinese Molecular Epidemiology Consortium. Molecular evolution of the SARS coronavirus during the course of the SARS epidemic in China. 2004. *Science*, 303(5664): 1666-1669.**（分子进化分析方法及应用代表性论文）

103. He, S. P., Liu, H. Z., Chen, Y. Y., Kuwahara, M., Nakajima, T., Zhong, Y. Molecular phylogenetic relationships of Eastern Asian Cyprinidae (Pisces: Cypriniformes) inferred from cytochrome *b* sequences. 2004. *Science in China Series C*, 47(2): 130-138.

104. Jian, S. G., Tang, T., Zhong, Y., Shi, S. H. Variation in inter-simple sequence repeat (ISSR) in mangrove and non-mangrove populations of *Heritiera littoralis* (Sterculiaceae) from China and Australia. 2004. *Aquatic Botany*, 79(1): 75-86.

105. Lei, Y. D., Nan, P., Tashi, T., Wang, L., Liu, S. P., Zhong, Y. Interpopulation variability of rhizome essential oils in *Rhodiola crenulata* from Tibet and Yunnan, China. 2004. *Biochemical Systematics and Ecology*, 32(6): 611-614.

106. Lei, Y. D., Tang, X. H., Tashi, T., Shi, S. H., Zhong, Y. A preliminary phylogenetic analysis of some *Rhodiola* species based on nrDNA ITS sequences. 2004. *Journal of Genetics and Molecular Biology*, 15(3-4): 190-197.

107. Nan, P., Hu, Y. M., Zhao, J. Y., Feng, Y., Zhong, Y. Chemical composition of the essential oils of two *Alpinia* species from Hainan Island, China. 2004. *Zeitschrift fur Naturforschung C*, 59(3-4): 157-160.

108. Qin, Z. Q., Zhong, Y., Zhang, J., He, Y. Y., Wu, Y., Jiang, J., Chen, J. M., Luo, X. M., Qu, D. Bioinformatics analysis of two-component regulatory systems in *Staphylococcus epidermidis*. 2004. *Chinese Science Bulletin*, 49(12): 1267-1271.

109. Ren, Z. M., Ma, E. B., Guo, Y. P., Zhong, Y. A molecular phylogeny of *Oxya* (Orthoptera: Acridoidea) in China inferred from partial cytochrome *b* gene sequences. 2004. *Molecular Phylogenetics and Evolution*, 33(2): 516-521.

110. Tang, T., Huang, J. Z., Zhong, Y., Shi, S. H. High-throughput S-SAP by fluorescent multiplex

PCR and capillary electrophoresis in plants. 2004. *Journal of Biotechnology*, 114(1-2): 59-68.

111. Tian, C. J., Lei, Y. D., Shi, S. H., Nan, P., Chen, J. K., Zhong, Y. Genetic diversity of sea buckthorn (*Hippophae rhamnoides*) populations in northeastern and northwestern China as revealed by ISSR markers. 2004. *New Forests*, 27(3): 229-237.

112. Tian, C. J., Nan, P., Chen, J. K., Zhong, Y. Volatile composition of Chinese *Hippophae rhamnoides* and its chemotaxonomic implications. 2004. *Biochemical Systematics and Ecology*, 32(4): 431-441.

113. Tian, C. J., Nan, P., Shi, S. H., Chen, J. K., Zhong, Y. Molecular genetic variation in Chinese populations of three subspecies of *Hippophae rhamnoides*. 2004. *Biochemical Genetics*, 42(7-8); 259-267.

114. Wang, L., Jian, S. G., Nan, P., Zhong, Y. Chemical composition of the essential oil of *Elephantopus scaber* from southern China. 2004. *Zeitschrift fur Naturforschung C*, 59(5-6): 327-329.

115. Zhang, L., Wu, G. W., Xu, Y. F., Li, W., Zhong, Y. Multilingual collection retrieving via ontology alignment. 2004. *Lecture Notes in Computer Science*, 3334: 510-514.

116. Zhang, Y., Zheng, N., Hao, P., Zhong, Y. Reconstruction of the most recent common ancestor sequences of SARS-Cov S gene and detection of adaptive evolution in the spike protein. 2004. *Chinese Science Bulletin*, 49(12): 1311-1313.

117. Zhao, J. Y., Nan, P., Zhong, Y. Chemical composition of the essential oils of *Clausena lansium* from Hainan Island, China. 2004. *Zeitschrift fur Naturforschung C*, 59(3-4): 153-156.

118. 陈永燕，钟扬，田波，杨继，李德铢. 被子植物SQUA类基因适应性进化的统计检测.《自然科学进展》2004年第14卷第9期，1063-1066.

119. 何舜平，刘焕章，陈宜瑜，钟扬. 基于细胞色素b基因序列的鲤科鱼类系统发育研究（鱼纲：鲤形目）.《中国科学C辑：生命科学》2004年第34卷第1期，96-104.

120. 李涛，赖旭龙，钟扬. 利用DNA序列构建系统树的方法.《遗传》2004年第26卷第2期，205-210.

121. 秦智强, 钟扬, 张健, 何有裕, 吴旸, 江娟, 陈洁敏, 罗小民, 瞿涤. 表皮葡萄球菌双组分调控系统的生物信息学分析.《科学通报》2004年第49卷第10期, 948-952.

122. 唐绍清, 施苏华, 钟扬, 王燕. 基于ITS序列探讨山茶属金花茶组的系统发育.《广西植物》2004年第24卷第6期, 488-492.

123. 张原, 郑楠, 郝沛, 钟扬. SARS冠状病毒S基因的最近共同祖先序列重建及Spike蛋白的适应性进化检测.《科学通报》2004年第49卷第11期, 1121-1122.

124. Bai, Z. K., Nan, P., Zhong, Y. Chemical composition of the essential oil of *Rhodiola quadrifida* from Xinjiang, China. 2005. *Chemistry of Natural Compounds*, 41(4): 418-419.

125. Chen, X., Zheng, J., Fu, Z., Nan, P., Zhong, Y., Lonardi, S., Jiang, T. Assignment of orthologous genes via genome rearrangement. 2005. *IEEE-ACM Transactions on Computational Biology and Bioinformatics*, 2(4): 302-315.

126. Chen, X., Zheng, J., Fu, Z., Nan, P., Zhong, Y., Lonardi, S., Jiang, T. Computing the assignment of orthologous genes via genome rearrangement. 2005. *Asia-pacific Bioinformatics Conference*, 17(4): 363-378.

127. Chen, Y. Y., Zhong, Y., Tian, B., Yang, J., Li, D. Z. Statistical analysis on adaptive evolution of SQUA genes in angiosperms. 2005. *Progress in Natural Science*, 15(1): 93-96.

128. Fu, C. Z., Luo, J., Wu, J. H., Lopez, J. A., Zhong, Y., Lei, G. C., Chen, J. K. Phylogenetic relationships of salangid fishes (Osmeridae, Salanginae) with comments on phylogenetic placement of the salangids based on mitochondrial DNA sequences. 2005. *Molecular Phylogenetics and Evolution*, 35(1): 76-84.

129. **Hao, P., He, W. Z., Huang, Y., Ma, L. X., Xu, Y., Xi, H., Wang, C., Liu, B. S., Wang, J. M., Li, Y. X., Zhong, Y. MPSS: an integrated database system for surveying a set of proteins. 2005. *Bioinformatics*, 21(9): 2142-2143.**（生物数据模型与信息系统代表性论文）

130. Li, Z. F., Liu, Q., Song, M. G., Zheng, Y., Nan, P., Cao, Y., Chen, G. Q., Li, Y. X., Zhong, Y. Detecting correlation between sequence and expression divergences in a comparative analysis of human serpin genes. 2005. *Biosystems*, 82(3): 226-234.

131. Li, Z. F., Wang, L., Zhong, Y. Detecting horizontal gene transfer with T-REX and RHOM

programs. 2005. *Briefings in Bioinformatics*, 6(4): 394-401.

132. Wang, L., Jian, S. G., Nan, P., Liu, J. M., Zhong, Y. Chemotypical variability of leaf oils in *Elephantopus scaber* from 12 locations in China. 2005. *Chemistry of Natural Compounds*, 41(5): 491- 493.

133. Wang, Q. B., Wang, L., Zhou, R. C., Zhao, X. M., Shi, S. H., Yang, Y., Zhong, Y. Phylogenetic position of *Ephedra rhytidosperma*, a species endemic to China: evidence from chloroplast and ribosomal DNA sequences. 2005. *Chinese Science Bulletin*, 50(24): 2901-2904.

134. Xiong, B. H., Hou, H. B., Zhong, Y. The effect of water depth on seedling emergence and early growth of *Vallisneria natans* in a eutrophic lake with reduced transparency. 2005. *Journal of Freshwater Ecology*, 20(1): 123-127.

135. Zhang, M., Wang, J. D., Li, Z. F., Xie, J., Yang, Y. P., Zhong, Y., Wang, H. H. Expression and characterization of the carboxyl esterase Rv3487c from *Mycobacterium tuberculosis*. 2005. *Protein Expression and Purification*, 42(1): 59-66.

136. Zhang, Y., Zheng, N., Hao, P., Cao, Y., Zhong, Y. A molecular docking model of SARS-CoV S1 protein in complex with its receptor, human ACE2. 2005. *Computational Biology and Chemistry*, 29(3): 254-257.

137. Zhao, B., Li, B., Zhong, Y., Nakagoshi, N., Chen, J. K. Estimation of ecological service values of wetlands in Shanghai, China. 2005. *Chinese Geographical Science*, 15(2): 151-156.

138. Cheng, G., Chen, W. Z., Li, Z. F., Yan, W. Y., Zhao, X., Xie, J., Liu, M. Q., Zhang, H., Zhong, Y., Zheng, Z. X. Characterization of the porcine alpha interferon multigene family. 2006. *Gene*, 382: 28-38.

139. Fu, Z., Chen, X., Vacic, V., Nan, P., Zhong, Y., Jiang, T. A parsimony approach to genome-wide ortholog assignment. 2006. *Lecture Notes in Computer Science*, 3909: 578-594.

140. Guo, S. C., Savolainen, P., Su, J. P., Zhang, Q., Qi, D. L., Zhou, J., Zhong, Y., Zhao, X. Q., Liu, J. Q. Origin of mitochondrial DNA diversity of domestic yaks. 2006. *BMC Evolutionary Biology*, 6: 73.

141. Jian, S. G., Zhong, Y., Liu, N., Gao, Z. Z., Wei, Q., Xie, Z. H., Ren, H. Genetic variation in the endangered endemic species *Cycas fairylakea* (Cycadaceae) in China and implications for conservation. 2006. *Biodiversity and Conservation*, 15(5): 1681-1694.

142. Lei, Y. D., Gao, H., Tashi, T., Shi, S. H., Zhong, Y. Determination of genetic variation in *Rhodiola crenulata* from the Hengduan Mountains Region, China using inter-simple sequence repeats. 2006. *Genetics and Molecular Biology*, 29(2): 339-344.

143. Li, R., Cao, S. L., Li, Y. Y., Tan, H., Zhu, Y. Y., Zhong, Y., Li, Y. X. A measure of semantic similarity between gene ontology terms based on semantic pathway covering. 2006. *Progress in Natural Science*, 16(7): 721-726.

144. Liu, J. M., Nan, P., Qiong, T., Tashi, T., Bai, Z. K., Wang, L., Liu, Z. J., Zhong, Y. Volatile constituents of the leaves and flowers of *Salvia przewalskii* Maxim. from Tibet. 2006. *Flavour and Fragrance Journal*, 21(3): 435-438.

145. Liu, J. M., Nan, P., Wang, L., Wang, Q. B., Tashi, T., Zhong, Y. Chemical variation in lipophilic composition of *Lamiophlomis rotata* from the Qinghai-Tibetan Plateau. 2006. *Chemistry of Natural Compounds*, 42(5): 525-528.

146. **Liu, J. M., Wang, L., Geng, Y. P., Wang, Q. B., Luo, L. J., Zhong, Y. Genetic diversity and population structure of *Lamiophlomis rotata* (Lamiaceae), an endemic species of Qinghai-Tibet Plateau. 2006. *Genetica*, 128(1-3): 385-394.**（生物多样性与植物基因组分析代表性论文）

147. Ren, W. Z., Li, W. D., Yu, M., Hao, P., Zhang, Y., Zhou, P., Zhang, S. Y., Zhao, G. P., Zhong, Y., Wang, S. Y., Wang, L. F., Shi, Z. L. Full-length genome sequences of two SARS-like coronaviruses in horseshoe bats and genetic variation analysis. 2006. *Journal of General Virology*, 87: 3355-3359.

148. Tang, S. Q., Bin, X. Y., Wang, L., Zhong, Y. Genetic diversity and population structure of yellow camellia (*Camellia nitidissima*) in China as revealed by RAPD and AFLP markers. 2006. *Biochemical Genetics*, 44(9-10): 449-461.

149. Wang, L., Liu, J. M., Jian, S. G., Zhang, W. J., Wang, Q. B., Zhao, X. M., Liu, N., Zhong, Y. Genetic diversity and population structure in *Elephantopus scaber* (Asteraceae) from South

China as revealed by ISSR markers. 2006. *Plant Biosystems*, 140(3): 273-279.

150. Wang, Q. B., Yang, Y., Zhao, X. M., Zhu, B., Nan, P., Zhao, J. Y., Wang, L., Chen, F., Liu, Z. J., Zhong, Y. Chemical variation in the essential oil of *Ephedra sinica* from northeastern China. 2006. *Food Chemistry*, 98(1): 52-58.

151. Wang, Z. H., Zou, Y. J., Li, X. Y., Zhang, Q. Y., Chen, L., Wu, H., Su, D. H., Chen, Y. L., Guo, J. X., Luo, D., Long, Y. M., Zhong, Y., Liu, Y. G. Cytoplasmic male sterility of rice with Boro II cytoplasm is caused by a cytotoxic peptide and is restored by two related PPR motif genes via distinct modes of mRNA silencing. 2006. *Plant Cell*, 18(3): 676-687.

152. **Xu, J. J., Zhang, J. H., Wang, L., Zhou, J., Huang, H. D., Wu, J. H., Zhong, Y., Shi, Y. Y. Solution structure of Urm1 and its implications for the origin of protein modifiers. 2006. *Proceedings of the National Academy of Sciences of the USA*, 103(31): 11625-11630.**（分子进化分析方法及应用代表性论文）

153. Zeng, H. Z., Zhong, Y., Luo, L. J. Drought tolerance genes in rice. 2006. *Functional & Integrative Genomics*, 6(4): 338-341.

154. Zhang, W. J., Zhang, Y., Zhong, Y. Using maximum likelihood method to detect adaptive evolution of HCV envelope protein-coding genes. 2006. *Chinese Science Bulletin*, 51(18): 2236-2242.

155. Zhao, J. Y., Liu, J. M., Zhang, X. Y., Liu, Z. J., Tashi, T., Zhong, Y., Nan, P. Chemical composition of the volatiles of three wild *Bergenia* species from western China. 2006. *Flavour and Fragrance Journal*, 21(3): 431-434.

156. Zhao, X. M., Wang, Q. B., Zhou, J., Zhong, Y. Phylogenetic relationships of Taxaceae and its related groups inferred from nuclear 18S rRNA, chloroplast *mat*K, *rbc*L, *rps*4, 16S rRNA and mitochondrial *cox* I sequences. 2006. *Journal of Genetics and Molecular Biology*, 17(2): 81-93.

157. Zheng, X. W., Wei, X. D., Nan, P., Zhong, Y., Chen, J. K. Chemical composition and antimicrobial activity of the essential oil of *Sagittaria trifolia*. 2006. *Chemistry of Natural Compounds*, 42(5): 520-522.

158. Zheng, X. Y., Zhong, Y., Duan, Y. H., Li, C. X., Dang, L., Guo, Y. P., Ma, E. B. Genetic

variation and population structure of oriental migratory locust, *Locusta migratoria manilensis*, in China by allozyme, SSRP-PCR, and AFLP markers. 2006. *Biochemical Genetics*, 44(7-8): 333-347.

159. Zhu, B., Wang, Q. B., Roge, E. F., Nan, P., Liu, Z. J., Zhong, Y. Chemical variation in leaf oils of *Pistacia chinensis* from five locations in China. 2006. *Chemistry of Natural Compounds*, 42(4): 422-425.

160. 李荣，曹顺良，李园园，谭灏，朱扬勇，钟扬，李亦学. 基于语义路径覆盖的Gene Ontology术语间语义相似性度量方法.《自然科学进展》2006年第16卷第7期，916-920.

161. 琼次仁，扎西次仁，索朗白珍，中普穷，德央，钟扬. 西藏红景天植物资源及其利用.《西藏大学学报》2006年第21卷第3期，80-84.

162. 王庆彪，王莉，周仁超，赵晓敏，施苏华，杨永，钟扬. 用叶绿体和核糖体DNA序列探讨中国特有植物斑子麻黄的系统位置.《科学通报》2006年第51卷第1期，110-113.

163. 向筑，张竞男，宋平，胡珈瑞，钟扬. 核糖体蛋白rpl15 cDNA序列在真骨鱼类系统进化研究中的价值.《遗传》2006年第28卷第2期，171-178.

164. 张文娟，张原，钟扬. 应用最大似然法检测丙肝病毒包膜蛋白编码基因的适应性进化.《科学通报》2006年第51卷第16期，1894-1899.

165. Fu, Z., Chen, X., Vacic, V., Nan, P., Zhong, Y., Jiang, T. MSOAR: a high-throughput ortholog assignment system based on genome rearrangement. 2007. *Journal of Computational Biology*, 14(9): 1160-1175.

166. Tang, S. Q., Bin, X. Y., Peng, Y. T., Zhou, J. Y., Wang, L., Zhong, Y. Assessment of genetic diversity in cultivars and wild accessions of Luohanguo (*Siraitia grosvenorii* [Swingle] A. M. Lu et Z. Y. Zhang), a species with edible and medicinal sweet fruits endemic to southern China, using RAPD and AFLP markers. 2007. *Genetic Resources and Crop Evolution*, 54(5): 1053-1061.

167. Tang, S. Q., Li, Y., Geng, Y. P., Zhang, G. R., Wang, L., Zhong, Y. Clonal and spatial genetic structure in natural populations of Luohanguo (*Siraitia grosvenorii*), an economic species endemic to South China, as revealed by RAPD markers. 2007. *Biochemical Systematics and*

Ecology, 35(9): 557-565.

168. Tashi, T., Bai, Z. K., Nan, P., Qiong, T., Lei, Y. D., Liu, J. M., Wang, L., Zhong, Y. Chemical composition of the essential oils of three *Rhodiola* species from Tibet. 2007. *Chemistry of Natural Compounds*, 43(6): 716-718.

169. **Zeng, H. Z., Luo, L. J., Zhang, W. X., Zhou, J., Li, Z. F., Liu, H. Y., Zhu, T. S., Feng, X. Q., Zhong, Y. PlantQTL-GE: a database system for identifying candidate genes in rice and *Arabidopsis* by gene expression and QTL information. 2007. *Nucleic Acids Research*, 35: D879-D882.**（生物数据模型与信息系统代表性论文）

170. Zhang, W. J., Zhou, J., Li, Z. F., Wang, L., Gu, X., Zhong, Y. Comparative analysis of codon usage patterns among mitochondrion, chloroplast and nuclear genes in *Triticum aestivum* L. 2007. *Journal of Integrated Plant Biology*, 49(2): 246-254.

171. Zhang, Y., Zhang, N., Nan, P., Cao, Y., Hasegawa, M., Zhong, Y. Computational simulation of interactions between SARS coronavirus spike mutants and host species-specific receptors. 2007. *Computational Biology and Chemistry*, 31(2): 134-137.

172. Zhang, Y., Zheng, N., Zhong, Y. Computational characterization and design of SARS coronavirus receptor recognition and antibody neutralization. 2007. *Computational Biology and Chemistry*, 31(2): 129-133.

173. Zhu, B., Ren, Z. M., Nan, P., Jiang, M. X., Zhao, J. Y., Zhong, Y. Chemical variation in leaf essential oils of *Rhus chinensis* from eight locations in southern and eastern China. 2007. *Chemistry of Natural Compounds*, 43(6): 741-743.

174. Zhu, Y., Liu, N., Wang, Q. B., Tang, S. Q., Zhong, Y. Phylogenetic position of *Nothotsuga* inferred from the sequences of nuclear 18S rRNA, chloroplast *rbc*L, and mitochondria *cox* I genes. 2007. *Journal of Genetics and Molecular Biology*, 18(1): 23-28.

175. Cheng, Q. Q., Cheng, H., Wang, L., Zhong, Y., Lu, D. A preliminary genetic distinctness of four *Coilia* fishes (Clupeiformes: Engraulidae) inferred from mitochondrial DNA sequences. 2008. *Russian Journal of Genetics*, 44(3): 339-343.

176. Cram, W. J., Zhong, Y., Tashi, T., Cai, J. Tibet's seeds must be stored as climate changes. 2008. *Nature*, 452(7183): 28. (Correspondence)

177. Ren, Z. M., Zhu, B., Wang, D. J., Ma, E. B., Su, D. M., Zhong Y. Comparative population structure of Chinese sumac aphid *Schlechtendalia chinensis* and its primary host-plant *Rhus chinensis*. 2008. *Genetica*, 132(1): 103-112.

178. **Sun, L., Jin, K., Liu, Y. M., Yang, W. W., Xie, X., Ye, L., Wang, L., Zhu, L., Ding, S., Su, Y., Zhou, J., Han, M., Zhuang, Y., Xu, T., Wu, X. H., Gu, N., Zhong, Y. PBmice: an integrated database system of *piggyBac* (PB) insertional mutations and their characterizations in mice. 2008. *Nucleic Acids Research*, 36: D729-D734.**（生物数据模型与信息系统代表性论文）

179. Tang, S. Q., Dai, W. J., Li, M. S., Zhang, Y., Geng, Y. P., Wang, L., Zhong, Y. Genetic diversity of relictual and endangered plant *Abies ziyuanensis* (Pinaceae) revealed by AFLP and SSR markers. 2008. *Genetica*, 133(1): 21-30.

180. Zhang, M., Cao, T. W., Jin, K., Ren, Z. M., Guo, Y. P., Shi, J., Zhong, Y., Ma, E. B. Estimating divergence times among subfamilies in Nymphalidae. 2008. *Chinese Science Bulletin*, 53(17): 2652-2658.

181. Zhang, M., Cao, T. W., Zhong, Y., Ren, Z. M., Guo, Y. P., Ma, E. B. Molecular phylogenetic analysis of the main lineages of Nymphalinae (Nymphalidae: Lepidoptera) based on the partial mitochondrial *COI* gene. 2008. *Agricultural Sciences in China*, 7(6): 731-739.

182. Zhang, M., Zhong, Y., Cao, T. W., Geng, Y. P., Zhang, Y., Jin, K., Ren, Z. M., Zhang, R., Guo, Y. P., Ma, E. B. Phylogenetic relationship and morphological evolution in the subfamily Limenitidinae (Lepidoptera: Nymphalidae). 2008. *Progress in Natural Science*, 18(11): 1357-1364.

183. Zhang, Z. H., Zhou, R. C., Tang, T., Huang, Y. L., Zhong, Y., Shi, S. H. Genetic variation in central and peripheral populations of *Excoecaria agallocha* from Indo-West Pacific. 2008. *Aquatic Botany*, 89(1): 57-62.

184. Zheng, S. Y., Sheng, J., Wang, C., Wang, X. J., Yu, Y., Li, Y., Michie, A., Dai, J. L., Zhong, Y., Hao, P., Liu, L., Li, Y. X. MPSQ: a web tool for protein-state searching. 2008. *Bioinformatics*, 24(20): 2412-2413.

185. 郝沛，余曜，张晓艳，屠康，范海威，钟扬. 顺式调控元件对"头对头"基因对共表达的影响.《中国科学C辑：生命科学》2008年第38卷第11期，1007-1012.

186. 昝启杰，任竹梅，李后魂，李罡，王勇军，王莉，钟扬. 用mtDNA *COII*基因序列确定我国北部湾红树植物白骨壤虫灾虫源.《自然科学进展》2008年第19卷第2期，1380-1385.

187. 王定江，杨汉远，钟扬，任竹梅. 贵州省八个种群角倍蚜ISSR遗传多样性.《生态学杂志》2008年第27卷第10期，1729-1733.

188. 张敏，曹天文，金科，任竹梅，郭亚平，施婧，钟扬，马恩波. 蛱蝶科亚科间的分歧时间估计.《科学通报》2008年第53卷第15期，1809-1814.

189. Bao, H., Xiong, Y. Y., Guo, H., Zhou, R. C., Lu, X. M., Yang, Z., Zhong, Y., Shi, S. H. MapNext: a software tool for spliced and unspliced alignments and SNP detection of short sequence reads. 2009. *BMC Genomics*, 10: S13.

190. Cheng, G., Zhao, X., Li, Z. F., Liu, X. Y., Yan, W. Y., Zhang, X. Y., Zhong, Y., Zheng, Z. X. Identification of a putative invertebrate helical cytokine similar to the ciliary neurotrophic factor/leukemia inhibitory factor family by PSI-BLAST-based approach. 2009. *Journal of Interferon and Cytokine Research*, 29(8): 461-468.

191. Cheng, Q. Q., Su, Z. X., Zhong, Y., Gu, X. Effect of site-specific heterogeneous evolution on phylogenetic reconstruction: a simple evaluation. 2009. *Gene*, 441(1-2): 156-162.

192. Consortium for Influenza Study at Shanghai. Subtyping of type A influenza by sequencing the variable regions of HA gene specifically amplified with RT-PCR. 2009. *Chinese Science Bulletin*, 54(13): 2164-2167.

193. Consortium for Influenza Study at Shanghai. The mutation network for the hemagglutinin gene from the novel influenza A (H1N1) virus. 2009. *Chinese Science Bulletin*, 54(13): 2168-2170.

194. Consortium for Influenza Study at Shanghai. Structure modeling and spatial epitope analysis for HA protein of the novel H1N1 influenza virus. 2009. *Chinese Science Bulletin*, 54(13): 2171-2173.

195. Geng, Y. P., Tang, S. Q., Tashi, T., Song, Z. P., Zhang, G. R., Zeng, L. Y., Zhao, J. Y., Wang, L., Shi, J., Chen, J. K., Zhong, Y. Fine- and landscape-scale spatial genetic

structure of cushion rockjasmine, *Androsace tapete* (Primulaceae), across southern Qinghai-Tibetan Plateau. 2009. ***Genetica***, **135(3): 419-427.**（生物多样性与植物基因组分析代表性论文）

196. Han, L. P., Huang, Q., Nan, P., Zhong, Y. Prediction of potential antimalarial targets of artemisinin based on protein information from whole genome of *Plasmodium falciparum*. 2009. *Chinese Science Bulletin*, 54(22): 4234-4240.

197. Hao, P., Yu, Y., Zhang, X. Y., Tu, K., Fan, H. W., Zhong, Y. The contribution of *cis*-regulatory elements to head-to-head gene pairs' co-expression pattern. 2009. *Science in China Series C: Life Science*, 52(1): 74-79.

198. Hao, P., Zheng, S. Y., Ping, J., Tu, K., Gieger, C., Wang-Sattler, R., Zhong, Y., Li, Y. X. Human gene expression sensitivity according to large scale meta-analysis. 2009. *MBC Bioinformatics*, 10: S56.

199. Hasegawa, M., Zhong, B. J., Zhong, Y. Adaptive evolution of chloroplast genomes in ancestral grasses. 2009. *Plant Signaling &Behavior*, 4(7): 623-624.

200. Li, Y. Y., Li, Z. F., He, Y. F., Kang, Y., Zhang, X. Y., Cheng, M. J., Zhong, Y., Xu, C. J. Phylogeographic analysis of human papillomavirus 58. 2009. *Science in China Series C: Life Science*, 52(12): 1164-1172.

201. Liu, N., Zhu, Y., Wei, Z. X., Chen, J., Wang, Q. B., Jian, S. G., Zhou, D. W., Shi, J., Yang, Y., Zhong, Y. Phylogenetic relationships and divergence times of the family Araucariaceae based on the DNA sequences of eight genes. 2009. *Chinese Science Bulletin*, 54(15): 2648-2655.

202. **Luo, R. Y., Wei, H. B., Ye, L., Wang, K. K., Chen, F., Luo, L. J., Liu, L., Li, Y. Y., Crabbe, M. J. C., Jin, L., Li, Y. X., Zhong, Y. Photosynthetic metabolism of C-3 plants shows highly cooperative regulation under changing environments: a systems biological analysis. 2009.** *Proceedings of the National Academy of Sciences of the USA*, **106(3): 847-852.**（分子进化分析方法及应用代表性论文）

203. Ren, Z. M., Zhu, B., Ma, E. B., Wen, J., Tu, T. Y., Cao, Y., Hasegawa, M., Zhong, Y. Complete nucleotide sequence and gene arrangement of the mitochondrial genome of the

crab-eating frog *Fejervarya cancrivora* and evolutionary implications. 2009. *Gene*, 441(1-2): 148-155.

204. Omirshat T., Geng, Y. P., Zeng, L. Y., Dong, S. S., Chen, F., Chen, J., Song, Z. P., Zhong, Y. Assessment of genetic diversity and population structure of Chinese wild almond, *Amygdalus nana*, using EST- and genomic SSRs. 2009. *Biochemical Systematics and Ecology*, 37(3): 146-153.

205. The Schistosoma japonicum Genome Sequencing and Functional Analysis Consortium. The *Schistosoma japonicum* genome reveals features of host-parasite interplay. 2009. *Nature*, 460(7253): 345-352. (as a Principal Investigator)（分子进化分析方法及应用代表性论文）

206. Tang, S. Q., Wei, F., Zeng, L. Y., Li, X. K., Tang, S. C., Zhong, Y., Geng, Y. P. Multiple introductions are responsible for the disjunct distributions of invasive *Parthenium hysterophorus* in China: evidence from nuclear and chloroplast DNA. 2009. *Weed Research*, 49(4): 373-380.

207. Tang, X. C., Li, G., Vasilakis, N., Zhang, Y., Shi, Z. L., Zhong, Y., Wang, L. F., Zhang, S. Y. Differential stepwise evolution of SARS coronavirus functional proteins in different host species. 2009. *BMC Evolutionary Biology*, 9: 52.

208. Tang, X. C., Vasilakis, N., Shi, Z. L., Zhong, Y., Wang, L. F., Zhang, S. Y. SARS coronavirus adaptation to human is partially constrained by host alteration. 2009. *American Journal of Tropical Medicine and Hygiene*, 81(5): 196.

209. Yang, W. W., Jin, K., Xie, X., Li, D. S., Yang, J. G., Wang, L., Gu, N., Zhong, Y., Sun, L. V. Development of a database system for mapping insertional mutations onto the mouse genome with large-scale experimental data. 2009. *BMC Genomics*, 10: S7

210. Zhang, H. M., Zhong, Y., Hao, B. L., Gu, X. A simple method for phylogenomic inference using the information of gene content of genomes. 2009. *Gene*, 441(1-2): 163-168.

211. Zhao, G. P., Zhong, Y. Learn to fight from the war. 2009. *Chinese Science Bulletin*, 54(13): 2157-2158. (Editor's note)

212. Zhong, B. J., Yonezawa, T., Zhong, Y., Hasegawa, M. Episodic evolution and adaptation of

chloroplast genomes in ancestral grasses. 2009. *PLoS ONE*, 4(4): e5297.

213. Zhou, D. W., Zhou, J., Meng, L. H., Wang, Q. B., Xie, H., Guan, Y. C., Ma, Z. Y., Zhong, Y., Chen, F., Liu, J. Q. Duplication and adaptive evolution of the *COR15* genes within the highly cold-tolerant *Draba* lineage (Brassicaceae). 2009. *Gene*, 441(1-2): 36-44.

214. Zhu, Y., Geng, Y. P., Tashi, T., Liu, N., Wang, Q. B., Zhong, Y. High genetic differentiation and low genetic diversity in *Incarvillea younghusbandii*, an endemic plant of Qinghai-Tibetan Plateau, revealed by AFLP markers. 2009. *Biochemical Systematics and Ecology*, 37(5): 589-596.

215. 韩利平, 黄强, 南蓬, 钟扬. 基于恶性疟原虫全基因组蛋白信息预测青蒿素的潜在抗疟靶点. 《科学通报》2009年第54卷第18期, 2806-2812.

216. 韩利平, 黄强, 曾丽艳, 钟扬, 卫海滨, 南蓬. 青蒿素及其衍生物的抗疟机制研究进展. 《自然科学进展》2009年第19卷第1期, 25-32.

217. 李燕云, 李作峰, 何以丰, 康玉, 张晓燕, 程明军, 钟扬, 徐丛剑. 人乳头瘤病毒58型的系统地理学分析. 《中国科学C辑: 生命科学》2009年第39卷第7期, 654-661.

218. 刘念, 朱勇, 魏宗贤, 陈婕, 王庆彪, 简曙光, 周党卫, 施婧, 杨永, 钟扬. 利用8个基因的DNA序列探讨南洋杉科的系统发育关系和起源时间. 《科学通报》2009年第54卷第14期, 2089-2095.

219. 张晓鹏, 黄智星, 周杰, 修乃华, 钟扬, 陈凡. 最短路径法在水稻ABA和环境胁迫条件下基因应答网络研究中的应用. 《植物学报》2009年第44卷第2期, 159-166.

220. Fang, F., Yu, L., Zhong, Y., Yao, L. TGFB1 509 C/T polymorphism and colorectal cancer risk: a meta-analysis. 2010. *Medical Oncology*, 27(4): 1324-1328.

221. Jian, S. G., Tang, T., Zhong, Y., Shi, S. H. Conservation genetics of *Heritiera littoralis* (Sterculiaceae), a threatened mangrove in China, based on AFLP and ISSR markers. 2010. *Biochemical Systematics and Ecology*, 38(5): 924-930.

222. Li, T., Geng, Y. P., Zhong, Y., Zhang, M., Ren, Z. M., Ma, J., Guo, Y. P., Ma, E. B. Host-associated genetic differentiation in rice grasshopper, *Oxya japonica*, on wild vs. cultivated rice. 2010. *Biochemical Systematics and Ecology*, 38(5): 958-963.

223. Liu, J. M., Nan, P., Qiong, T., Tashi, T., Bai, Z. K., Wang, L., Liu, Z. J., Zhong, Y. Volatile

constituents of the leaves and flowers of *Salvia przewalskii* Maxim. from Tobet. 2010. *Flavours and Frangrance Journal*, 21: 435-438.

224. Liu, N., Wang, Q. B., Chen, J., Zhu, Y., Tashi, T., Hu, Y. Y., Chen, F., Zhong, Y. Adaptive evolution and structure modeling of *rbc*L gene in *Ephedra*. 2010. *Chinese Science Bulletin*, 55(22): 2341-2346.

225. Tang, S. Q., Zhang, Y., Zeng, L. Y., Luo, L. J., Zhong, Y., Geng, Y. P. Assessment of genetic diversity and relationships of upland rice accessions from Southwest China using microsatellite markers. 2010. *Plant Biosystems*, 144(1): 85-92.

226. Yang, Z., Ren, F., Liu, C. N., He, S. M., Sun, G., Gao, Q., Yao, L., Zhang, Y. D., Miao, R. Y., Cao, Y., Zhao, Y., Zhong, Y., Zhao, H. T. dbDEMC: a database of differentially expressed miRNAs in human cancers. 2010. *BMC Genomics*, 11: S5.

227. Yao, L., Fang, F., Wu, Q., Yang, Z., Zhong, Y., Yu, L. No association between CYP17 T-34C polymorphism and breast cancer risk: a meta-analysis involving 58, 814 subjects. 2010. *Breast Cancer Research and Treatment*, 122(1): 221-227.

228. Yao, L., Fang, F., Wu, Q., Zhong, Y., Yu, L. No association between CYP1B1 Val432Leu polymorphism and breast cancer risk: a meta-analysis involving 40, 303 subjects. 2010. *Breast Cancer Research and Treatment*, 122(1): 237-242.

229. Yao, L., Qiu, L. X., Yu, L., Yang, Z., Yu, X. J., Zhong, Y., Yu, L. The association between TA-repeat polymorphism in the promoter region of UGT1A1 and breast cancer risk: a meta-analysis. 2010. *Breast Cancer Research and Treatment*, 122(3): 879-882.

230. Yao, L., Qiu, L. X., Yu, L., Yang, Z., Yu, X. J., Zhong, Y., Hu, X. C., Yu, L. The association between ERCC2 Asp312Asn polymorphism and breast cancer risk: a meta-analysis involving 22,766 subjects. 2010. *Breast Cancer Research and Treatment*, 123(1): 227-231.

231. Yao, L., Fang, F., Zhong, Y., Yu, L. The association between two polymorphisms of eNOS and breast cancer risk: a meta-analysis. 2010. *Breast Cancer Research and Treatment*, 124(1): 223-227.

232. Ye, L., Zhang, Y., Mei, Y., Nan, P., Zhong, Y. Detecting putative recombination events of hepatitis B virus: an updated comparative genome analysis. 2010. *Chinese Science Bulletin*,

55(22): 2373-2379.

233. Zeng, L. Y., Li, Z. H., Tashi, T., Chen, J., Zhong, Y., Geng, Y. P. Microsatellite markers for the cushion rock Jasmine, *Androsace tapete* (Primulaceae), a species endemic to the Qinghai-Tibetan Plateau. 2010. *American Journal of Botany*, 97(10): E94-E96.

234. Zeng, L. Y., Xu, L. L., Tang, S. Q., Tashi, T., Geng, Y. P., Zhong, Y. Effect of sampling strategy on estimation of fine-scale spatial genetic structure in *Androsace tapete* (Primulaceae), an alpine plant endemic to Qinghai-Tibetan Plateau. 2010. *Journal of Systematics and Evolution*, 48(4): 257-264.

235. Zhong, B. J., Yonezawa, T., Zhong, Y., Hasegawa, M. The position of Gnetales among seed plants: overcoming pitfalls of chloroplast phylogenomics. 2010. *Molecular Biology and Evolution*, 27(12): 2855-2863.

236. 刘念, 王庆彪, 陈婕, 朱勇, 扎西次仁, 胡祎瑶, 陈凡, 钟扬. 麻黄属*rbc*L基因的适应性进化检测与结构模建.《科学通报》2010年第55卷第14期, 1341-1346.

237. 沈灵犀, 扎西次仁, 耿宇鹏, 钟扬, 南蓬. 西藏蒿属六种植物精油化学成分分析及抑菌效果.《复旦学报（自然科学版）》2010年第49卷第1期, 73-80.

238. Chen, H. K., Zhu, L. L., Cao, X. H., Song, D. D., Zhong, Y. Development and characterization of polymorphic microsatellite primers in *Reaumuria soongorica* (Tamaricaceae). 2011. *American Journal of Botany*, 98(8): E221-E223.

239. Jin, K., Xue, C. Y., Wu, X. L., Qian, J. Y., Zhu, Y., Yang, Z., Yonezawa, T., Crabbe, M. J. C., Cao, Y., Hasegawa, M., Zhong, Y., Zheng, Y. F. Why does the giant panda eat bamboo? A comparative analysis of appetite-reward-related genes among mammals. 2011. *PLoS ONE*, 6(7): e22602.

240. Li, T., Zhang, M., Qu, Y. H., Ren, Z. M., Zhang, J. Z., Guo, Y. P., Heong, K. L., Villareal, B., Zhong, Y., Ma, E. B. Population genetic structure and phylogeographical pattern of rice grasshopper, *Oxya hyla intricata*, across Southeast Asia. 2011. *Genetica*, 139(4): 511-524.

241. Min, Y., Jin, X. G., Chang, J., Peng, C. H., Gu, B. J., Ge, Y., Zhang, Y. Weak indirect effects inherent to nitrogen biogeochemical cycling within anthropogenic ecosystems: a network environ analysis. 2011. *Ecological Modelling*, 222(17): 3277-3284.

242. Yang, Z., Yu, Y., Yao, L., Li, G. G., Wang, L., Hu, Y. Y., Wei, H. B., Wang, L., Hammami, R., Razavi, R., Zhong, Y., Liang, X. F. DetoxiProt: an integrated database for detoxification proteins. 2011. *BMC Genomics*, 12: S2.

243. Zhang, L. P., Zheng, Y. F., Li, D. F., Zhong, Y. Self-organizing map of gene regulatory networks for cell phenotypes during reprogramming. 2011. *Computational Biology and Chemistry*, 35(4): 211-217.

244. Zhang, M., Cao, T. W., Zhong, Y., Guo, Y. P., Ma, E. B. Phylogeny of Limenitidinae butterflies (Lepidoptera: Nymphalidae) inferred from mitochondrial cytochrome oxidase I gene sequences. 2011. *Agricultural Sciences in China*, 10(4): 566-575.

245. Zhang, Y., Zhang, H., Zhou, T. Y., Zhong, Y., Jin, Q. Genes under positive selection in *Mycobacterium tuberculosis*. 2011. *Computational Biology and Chemistry*, 35(5): 319-322.

246. 张君诚，黄晖，张杭颖，钟扬. 武夷山脉石杉科植物石杉碱甲含量的研究.《植物遗传资源学报》2011年第12卷第4期，629-633.

247. 张君诚，宋育红，朱勇，张杭颖，钟扬. 长柄石杉居群遗传多样性和遗传结构的AFLP分析.《应用与环境生物学报》2011年第17卷第1期，18-23.

248. 赵佳媛，薛晨轶，金科，王合瑞，王爱善，范敏玉，孙璘，王莉，钟扬. 节尾狐猴二氧化碳感知相关基因CAII的序列分析及其进化意义.《科学通报》2011年第56卷第3期，210-214.

249. Dong, D., Jin, K., Wu, X. L., Zhong, Y. CRDB: database of chemosensory receptor gene families in vertebrate. 2012. *PLoS ONE*, 7(2): e31540.

250. Qiu, Q., Zhang, G. J., Ma, T., Qian, W. B., Wang, J. Y., Ye, Z. Q., Cao, C. C., Hu, Q. J., Kim, J., Larkin, D. M., Auvil, L., Capitanu, B., Ma, J., Lewin, H. A., Qian, X. J., Lang, Y. S., Zhou, R., Wang, L. Z., Wang, K., Xia, J. Q., Liao, S. G., Pan, S. K., Lu, X., Hou, H. L., Wang, Y., Zang, X. T., Yin, Y., Ma, H., Zhang, J., Wang, Z. F., Zhang, Y. M., Zhang, D. W., Yonezawa, T., Hasegawa, M., Zhong, Y., Liu, W. B., Zhang, Y., Huang, Z. Y., Zhang, S. X., Long, R. J., Yang, H. M., Wang, J., Lenstra, J. A., Cooper, D. N., Wu, Y., Wang, J., Shi, P., Wang, J., Liu, J. Q. The yak genome and adaptation to life at high altitude. 2012. *Nature Genetics*, 44(8): 946-949.

251. Yang, Z., Dong, D., Zhang, Z. L., Crabbe, M. J. C., Wang, L., Zhong, Y. Preferential regulation of stably expressed genes in the human genome suggests a widespread expression buffering role of microRNAs. 2012. *BMC Genomics*, 13: S14.

252. Zeng, L. Y., Lhag, C., Tashi, T., Nor, B., Zhao, J. Y., Zhong, Y. Microsatellite markers for *Saussurea gnaphalodes* (Asteraceae), a native Himalayan mountain species. 2012. *American Journal of Botany*, 99(8): E326-E329.

253. Zhang, M. L., Kang, Y., Zhong, Y., Sandersone, S. C. Intense uplift of the Qinghai-Tibetan Plateau triggered rapid diversification of *Phyllolobium* (Leguminosae) in the Late Cenozoic. 2012. *Plant Ecology & Diversity*, 5(4): 491-499.

254. Zhang, M., Nie, X. P., Cao, T. W., Wang, J. P., Li, T., Zhang, X. N., Guo, Y. P., Ma, E. B., Zhong, Y. The complete mitochondrial genome of the butterfly *Apatura metis* (Lepidoptera: Nymphalidae). 2012. *Molecular Biology Reports*, 39(6): 6529-6536.

255. Zhang, T. C., Qiao, Q., Zhong, Y. Detecting adaptive evolution and functional divergence in aminocyclopropane-1-carboxylate synthase (ACS) gene family. 2012. *Computational Biology and Chemistry*, 38: 10-16.

256. 李汐, 祝铭, 孙延霞, 钟扬, 李建强. 基于叶绿体 rps16 基因和核基因 ITS 片段研究肉苁蓉属系统位置.《植物科学学报》2012年第30卷第5期, 431-436.

257. Li, X., Zhang, T. C., Qiao, Q., Ren, Z. M., Zhao, J. Y., Yonezawa, T., Hasegawa, M., Crabbe, M. J. C., Li, J. Q., Zhong, Y. Complete chloroplast genome sequence of holoparasite *Cistanche deserticola* (Orobanchaceae) reveals gene loss and horizontal gene transfer from its host *Haloxylon ammodendron* (Chenopodiaceae). 2013. *PLoS ONE*, 8(3): e58747.

258. Li, Y. L., Ren, Z. M., Shedlock, A. M., Wu, J. Q., Sang, L., Tashi, T., Hasegawa, M., Yonezawa, T., Zhong, Y. High altitude adaptation of the schizothoracine fishes (Cyprinidae) revealed by the mitochondrial genome analyses. 2013. *Gene*, 517(2): 169-178.

259. Ma, W. L., Wu, M., Wu, Y., Ren, Z. M., Zhong, Y. Cloning and characterisation of a phenylalanine ammonia-lyase gene from *Rhus chinensis*. 2013. *Plant Cell Reports*, 32(8): 1179-1190.

260. Ren, Z. M., Zhong, Y., Kurosu, U., Aoki, S., Ma, E. B., von Dohlen, C. D., Wen, J. Historical

biogeography of Eastern Asian-Eastern North American disjunct Melaphidina aphids (Hemiptera: Aphididae: Eriosomatinae) on *Rhus* hosts (Anacardiaceae). 2013. *Molecular Phylogenetics and Evolution*, 69(3): 1146-1158.

261. Yang, Z., Miao, R. Y., Li, G. B., Wu, Y., Robson, S. C., Yang, X. B., Zhao, Y., Zhao, H. T., Zhong, Y. Identification of recurrence related microRNAs in hepatocellular carcinoma after surgical resection. 2013. *International Journal of Molecular Sciences*, 14(1): 1105-1118.

262. Zhao, J. Y., Zheng, X. X., Newman, R. A., Zhong, Y., Liu, Z. J., Nan, P. Chemical composition and bioactivity of the essential oil of *Artemisia anomala* from China. 2013. *The Journal of Essential Oil Research*, 25(6): 520-525.

263. Zheng, W. W., Wang, X., Tian, D. J., Jiang, S. H., Andersen, M. E., He, G. S., Crabbe, M. J. C., Zheng, Y. X., Zhong, Y., Qu, W. D. Water pollutant fingerprinting tracks recent industrial transfer from coastal to inland China: a case study. 2013. *Scientific Reports*, 3: 1031.

264. 吕帝瑾，赵佳媛，陈婧，钟扬，南蓬．植物microRNA的研究进展．《植物生理学报》2013年第49卷第9期，847-854.

265. 索朗次仁，格桑卓玛，钟扬，扎西次仁．西藏尼洋河流域陆生脊椎动物的夏季调查．《西藏科技》2013年第8期（总第245期），71-75.

266. 张体操，乔琴，钟扬．青藏高原生物资源开发的现状与前景．《生命科学》2013年第25卷第5期，446-450.

267. Chen, H. K., Zhong, Y. Microsatellite markers for *Lycium ruthenicum* (Solananeae). 2014. *Molecular Biology Reports*, 41(9): 5545-5548.

268. Chen, H. K., Zeng, L. Y., Yonezawa, T., Ren, X., Zhong, Y. Genetic population structure of the desert shrub species *Lycium ruthenicum* inferred from chloroplast DNA. 2014. *Pakistan Journal of Botany*, 46(6): 2121-2130.

269. Chen, J., Sheng, J., Lv, D. J., Zhong, Y., Zhang, G. Q., Nan, P. The optimization of running time for a maximum common substructure-based algorithm and its application in drug design. 2014. *Computational Biology and Chemistry*, 48: 14-20.

270. Lin, P. C., Zeng, L. Y., Yang, Z., Liu, R. K., Zhong, Y. Development and characterization of polymorphic microsatellite marker for *Dactylorhiza hatagirea* (D. Don) Soo. 2014.

Conservation Genetics Resources, 6(1): 29-31.

271. Liu, C. C., Lei, C., Zhong, Y., Gao, L. X., Li, J. Y., Yu, M. H., Li, J., Hou, A. J. Novel grayanane diterpenoids from *Rhododendron principis*. 2014. *Tetrahedron*, 70(29): 4317-4322.

272. Liu, Z., Zheng, G. Y., Dong, X., Wang, Z., Ying, B. L., Zhong, Y., Li, Y. X. Investigating co-evolution of functionally associated phosphosites in human. 2014. *Molecular Genetics and Genomics*, 289: 1217-1223.

273. Wen, J., Zhang, J. Q., Nie, Z. L., Zhong, Y., Sun, H. Evolutionary diversifications of plants on the Qinghai-Tibetan Plateau. 2014. *Frontiers in Genetics*, 5: Article 4.

274. Wu, J. Q., Hasegawa, M., Zhong, Y., Yonezawa, T. Importance of synonymous substitutions under dense taxon sampling and appropriate modeling in reconstructing the mitogenomic tree of Eutheria. 2014. *Genes & Genetic Systems*, 89(5): 237-251.

275. Xu, L., Wang, H., Lhag, Q., Lu, F., Sun, K., Fang, Y., Yang, M., Zhong, Y., Wu, Q. H., Chen, J. J., Birks, H. J. B., Zhang, W. J. Microrefugia and shifts of *Hippophae tibetana* (Elaeagnaceae) on the north side of Mt. Qomolangma (Mt. Everest) during the last 25000 years. 2014. *PLoS ONE*, 9(5): e97601.

276. Xu, L. L., Zeng, L. Y., Liao, B. W., Zhong, Y. Microsatellite markers for a mangrove species, *Cerbera manghas*, from South China. 2014. *Conservation Genetics Resources*, 6(1): 45-48.

277. Yang, Z., Chen, H. K., Yang, X. B., Wan, X. S., He, L., Miao, R. Y., Yang, H. Y., Zhong, Y. Wang, L., Zhao, H. T. A phylogenetic analysis of the ubiquitin superfamily based on sequence and structural information. 2014. *Molecular Biology Reports*, 41(9): 6083-6088.

278. Yang, Z., Wan, X. S., Gu, Z. Y., Zhang, H. H., Yang, X. B., He, L., Miao, R. Y., Zhong, Y., Zhao, H. T. Evolution of the mir-181 microRNA family. 2014. *Computers in Biology and Medicine*, 52: 82-87.

279. Yonezawa, T., Hasegawa, M., and Zhong, Y. Polyphyletic origins of schizothoracine fish (Cyprinidae, Osteichthyes) and adaptive evolution in their mitochondrial genomes. 2014. *Genes & Genetic Systems*, 89(4): 187-191.

280. 胡祎瑶，钟扬. 基因组比较数据可视化的快速实现.《复旦学报（自然科学版）》2014

年第53卷第1期，107-112.

281. 拉琼，扎西次仁，朱卫东，许敏，钟扬. 雅鲁藏布江河岸植物物种丰富度分布格局及其环境解释.《生物多样性》2014年第22卷第3期，337-347.

282. 吕帝瑾，陈婧，钟扬，南蓬. 两种提取蒙古黄芪miRNA方法的比较.《植物生理学报》2014年第50卷第8期，1255-1258.

283. An, M., Zeng, L. Y., Zhang, T. C., Zhong, Y. Phylogeography of *Thlaspi arvense* (Brassicaceae) in China inferred from chloroplast and nuclear DNA sequences and ecological niche modeling. 2015. *International Journal of Molecular Sciences*, 16(6): 13339-13355.

284. Chen, H. K., Feng, Y., Wang, L. N., Yonezawa, T., Crabbe, M. J. C., Zhang, X., Zhong, Y. Transcriptome profiling of the UV-B stress response in the desert shrub *Lycium ruthenicum*. 2015. *Molecular Biology Reports*, 42(3): 639-649.

285. Chen, J., Wu, X. T., Xu, Y. Q., Zhong, Y., Li, Y. X., Chen, J. K., Li, X., Nan, P. Global transcriptome analysis profiles metabolic pathways in traditional herb *Astragalus membranaceus* Bge. var. *mongolicus* (Bge.) Hsiao. 2015. *BMC Genomics*, 16: S15.

286. Chen, J., Yuan, H., Zhang, L., Pan, H. Y., Xu, R. Y., Zhong, Y., Chen, J. K., Nan, P. Cloning, expression and purification of isoflavone-2-hydroxylase from *Astragalus membranaceus* Bge. var. *mongolicus* (Bge.) Hsiao. 2015. *Protein Expression and Purification*, 107: 83-89.

287. Lin, L., Cai, Q. Q., Zhang, X. Y., Zhang, H. W., Zhong, Y., Xu, C. J. Two less common human microRNAs miR-875 and miR-3144 target a conserved site of E6 oncogene in most high-risk human papillomavirus subtypes. 2015. *Protein & Cell*, 6(8): 575-588.

288. Ma, W. L., Wu, Y., Wu, M., Ren, Z. M., Zhong, Y. Cloning, characterization and expression of chalcone synthase from medicinal plant *Rhus chinensis*. 2015. *Journal of Plant Biochemistry and Biotechnology*, 24(1): 18-24.

289. Meng, Y. H., Zhang, W. L., Zhou, J. H., Liu, M. Y., Chen, J. H., Tian, S., Zhuo, M., Zhang, Y., Zhong, Y., Du, H. L., Wang, X. N. Genome-wide analysis of positively selected genes in seasonal and non-seasonal breeding species. 2015. *PLoS ONE*, 10(5): e0126736.

290. Razavi, N. R., Qu, M. Z., Chen, D. M., Zhong, Y., Ren, W. W., Wang, Y. X., Campbell, L. M. Effect of eutrophication on mercury (Hg) dynamics in subtropical reservoirs from a high Hg

deposition ecoregion. 2015. *Limnology and Oceanography*, 60: 385-401.

291. 罗天龙，钟扬. 人类干细胞线粒体外膜蛋白PPI网络构建与分析.《复旦学报（自然科学版）》2015年第54卷第4期，491-497.

292. 龙毅，孟凡栋，王常颖，白玲，钟扬，汪诗平. 高寒草甸主要植物地上地下生物量分布及退化对根冠比和根系表面积的影响.《广西植物》2015年第35卷第4期，532-538.

293. Gu, Z. Y., Jin, K., Crabbe, M. J. C., Zhang, Y., Liu, X. L., Huang, Y. Y., Hua, M. Y., Nan, P., Zhang, Z. L., Zhong, Y. Enrichment analysis of Alu elements with different spatial chromatin proximity in the human genome. 2016. *Protein & Cell*, 7(4): 250-266.

294. Huang, C. H., Sun, R. R., Hu, Y., Zeng, L. P., Zhang, N., Cai, L. M., Zhang, Q., Koch, M. A., Al-Shehbaz, I., Edger, P. P., Pires, J. C., Tan, D. Y., Zhong, Y., Ma, H. Resolution of Brassicaceae phylogeny using nuclear genes uncovers nested radiations and supports convergent morphological evolution. 2016. *Molecular Biology and Evolution*, 33(2): 394-412.

295. Li, X. C., Liu, C. L., Huang, T., Zhong, Y. The occurrence of genetic alterations during the progression of breast carcinoma. 2016. *BioMed Research International*, 2016: 5237827.

296. **Qiao, Q., Huang, Y. Y., Qi, J., Qu, M. Z., Jiang, C., Lin, P. C., Li, R. H., Song, L. R., Yonezawa, T., Hasegawa, M., Crabbe, M. J. C., Chen, F., Zhang, T. C., Zhong, Y. The genome and transcriptome of *Trichormus* sp. NMC-1: insights into adaptation to extreme environments on the Qinghai-Tibet Plateau. 2016. *Scientific Reports*, 6: 29404.**
（生物多样性与植物基因组分析代表性论文）

297. Qiao, Q., Wang, Q., Han, X., Guan, Y. L., Sun, H., Zhong, Y., Huang, J. L., Zhang, T. C. Transcriptome sequencing of *Crucihimalaya himalaica* (Brassicaceae) reveals how *Arabidopsis* close relative adapt to the Qinghai-Tibet Plateau. 2016. *Scientific Reports*, 6: 21729.

298. Qiao, Q., Xue, L., Wang, Q., Sun, H., Zhong, Y., Huang, J. L., Lei, J. J., Zhang, T. C. Comparative transcriptomics of strawberries (*Fragaria* spp.) provides insights into evolutionary patterns. 2016. *Frontiers in Plant Science*, 7: 1839.

299. Shi, M. F., Xu, M., Zhang, Y. Q., Zhong, Y. Composition and antibacterial activity of the

essential oil of *Chenopodium foetidum*. 2016. *Chemistry of Natural Compounds*, 52(5): 930-931.

300. 鲍武印，张阳，林鹏程，南蓬，黄艳燕，靳浩飞，钟扬. 青藏高原植物手参的谱系地理学研究.《生物技术通报》2016年第32卷第12期，96-102.

301. 李信，徐翌钦，郑煜芳，钟扬. Unc5基因家族多样性进化的研究.《生命科学研究》2016年第20卷第4期，301-308.

302. De, J., Lu, Y., Ling, L. J., Nan, P., Zhong, Y. Essential oil composition and bioactivities of *Waldheimia glabra* (Asteraceae) from Qinghai-Tibet Plateau. 2017. *Molecules*, 22(3): 460.

303. De, J., Zhu, W. D., Liu, T. M., Wang, Z., Zhong, Y. Development of microsatellite markers using Illumina MISEQ sequencing to characterize *Ephedra gerardiana* (Ephedraceae). 2017. *Applications in Plant Sciences*, 5(3): 1600104.

304. Jin, H. F., Yonezawa, T., Zhong, Y., Kishino, H., Hasegawa, M. Cretaceous origin of giant rhinoceros beetles (Dynastini; Coleoptera) and correlation of their evolution with the Pangean breakup. 2017. *Genes & Genetic Systems*, 91(4): 209-215.

305. **Lhag, C., Zhang, W. J., Wang, H., Zeng, L. Y., Birks, H. J. B., Zhong, Y. Testing the effect of the Himalayan mountains as a physical barrier to gene flow in *Hippophae tibetana* Schlect. (Elaeagnaceae). 2017. *PLoS ONE*, 12(5): e0172948.**（生物多样性与植物基因组分析代表性论文）

306. Li, H. L., Pu, J. Y., Zeng, L. Y., Zhong, Y., Xu, F. L., Nan, P. Chemical composition of essential oil of *Artemisia sieversiana* from Tibet. 2017. *Journal of Essential Oil Bearing Plants*, 20(5): 1407-1412.

307. Lou, Q. J., Chen, L., Mei, H. W., Xu, K., Wei, H. B., Feng, F. J., Li, T. M., Pang, X. M., Shi, C. P., Luo, L. J., Zhong, Y. Root transcriptomic analysis revealing the importance of energy metabolism to the development of deep roots in rice (*Oryza sativa* L.). 2017. *Frontiers in Plant Science*, 8: 1314.

308. Ren, Z. M., Harris, A. J., Dikow, R. B., Ma, E. B., Zhong, Y., Wen, J. Another look at the phylogenetic relationships and intercontinental biogeography of eastern Asian-North American *Rhus* gall aphids (Hemiptera: Aphididae: Eriosomatinae): evidence

from mitogenome sequences via genome skimming. 2017. *Molecular Phylogenetics and Evolution*, 117: 102-110.

309. Yuan, H., Wu, J. Q., Wang, X. Q., Chen, J. K., Zhong, Y., Huang, Q., Nan, P. Computational identification of amino-acid mutations that further improve the activity of a chalcone-flavonone isomerase from *Glycine max*. 2017. *Frontiers in Plant Science*, 8: 1-8.

310. Zeng, L. Y., Gu, Z. Y, Xu, M., Zhao, N., Zhu, W. D., Yonezawa, T., Liu, T. M., Lhag, Q., Tashi, T., Xu, L. L., Zhang, Y., Xu, R. Y., Sun, N. Y., Huang, Y. Y., Lei, J. K., Zhang, L., Xie, F., Zhang, F., Gu, H. Y., Geng, Y. P., Hasegawa, M., Yang, Z. H., Crabbe, M. J. C., Chen, F., Zhong, Y. Discovery of a high-altitude ecotype and ancient lineage of *Arabldopsis thaliana* from Tibet. 2017. *Science Bulletin*, 62(24): 1628-1630.（生物多样性与植物基因组分析代表性论文）

311. Zhang, Y., Su, X., Harris, A. J., Caraballo-Ortiz, M. A., Ren, Z. M., Zhong, Y. Genetic structure of the bacterial endosymbiont *Buchnera aphidicola* from its host aphid *Schlechtendalia chinensis* and evolutionary implications. 2017. *Current Microbiology*, DOI: 10.1007/s00284-017-1381-0 (online)

312. Wang, Q. B., Xu, Y. Q., Gu, Z. Y., Liu, N., Jin, K., Li, Y., Crabbe, M. J. C., Zhong, Y. Identification of new antibacterial targets in RNA polymerase of *Mycobacterium tuberculosis* by detecting positive selection sites. 2018. *Computational Biology and Chemistry*, 73: 25-30.

钟扬教授著作选摘

I. 专 业 著 作

《数量分类的方法与程序》前言

  分类学是一门古老的基础学科。迄今为止，在生物学研究和应用中仍占有极其重要的地位。然而，随着生物科学的不断发展，传统分类学中存在的不足之处显得十分突出。一方面，生物物种的不断发现和研究领域的逐渐扩展导致资料积累日益增多；另一方面，长期沿用的定性描述方法受人为因素的影响较大，加之繁琐低效的工作方式，不可避免地造成分类研究中的一些混乱。因此，有关分类方法的定量化和科学化的工作业已成为生物学界一项十分重要的任务。

  数量分类学（numerical taxonomy），或译数值分类学，是一门应用数学理论和电子计算机技术处理生物分类问题的边缘学科。它是本世纪50年代末期，由英国微生物学家史尼斯（P. H. A. Sneath）、动物与人类学家克恩（A. J. Cain）和美国生物统计学家索卡尔（R. R. Sokal）等人在总结前人对定量分类方法的研究基础上创立的。此后，越来越多的生物学工作者对这门科学的理论、方法和应用产生了浓厚的兴趣。70年代以后，数量分类方法已广泛应用于生物学及相关学科的众多研究领域中，其数学理论和分析模型也逐步得到发展。

国内数量分类学研究的开展也引起了普遍的重视。近年来，不少单位从事了这方面的工作，尤其是广大青年分类学工作者兴趣较高。为了结合生物系研究生"现代植物分类学技术"课程的教学实习及培训其他分类学工作者的需要，我们编写了本书。编写中主要参考了史尼斯博士等著《数值分类学：数值分类的原理和应用》和阳含熙教授等著《植物生态学的数量分类方法》两本著作，以及徐克学先生在沈阳等地的数量分类学讲习班教材。考虑到大多数生物学工作者在较短时间内全面掌握众多数量分类方法的困难，本书仅综合介绍了数量分类学中的基本概念和较为成熟的分析模型，也选择了一些正在发展中的方法。力求通俗易懂，简明实用。同时，还介绍了常用数量分类方法的BASIC程序，供实际应用中参考。尽管书中大部分例子都与植物分类有关，然而，这些分析方法对于其他分类学科也是适合的。

本书是"数量分类方法与程序"课题的部分总结。中国科学院植物研究所系统与进化植物学开放实验室给予了专项资助。在写作过程中，得到了我们所在单位中国科学院武汉植物研究所和武汉大学生物系的鼓励与支持。李群副教授精心审编本书。张晓艳同志参加了大量工作。李伟、邹洪才等同学给予了帮助。在此，我们一并表示衷心的感谢。

虽然本书的编写历时两年有余，但由于我们水平所限，仍会有不少错误和不足之处。敬请读者批评指正。

编著者
1989年10月于
武昌磨山、珞珈山

《分支分类的理论与方法》前言

　　1990年，我们在编著《数量分类的方法与程序》一书时，就萌生了撰写其姊妹篇《分支分类的理论与方法》的想法，并着手收集了一批资料。尽管当时已对该学科领域的主要原理和方法有了一定的了解，但仍感觉掌握的材料还不够多，理解也不够深，因此只是试写了几段就搁下来了，仅对当时手头上几本还算新鲜的原著（如 E. O. Wiley 的 *Phylogenetics* 一书和 T. Duncan 与 T. F. Stuessy 合编的 *Cladistics* 论文集等）作了一些翻译工作。1992年初，我到美国密西根州立大学（MSU）做访问学者，研究项目为分类信息系统（TIS）的设计，有关分支分类学的研究也算是"业余"课题。在MSU的图书馆，我开始为本书补充资料，却很快发现这一学科领域近几年有了突飞猛进的发展，浩如烟海的文献、眼花缭乱的方法和功能多样的程序使我感到必须对本书另起炉灶才行。到了1993年下半年，我们向美国国家自然科学基金会（NSF）申请新的课题，得到了将分支分类方法与分类信息系统相结合的建议，这就让我"名正言顺"地重新开始撰写本书。在MSU的植物标本馆，我和同事们关于分支学的多次讨论使我获益匪浅。回到武汉植物所后，我们植物计算生物学青年实验室的同志们团结奋斗，发挥集体的智慧和力量，终于在短短的半年时间内完成了本书的修改、定稿和计算机排版工作。

　　本书的编著过程在一定程度上反映了科学工作者面对新的学科领域所应具备的学习探索和协作攻关的精神。当然，犹豫和彷徨也属正常。我们承认自己是初学者，愿意与读者们一道从头学起。从这个观点出发，本书希望能用有限的篇幅来介绍有关分支分类的各种理论和尽可能多的方法，并辅以例子说明，最后附录了大量的参考文献，以使初学者能对该学科的基本原理和方法有一个初步的了解，也使研究者能根据讨论和文献目录作进一步的深入研究。

中国科学院分类区系特别支持费为我们撰写本书提供了有利条件。感谢中国科学院武汉植物研究所和密西根州立大学的同事们，特别是胡鸿钧所长和John H. Beaman教授，没有他们的鼓励、支持与帮助就不可能完成本书。感谢王恒玲、刘莹、雷一东、余清清和洪亚平等同志协助文字输入工作，特别感谢君安证券有限公司为本书出版印刷所给予的雪中送炭般的资助。我们还必须感谢各位编著者的家属们多年来不断的支持和极大的忍耐，使我们在无数个夜晚及节假日得以对这本"讨厌"的书进行写作、讨论和修改。

最后，我们殷切期待着读者对本书的缺点错误和可商榷之处批评指正。

钟扬
1994年10月于
武昌 磨山

《简明生物信息学》前言

生物信息学是生命科学、信息科学、数理科学等众多相关学科相互交融所形成的一门新兴边缘学科，它随人类基因组计划（HGP）的实施而诞生，已迅速发展成为当今生命科学的重大前沿领域之一。

生物信息学针对海量分子数据处理与分析的实际需求，综合运用计算机科学、数理科学和系统科学等相关知识与技术，开展大范围、深入的信息收集、整理、组织、检索和分析工作。这无论对生命科学本身还是众多相关学科都是一个极大的挑战，也是一次发展的机会。该学科领域涉及面广，技术性强，研究和学习的难度很大，迫切需要新生的研究力量。事实上，国外著名大学已在20世纪90年代后期纷纷开设生物信息学教育项目，培养研究与应用型人才。国内也开始认识到该领域的重要性，正奋力直追。

本书系复旦大学生命科学学院和计算科学系"生物多样性信息学"联合研究组学习、研究生物信息学的成果之一。本书在综合国外有关文献的基础上，结合我们的研究成果，概述了生物信息学的基本概念、必备的计算机基础知识，介绍了DNA序列分析、系统发育分析、基因组分析以及蛋白质组分析等原理与方法、关键技术以及常用软件。书中除列有阅读材料、参考文献和思考题外，还附录了最新生物信息学网址和相关刊物。本书在编写过程中主要参考的书籍有：Hooman Rashidi 和 Lukas Buehler（2000）的 *Bioinformatics Basics: Applications in Biological Science and Medicine*、Teresa K. Attwood 和 Dacid J. Parry-Smith（1999）的 *Introduction to Bioinformatics*、Andreas D. Baxevanis 和 B. F. Francis Ouellette（1998）的 *Bioinformatics: a practical guide to the analysis of genes and proteins*（李衍达、孙之荣等译）以及郝柏林和张淑誉（2000）的《生物信息学手册》等，书中分子与遗传学名词主要参考了赵寿元（1999）的《英汉遗传工程词典》（增订版）。本书的大部分内容已在复旦大学98级生物学专业本科生和部分研究生选修的"生物信息学"课程中试用，效果良好。

感谢联合研究组的全体师生日以继夜的工作，使本书能按时付梓；国家自然科学基金重大项目（39893360）、重点项目（69933010，69935020）和教育部博士点基金"植物分子系统发育分析的统计检验模型与计算机模拟技术研究"提供经费资助；复旦大学生命科学学院的教材编写基金给予了雪中送炭般的支持；挂靠单位"生物多样性与生态工程"教育部重点实验室和生物多样性科学研究所则提供了人力、物力和场地的便利；中山大学生命科学学院徐安龙教授审阅了文稿并提出宝贵意见；谈家桢院士欣然作序；特别感谢高等教育出版社的孙素青、刘丽两位编辑和高校生物学教学指导委员会的陈家宽和乔守怡两位教授，没有他们的鼓励、支持与帮助，本书就不可能面世。经佐琴、蒋如敏两位老师和王卿同学在本书编写和课程教学中也付出了辛勤劳动，在此一并致谢。

由于生物信息学内容新、覆盖面广，我们的知识水平有限，加之时间仓促，书中难免有不少错误和遗漏之处，欢迎大家提出宝贵意见。

复旦大学生命科学学院、计算科学系
"生物多样性信息学"联合研究组
2001年8月于上海

II. 专 业 译 著

《水生植被研究的理论与方法》编译前言

在我们编译本书的时候，我国南方刚刚遭受了一场罕见的洪涝灾害，又很快进入了酷暑季节。水灾和旱灾不仅使人民生命财产损失巨大，而且给农业生产造成十分不利的局面。目前，水自然是一个最热门的话题。

人类对于水及水生植物的认识与研究已有了很长的历史。本世纪初以来，随着人们对生态环境问题的日益重视，有关水生植被的研究也有了较大的发展，特别是在水生植物群落的描述与分类、结构与功能、演替趋势以及对水生生态系统的影响等方面建立了若干种研究体系，并结合了物理、化学、遥感、统计学和计算机等技术，形成了不少新的理论与方法。我国是地表水最丰富的国家之一，有着开展水生植被研究的优越条件，长期以来进行了许多富有特色的研究工作。然而，由于种种原因，国内还比较缺乏有关该领域国际研究进展及发展趋势的专题文献，不利于今后深入的研究。鉴此，近年来我们开始系统地收集和整理了一批国内外研究资料。1990年7月，国际知名的水生植物学家、瑞士苏黎世大学的 C. D. K. Cook 教授来武汉大学讲学期间带来了许多新的文献。荷兰奈梅亨天主教大学的 C. den Hartog 教授和美国西南得克萨斯州立大学的 E. L. Schneider 教授等也先后给我们寄来了他们最新的研究成果，并希望能尽快将其介绍到国内学术界。

本书编译了国外近年来有关水生植被研究理论与方法的综述与专论10篇，这些文章分别是：1. 植物生活的水环境（R. G. Wetzel）；2. 水生植被环境的调查与分析方法（H. L. Golterman 等）；3. 水生维管束植物的生活型与生态学分类（G. E. Hutchinson）；4. 水生植物群落的结构概况（C. den Hartog 等）；5. 水生植被描述与分类的植物社会学途径（E. P. H. Best 等）；6. 水生植物群落的群落生态学分类（C. den Hartog）；7. 水生植物的初级生产力（D. F. Westlake）；8. 水生植物的生活策略，对进一步研究的评论与建议（J. T. A. Verhoeven 等）；9. 水生植被演替过程中结构发展的一般趋势（S. Segal）；10. 大型沉水植物对生态系统过程的影响（S. R. Carpenter 等）。它们分别出自下列图书、文集和杂志。J. J. Symoens 主编：*Vegetation of Inland Waters*；G. E. Hutchinson：*A Treatise on Limnology III*, *Limnological Botany*；*Colloq. Phytosoc. X, Aquatiques*。J. J. Symoens 等主编：*Studies on Aquatic Vascular Plants*；*Aquatic Botany*, Vol.26。在编译过程中，一般是一个专题以一篇主要的文章为蓝本，再辅以其他材料加以增删。值得说明的是，各专题之间可能出现一些重叠；作者的写作风格各异，学术观点纷呈甚至相左；研究区域和结果差别较大，等等。我们则力求保持原有特色，突出其重点，尽可能以有限的篇幅提供更多的信息量。此外，还附录了一篇有关我国内陆水生植被研究概况的专题综述，供读者参考。

本书的编译出版得到了中国科学院水生生物研究所"淡水生态与生物技术"国家重点实验室的专项资助。我们所在单位中国科学院武汉植物研究所和华中师范大学出版社给予了鼓励与支持。武汉大学生物系陈家宽教授审阅了本书并欣然作序，廖矛川、南蓬、周进等同志对部分章节提出了修改意见，张敏华、张晓艳、彭学军三位同志在酷暑中誊抄文稿、绘制插图和打印文献，部分插图由王莉娟、陈宝联同志精心绘制，在此，我们一并致以衷心的感谢。

由于时间仓促，加之水平有限，本书仍会有不少的缺点，敬请读者批评指正。

编译者

1991年8月于武昌磨山

《延续生命》译者的话

"你知道生物多样性吗？"如果一个电视节目主持人这么询问路人，相信大多数人能给出肯定答案。倘若再追问，"世界上哪里的生物多样性最丰富？""生物多样性与人类健康有什么关系？"，以及"深海中某种微生物灭绝会影响药物开发吗？"，恐怕大多数人就只能摇头微笑了。

的确，民众对"生物多样性"一词似乎耳熟能详但却知之甚少。大量有关生物多样性的信息来自教科书、标语或新闻，但有时只是一些刻板的概念而已；充斥于电视的探险、综艺、旅游节目，虽带领人们身临生命世界的奇境，但真假混淆、效果夸张，让人怀疑其科学性。本书正好充当两者之间的桥梁——由一线科学家撰写，旁征博引大量史料，科学性上不亚于任何综述论文，却又是贴近生活的案例汇集，犹如故事书一般引人入胜。

本书的独特之处在于聚焦生物多样性与人类健康这一较少被讲述的主题。一方面，它传播新鲜及时的生物多样性知识，可以代替老生常谈的教科书。例如，"为什么熊不得糖尿病？""鲨鱼的软骨真的能抗癌吗？""青蒿素是怎样被发现的？""除蛇毒之外还有哪些生物毒素能入药？"……这些有趣问题所对应的科学知识，令人们深切认识到生物多样性丧失将使我们错失宝贵的生物药物开发良机。另一方面，本书作者秉持批判性思维，对诸如食品安全和有机农业之类的问题不作简单肯定与否定，而是通过各色案例提出多种看法。鉴于培养批判性思维几乎是所有通识教育的难点，本书可作为一本不可多得的通识教育教材。正因为这些独特之处，使我们在翻译本书过程中丝毫不觉枯燥，并收获了大量新知。当然，我们更希望我们的"转述"能给读者带来同样的感受。

本书翻译始于2010年——国际生物多样性年，但断断续续达六年之久，译稿也随我们遍访了世界多地，尤其是主译者在2015年突发重病，险些中断了译校工作。如果没有牛津大学出版社和科学出版社极大的宽容和耐心，这本译著就不可能面世。在本书付梓之际，我们衷心感谢大力推动本书翻译出版的牛津大学出版社和科学出版社，以及相关编辑的辛勤劳动；感谢参与本书翻译和校对的乔琴、张体操、臧婧泽、金燕和顾卓雅等，以及为编制索引提供帮助的陈科元和孙宁宇，他们既积极投身于青藏高原生物多样性与进化研究，也为科普教育奉献了宝贵的时间和精力；感谢陈新、耿宇鹏和郑煜芳对本书翻译所提出的学术上和语言上的中肯意见和建议；感谢欧珠罗布、琼次仁（已故）、卢宝荣、南蓬、黄艳燕、扎西次仁、吾买尔夏提·塔汉、拉琼和德吉等与我们在上海和拉萨进行了不计其数的有关西藏生物多样性与民族生物医学的讨论。

西藏大学理学院生物多样性与地生物学研究所、医学院高原医学研究中心、"青藏高原生物多样性与分子进化研究"教育部创新团队和国家自然科学基金"微进化过程的多基因作用机制"重大研究计划为本书的翻译出版提供了资助，白玛多吉、旦增罗布、徐宝慧、白玲等院领导和复旦大学生命科学学院生物多样性与生态工程教育部重点实验室也提供了便利条件，在此一并致谢。

西藏大学生物多样性与地生物学研究所/高原医学研究中心　索　顿
复旦大学生物多样性与生态工程教育部重点实验室　赵佳媛

III. 科 普 著 作

《基因计算》后记

美国未来学家、麻省理工学院教授尼葛洛庞帝（Nicholas Negroponte）在其名著《数字化生存》一书中曾经说道：计算不再只和计算机有关，它决定了我们的生存。DNA 计算机的出现，预示着这一论断开始走向了生命的本质。生命的基本组成——DNA 或 RNA，本身就具有超乎想象的巨大的计算能力，这种能力一旦被人类发掘，社会将发生翻天覆地的进步，也必将揭开人类文明史上的新篇章。

从婴儿的牙牙学语、蹒跚学步，到青年的聪明强壮，人类每一时、每一天都在历经这样的成长，但需要耐心等待数十年。对于刚刚起步的 DNA 计算机领域而言，更需要这样的等待。我们的科学家们正在努力缩短这一进程。就在十年前，DNA 计算机还被认为是科学上的想象，人们只能从科幻小说中去寻觅它的影子。1994 年，具有丰富想象力和实干精神的阿勒德曼教授敢于第一个"吃螃蟹"，一种简陋但具有真正计算能力的 DNA 计算机雏形终于在他的实验室中变成了现实。作为前所未有的尝试，DNA 计算机一方面遭遇了许许多多的质疑，另一方面又令更多具有卓越眼光的科学家看到了未来，他们不仅小心呵护着这一新兴技术的成长，而且自己也满腔热情地投身其中。自从 2000 年人类基因组计划取得重大进展以来，一批非生物专业出身的科技人员也迅速加入到生命科学领域，这支队伍中有计算机科学、数学、物理、化学、微电子技术、激光技术以及控制科学等诸多学科的科研工作者。尽管他们可能对生命科学不甚了解，甚至在此之前，对什么是 DNA 都一知半解，是大科学的召唤使得他们与生物学家们携手同行。他们中的大多数人擅长物理与数字技术，于是 DNA 研究也成了数学和物理研究中的题材，就是这些形形色色"半路出家"的科研人员大大加速了 DNA 计算机的发展。

在阿勒德曼教授的"新生儿"懵懵懂懂地来到这个世界之时，它的"大哥"——电子计算机已经发展了几十年。面对同一个经典问题——旅行推销员问题,目前一般的电子计算机只要花费数秒的时间就可以轻松地解决,即使单凭脑力,一般人稍花几分钟也可以得出正确答案,但第一台DNA计算机却花费了整整七天才找到答案。这一结果也可能让那些对DNA计算机寄予厚望的人们哭笑不得,也使人不禁联想起1814年间那位沮丧的斯蒂芬逊(George Stephenson)。当时这个制造出世界第一辆蒸汽机车的英国人,也遭遇到火车跑不过马车的尴尬。

什么是真正的DNA计算机？ 2002年初,日本东京大学与奥林巴斯(Olympus)光学公司联合研制出一台可用于分析基因表达概形的DNA计算机,希望其在协助疾病分析、诊断和药物开发方面有所突破。这台机器引发了许多争论。《科学》杂志也报道了这场来自学科交叉的争鸣,不少学者认为,DNA计算机还是应该像传统的电子计算机一样能够进行逻辑计算,初步解决一些简单的教学、逻辑或棋局等方面的问题,而日本研制的这台DNA计算机只能算是比较先进的基因分析系统,并不具备计算能力。事实上,这套系统具有两个部分：分子计算元件和电子运算元件,前者负责计算DNA所结合的分子数、经历的化学反应,搜寻获得正确的DNA运算结果,而后者则用来执行处理的程序和分析相应的结果。这种结合了DNA反应和电子计算机分析的系统,在目前的DNA计算机研究者看来,只是专门针对基因分析应用而设计的,缺乏通用性,非大众所能接受。然而,它的研发者认为,应该从广义的角度去理解DNA计算机,无论如何,这场争论推动了DNA计算机的多元化发展。

著名的化学家、以色列第一任总统魏茨曼(Chaim Weizmann)在建立以色列国的同时也创建了一座世界水平的科学研究院。目前,魏茨曼研究院已是世界公认的高水平科研机构之一,共有13个研究中心,2 000多名科研人员,规模庞大、学科众多。研究院附近的魏茨曼工业区依靠研究院雄厚的科研实力,创建了几十家科技公司,专业范围涉及光学、电子学、遗传学等学科。在DNA计算机领域,魏茨曼研究院也一直走在世界前列。2001年,他们已率先完成基于DNA分子的自动机模型,并以世界上最小的生物计算机入选吉尼斯纪录。2004年,《自然》杂志报道了魏茨曼研究院的科学家们在DNA计算机领域的新突破,他们利用反义技术,在试管中制造出可以进行初步的疾病诊断和药物释放的DNA计算机。这台特殊的DNA计算机可以进行逻辑"思考",分析"输入数据"中若干种信使RNA的丰度,并根据分析结果释放出能够影响基因表达的相应分子。其中,包括三个程式化模块:充当随机分子自动机角色的运算模块、以特定的信使RNA的丰度或者调控点突变的因子浓度为数据的输入模块(其中控制点突变的调控因子的作用类似于自动机转换频率),以及控制一条短的单链DNA分子释放的输出模块。为了检测这台DNA计算机的性能,研究人员以小细胞肺癌和前列腺癌作为研究对象,将其相关基因的信使RNA丰度作为输入,在一系列DNA自身运算之后,释放出利用抗癌药物模拟的单链DNA分子。

继魏茨曼研究院发明了DNA计算机之后,2005年,以色列理工学院的克南(Ehud Keinan)教授等又开发了一台可以在表面实现并行分子计算的DNA计算机。比起几年前仅能进行765个程序运算并且仍需要人工监督的DNA计算机来,这台DNA计算机能够同时进行10亿次运算而只需少量DNA和酶。被《科学》杂志称为目前世界上最快的DNA计算机。

在DNA计算机的国际竞赛中,我国科学家也不甘落后。2004年初,上海交通大学和中国科学院上海生命科学研究院联合宣布,已经在试管中完成了DNA计算机雏形的研制工作,通过试验把自动机与表面DNA计算结合到一起。他们采用了双色荧光标记技术,对输入和输出的分子同时进行检测,用测序仪对自动运行的过程进行实时监测,而磁珠表面反应法等固化手段可以用来提高操作的可控性,完成了模拟电子计算机处理0、1信号的功能。诚然,这只是我们自主开发研制DNA计算机的一个小小的开端,然而意义却十分重大。

DNA计算会有什么用吗?还是让我们回到那道旅行推销员问题吧。目前的世界纪录由3台Digital AlphaServer 4100s巨型机(包含了12个处理单元,每个节点有32台Pentium-II PC)保持,它们可以解决涉及13 509座城市的推销员难题。事实上,创造这个惊人的纪录并不仅仅是依靠电子计算机本身,更重要的原因是采用了非常有效的分支算法。因此，DNA中蕴藏的计算能量还有可能通过算法来提高。

DNA计算算法的革新来自何方？生命的最大奥秘来自于进化，生生不息的生命是以DNA为载体的，DNA计算就像是大自然在做算数，只不过加、减变成了剪切、连接、插入和删除等。DNA计算所蕴涵的理念也可使计算算法产生进化吗？目前，DNA计算理论是一个研究热点，有些困难通过新的数学模型和程序设计技术就可以解决，预示着分子计算在未来可以实际应用，而大自然可能就是我们最好的老师。

毋庸讳言，尽管国际学术界和技术领域不断传来有关DNA计算机的好消息，但它们都还处于理论研究和实验室试验阶段，离真正的实际应用还是距离很远。学术界自身也还存在很多不同的声音。美国麻省理工学院于2003年出版的《分子计算》一书中提到："DNA计算的优势仍然停留在理论。实验室的试验也只是探测到其最近的浅滩。在解决一些问题时它们优于传统方法，但在另一些问题上则相反。从这个角度来看，我认为与其说DNA计算取代了目前的计算方法，还不如说是一个补充。"同时，现阶段的DNA计算机错误率依然很高，要想真正超越电子计算机，还有很长一段路要走。然而，无论如何，有关基因计算传奇般的故事和DNA计算机十年来的飞速发展已经让我们看到了希望的曙光。

IV. 科 普 译 著

《林肯的DNA》译后记

从最初拿到《林肯的DNA》一书的原件到现在译稿成形几近付梓，算来也有一年了。书中由遗传学故事引发出的伦理争论和思索，虽并非短短的24个章节和一年的翻译修改可以穷尽，却也能深入浅出地加深人们对一些常见遗传学问题的了解。

这本书写成于分子生物学和遗传学蓬勃发展的21世纪之初，作者在序言中提到，自己投身于遗传学是受到了神秘的声音的召唤。抛开这种有些玄妙的说法，不如说是遗传学当时的感召力令许多人都对它产生了好奇。那些与人类本性、日常生活、疾病健康等息息相关的问题，看似都能在遗传学中找到答案，但又扑朔迷离。作者正是掌握了读者的好奇的心态，用说故事的轻松方式和自己特有的遗传学、医学和法律的综合背景，为我们提供了一册相当不错的遗传学科普读物，内容海阔天空，不一而足。读者既可以看到作者身为遗传学家用浅显的语言为大家揭示一个个遗传机制，又能看到他作为医生时在医学上面临的一些遗传学难题，当然还有他作为律师时狡黠地对相关法律进展进行引经据典。

当然，不可避免的是，由于遗传学的飞速发展，书内的一些内容现在看来终有其局限性。作者也无法面面俱到，对于某些问题的看法也可能因为了解不够全面而有失偏颇。比如第24章中，作者讲述了优生学和优生运动的简史及他对优生学的一些看法，并对中国的现行法律之一《中华人民共和国母婴保健法》的第十条、第十六条提出了质疑。但如果作者能够了解这是由于语言上的错用而造成的误解，想必也会收回自己的疑问。《母婴保健法》出台甫始的译名曾是"Eugenic Law"，而西方人普遍将"eugenics"用来表示高尔顿当年发起并影响到美国、德国生育政策的优生运动。但中文"优生"的意思则是"生一个好而健康的孩子"或"提高人口素质"，该词也已沿用了数十年。此外，正如作者在书中提到的，中国目前残疾人的总数已有6000万，而这个数字几乎与法国人口总数相当。伦理学上的争议和中国国情也是必须考虑的因素。不过，在1998年于北京举行的第十八届国际遗传学大会（北京曾因为《母婴保健法》引发的争议而险些被取消承办该届会议的资格）上，中国政府对此不当翻译进行了更正，并决定今后新闻部门将把"优生"一词译为"healthy birth"或"birth health"，并对上述存有争议的条文给予了解释和伦理讨论。会议之后的数年内，中国政府和相关部门也对该法进行了一些补充，提出了一些建设性意见。在翻译的过程中，我们为忠实原著，采取了直译的方式。在此，借译后记一角加以说明。

翻译中，我们只能兢兢业业力求不失作者的本意，难免在文采上稍有不足。所幸这些故事本身也着实有趣，不会因语言而有所逊色。只希望这本书，也能焕发出神秘的召唤，让更多对遗传学抱有兴趣的人能真正跨入遗传学的大门。

译者
2005年7月

《大流感》译后记

但凡著书，都希望好卖，对译书者而言，亦是此般心态，希望辛辛苦苦翻译出来的书——无论质量好坏——能够被读者认同、喜爱，而这又与原作本身的内容休戚相关。《大流感》一书是在2004年，时任美国辛辛那提大学医学院教授、甫任复旦大学生命科学学院院长的金力介绍给我们和出版社的。据他本人所言，他是在横跨太平洋的中美航班上将本书通读完毕。原版书的磨损程度证实金教授不止一次翻阅过它。普通人对此书最初的认知，是它得到了美国总统布什的青睐，带去度假。这大抵是这本书最肤浅的噱头了。再进一步说，当时恰逢禽流感、SARS，一波未平、一波又起，人们惊魂甫定之时，的确正是这本书热卖的好时机。然而，光抓住一哄而上的卖点来炒作的书，也许并没有称作"经典"的资格。今天看来，《大流感》并不是一本需要靠炒作来畅销的书。即便再过几年，它仍会是书架上不被冷落的那类。这让我们觉得，花费时间和精力来翻译这样一本书，介绍给读者，的确是件幸事。

持续三年的译书过程当然不那么轻松，仅从内容而言，翻译者就常因太过投入而陷入悲痛。因为翻译《大流感》一书，就好像在亲历那场堪称瘟疫的疾病流行：跟着流感的脚步，尾随其后，眼见其将魔爪伸向各处，虽然是多年之前的往事，但仿若历历在目，到处是呻吟的患者、无力回天的医生、焦头烂额的政府，机构、军队、平民、医生、科学家，无一不受到流感的侵害。人体、人心、城市、国家均被流感所蚕食。比起真正的战争，这场没有硝烟的战斗似乎更加令人殚精竭虑、死伤惨重。

每一章中，作者都浓墨刻画了一个又一个在这场人类同流感的殊死搏斗中发挥重要作用的人物，他们或是亲身投入或是调兵遣将。你能看到他们与流感斗争的同时也与自己的人性进行抗争或者妥协；看到他们对科学的执着或者偏见，对权威的崇拜或者质疑；看到他们现实生活中的痛苦，与

下　编

形形色色的人交往的苦恼,科研中久久未能突破的瓶颈……个人命运随着流感跌宕起伏。巨大的死亡数字,让人觉得悚然,而那一个个具体的人物,却更教人扼腕。或许读者尚不会为了那些近百年前因病而亡的美国民众或士兵感到悲恸;但那些已被作者描绘得鲜活的、原本可能该有更好的生活、更大发展空间的人,却因为一场流感而走向不归路,又怎能不令人为之动容?

尽管翻译这些时不时牵涉死亡的内容让我们不甚愉快,但查究典故,却令我们乐在其中。比如书中反复出现的"流感"一词,如何为其定名,我们颇费了一番脑筋。流行性感冒在英语中有好几种说法:influenza、grippe、flu,原书中不断交替出现。其中的flu一般看作是influenza的缩写,grippe在词典中就被解释为influenza,翻译成汉语之后,形态上的区分就无法移植,而作者是否有其微妙的用意,似乎也难以揣摩。于是我们不得不追究一番influenza的词源,才定下书中最后的译法。

而在知其意后,如何寻找恰当的措辞方面,有时候也需要在现有的资料上稍加斟酌。这场1918年的大流感以"西班牙大流感"著名,但正如书中所说,其并非起源于西班牙,只是因为当时未参与一战的西班牙的媒体对此进行了大量宣传而造成误解,《真理报》称作为"Spanish Lady"。这个词之前总是被译作"西班牙女士",但我们感觉该词所体现的温柔之意同那场流感的肆意大不搭调,于是考量之下译成"西班牙女郎",那种热辣的感觉兴许还能同流感的狂暴沾上点边。

由于作者的旁征博引,书中牵涉了不少文学作品,如歌德的《浮士德》、加缪的《鼠疫》、安妮·波特的《灰色马,灰色的骑士》,等等。翻译过程中,有的作品有着多个优秀译本可供选择,有的则是要翻遍图书馆的角落才能找到。在能找到现成译本的情况下,我们尽量参考采用现有翻译,当然也会根据情况做些小小改动。而少数诗歌的翻译则没有那么简单,"信"和"达"不容易,"雅"更是一个问题。如上所述,若有现成译本翻译起来可以轻松不少,倘若碰上一些原本就是打油诗,或是标语、谚语的部分,则无据可考,无译可参,我们只能硬着头皮反复琢磨,力求在信、达的基础上,能还原原文或雅或俗的意味和语感。

《大流感》并不是一本用来娱乐的书，读后反而让人感觉有些沉重，每个翻译过哪怕只是一个篇章的人都能体会到这一点。我们希望，这本名为"史诗"、实则讲述悲欢交加而悲痛更甚的"悲剧"的书，能让我们对过去的那场大流感有更多的了解，并明白疾病的强大和科研的艰难。的确，我们今天已经可以快速测定每一株流感病毒的基因（甚至全基因组）序列，但对其起源和进化规律仍未做到了如指掌，更遑论预测其爆发时间了。药物——我们与病毒或致病菌斗争的武器，也远不如它们上市时宣传的那么神奇。一方面，它们的确缓解了我们的痛苦并带给我们战胜病魔的希望，但另一方面，它们本身不是"常胜将军"，也不是书中提到的"神奇子弹"，而其副作用甚至还会增添新的麻烦。更令人惊讶的是，许多病毒和病菌具有极为高超的进化本领，能在人类活动和药物作用的巨大压力下快速生成新的抗药变异株，使我们开发药物的工作不可能一劳永逸，必须时刻迎接新的挑战。在进化意义上，人类与病毒、病菌的斗争可以说是一场永不停歇的"军备竞赛"。其实不仅流感病毒如此，肝炎病毒如此，就连我们原以为牢牢控制的结核杆菌也是如此，它们随时会卷土重来。面对这些问题时怎么办？读者也许会从《大流感》的字里行间得到有益的启示。

衷心感谢李作峰、王莉、赵晓敏、梅旖，他们在繁重的研究生学习的同时协助翻译了许多章节的初稿。感谢上海科技教育出版社的侯慧菊老师，没有她的鼓励、鞭策和耐心，本书不可能面世。

对于本书中大量的医学术语，我们虽一一借助字典、网络或是求教于有医学背景的同行以求精确，但仍可能有纰漏谬误之处，希望读者指正。

钟　扬　赵佳媛
2008 年 11 月于复旦

下 编

《DNA博士》译后记

作为现代社会的普通民众,我们一直将崇尚科学和崇拜科学家混为一谈。从小学作文描述"我的理想",到填写高考志愿专业,我们的脑海中时常闪现出古今中外某些大科学家的名字,他们的辉煌成就更是吸引无数青年学子踏上了科学研究之路。多少年后回首,方知这条崎岖山路的艰辛,能攀蟾折桂者又有几人?在科学的竞技场上,失败者自不待言,无人相信眼泪。成功者笑逐颜开,谁人闻其心声。即使在多元化开放的今天,大多数媒体在涉及与科学相关的新闻报道时,总是极力渲染科学成就的巨大价值,而有意无意地掩盖成就创造者真实的心路历程,无怪乎大众眼中的科学家逐渐形成了千人一面的固定模式。

于是,为大科学家作传似乎成为一项颇有意义的工作。事实也证明,《居里夫人传》的影响力绝不止于异国求学的女性,《哥德巴赫猜想》确实唤起了国人对摘取"数学皇冠上的明珠"的热情。不用提诺贝尔奖获得者,只要你手头有一份美国国家科学院院士名录,就可以从网上搜索到其中每一位的传记。当然,这些传记中包括自传和访谈录——这对科学界来说还是一种较为稀少的形式。一般来说,传记是真实素材加工后的主观产物,自传更是如此。在阅读科学家传记时,我们似乎比读文学家和政治家的传记更苛求其真实性,一如科学本身来不得半点虚假。因此,设想有某位成就非凡又乐于"高调"行事的科学家出版了几本传记,若将它们放在一起进行比较分析,不仅可以发现其中的差别,你甚至还会产生想去向科学家本人对质的冲动。

詹姆斯·沃森——DNA双螺旋结构的发现者之一,就是这样一位科学家。他的伟大成就、古怪行径、丰富故事以及有趣文风足以让他为自己写上好几部自传——事实上他也毫不客气地这么做了。在获得诺贝尔奖后不久,他就出版了风靡全球的自传——《双螺旋》[1]；在双螺旋发现50周年纪念之前，以其情感历程为主线的自传《基因·女郎·伽莫夫》[2]又得以问世。这些传记和学术界如潮的评论令他成为了一名现代"话题"科学家。沃森喜欢自己掌控故事，就像他在《双螺旋》的序中强调的："这是如何发现DNA结构的我的版本。"而对于接受他人的深度采访，沃森却一直小心谨慎。所幸最终他还是对"传记垄断"有了让步，才令本书成为沃森自传之外第一本经由别人讲述沃森其人的传记。

这本访谈录——《DNA博士》亦可说是对上述传记的"对质"之作。作者先对沃森的自传做足了功课，在同沃森一次次的谈话中，话题貌似天南海北，观点相和相悖，甚至提问和被问的角色互换，再加上采访角度的巧妙，力图向大家呈现一个多样的沃森，一个真实的人，而不仅仅是那个笼罩在"DNA结构发现者"光环下的科学家。即便是在谈及大家耳熟能详的DNA结构的发现之旅时，惯常争论的到底有多少人作出了切实贡献已不再是话题的关键，我们从对话中看到的是沃森的工作以及他和诸多同行在亲密合作、交流互动间迸发的火花，让这片断成为最终能组合出一个真实科学事件的"罗生门"拼图的一块，而不是简单重复。访谈时作者有意诱导抑或沃森无意流露出来的故事，也让沃森这个科学家的形象更加丰满起来。就像作者转述的、沃森自己不无得意说过的那样，"如果拿他和爱因斯坦及百年内的其他人比较，其生活的每个细节都应该有趣至极"。而这些有趣至极的故事，不单单有着沃森自己的主观版本，还有了其他作者的客观版本。

访谈录在受访者的成就之外，必然会涉及逸闻趣事，人们会在读罢这类访谈录之后得出一个结论：科学家都是差不多样子的普通人嘛！可《DNA博士》却不会用这样的方式给你一个"普通的"沃森。对话字里行间展示出

1　*The Double Helix: A Personal Account of the Discovery of the Structure of DNA*，1968 出版，2006 年科学出版社出版影印版；中文版《双螺旋：发现DNA结构的个人经历》，2001 年生活·读书·新知三联书店出版，田洺译。

2　*Genes, Girls, and Gamow: After the Double Helix*，2001 年出版；中文版《基因·女郎·伽莫夫：发现双螺旋之后》，2003 年上海科技教育出版社出版，钟扬等译。

下 编

沃森及其一些同行的思维方式和所具备的素质，绝不是用"普通"一词就能轻描淡写而过。对于诺贝尔奖遗珠、科学与伦理、女性科学家等话题的讨论，其观点的鲜明、尖锐、独到，都不是普通人可以达到的。还有那些故事中所谓的巧合，比如偶然看到DNA晶体衍射照片的沃森和克里克，比如在寻究聚苯丙氨酸性质时，"问到了世上唯一知晓这个问题答案的人"的尼伦伯格，与其说那是运气，不如说他们对所见所闻感知与常人不同，才能抓住那瞬息流逝的机遇之光。但更值得钦佩的是，沃森的天赋异禀和青年得志乃至这个社会的浮躁，并没有使他变得急功近利。如果没有为某篇论文做出贡献，他不会在文章上署名；如果有一项年轻科学家首创的科研项目，他就会呼吁为其拨给经费；如果他念念不忘的关于人脑的研究在未来被一一揭示的话，就更证明了他的目光高瞻远瞩。此外，值得一提的是作者本身也是一位颇有成就的科学家，他同许多科学家的交情匪浅，又有机会阅读大量文献资料，我们可以看到在对沃森访谈的实时记述中不时穿插着他的严谨评述。科学家作者同一般传记作家不同的写作风格和思考角度，也是我们从作者这方面来了解科学家思维方式的重要途径。如果你耐心读完这本访谈录，就会逐渐发现它的价值远不在于"对质"沃森的自传和反映科学家的生活多样化那么简单，它还触及到了科学家在现代社会中应有的地位问题。阿基米德、伽利略、牛顿、达尔文，这些历史上的科学大家，在很多人眼中俨然成了神话。科学家在大众心中不是深居简出、生活邋遢、无能的怪人，就是一呼百应、名利双收、万能的超人。20世纪中叶，以色列政府就这样如神一般地崇拜着爱因斯坦（或者只是爱因斯坦的名声），甚至想邀请他去当以色列的总统。爱因斯坦当然是拒绝了，但如此这般对科学家的盲目崇拜绝非个案。否则当今世界怎么能有那么多"学而优则仕"的科学家？事实上，很少有在出仕之后还能有科研突破的科学家，而科学家无所不能的神话该被打破了！纵然是沃森这样的天才，在某些方面也有其幼稚的一面。在这部访谈录中，作者甚至保留了一些沃森不小心爆出的粗口。这或者是沃森表达愤懑情绪的方式之一吧。不过，会吃光小朋友生日蛋糕的沃森，有什么做不出来、说不出来呢？尽管他自称（似乎还颇有些遗憾）出于冷泉港实验室负责人的立场，不能如克里克一样口无遮拦，但如果他真如其所言那般懂得缄默，就不会因那些对人种与先天智力的关系有失偏颇的言论而备受指责，以至于不得不辞去行政职务[1]。沃森常常从自以为客观而科学的角度来评价一些事物，比如他一方面夸赞着犹太人的智商的同时又对犹太人的宗教、传统和食物颇多微词。对于敏感的政治、宗教问题，他也是抱持这样的立场和价值观而有时语出惊人。遗憾的是，科学界有时是宽容的，地球村却不是。

钟　扬　赵佳媛

2009年7月于复旦

---

[1] 此处指沃森因其黑人智商不及白人的言论为2007年10月14日的《星期日时报》(The Sunday Times)所引用而被冷泉港实验室于18日解除了职务一事。

下 编

V. 科普译著审校

《造就适者》导读：证据的力量

自达尔文（Charles Darwin）于1859年出版《物种起源》以来，"进化"（evolution，或译为演化）已逐渐成为生物学界使用频率最高的词语之一，并渗透到自然科学与社会科学的众多领域。150年来，进化理论不断发展并广为传播，终成主流科学思想。连教皇约翰·保罗二世在1996年写给教皇科学院的信中也表示："新的发现引导我们承认进化论不只是一种假说。事实上，在不同科学领域一系列的发现之后，这个理论不可思议地对研究人员的心灵产生愈来愈大的影响。"不过，人们也注意到，分子生物学兴起的60年来，一些不能用达尔文进化论直接解释的科学现象开始涌现，进化理论似乎面临着新的挑战。

一个月前，我应邀为上海的一个公众科普活动——"科学咖啡馆"做了一场题为"生物进化与我们的未来"的报告。我在报告中除简要介绍进化生物学（尤其是分子进化）的基本概念和研究进展外，还列举了四个开放问题（open questions）：压力还是动力？缺失还是获得？数量还是质量？个体还是群体？目的是帮助听众了解自然选择的力量、性状进化的方向、延长寿命的意义以及长期进化的策略等当代进化生物学研究的热门领域。所谓开放问题一般都是没有标准答案的多向思维问题，这在国外学术讨论和科学普及活动中十分常见，但在我国还是一种较为罕见的形式。鉴此，我采用了若干实际案例而不是直接用学术界目前流行的理论来解释上述问题，这给习惯于只接受一种"正确"的理论，以及长期受熏陶于重科学结论而轻研究过程的教育模式的听众们带来了些许新鲜感。

现在，对当代进化生物学中开放问题感兴趣的读者可以从《造就适者——DNA 和进化的有力证据》一书中获取更多案例和进化证据了。比如，书中提到，布韦岛的冰鱼是一种完全丧失血红蛋白、没有红细胞的南极"无血"鱼，由于缺乏化石证据，因而很难从形态学或生理学上提供其起源与进化的明证，不过人们还是可以获得其现存种群的遗传物质——DNA。DNA分析结果清晰地表明，冰鱼在其进化过程中"舍弃"了两个合成血红蛋白中珠蛋白的基因，而在5亿年前这两个基因却是其生活于温暖水域的祖先不可或缺的。进一步比较冰鱼不同近亲及其他南极鱼类的DNA序列与结构，科学家们终于揭示了冰鱼由生活于温水、依赖血红蛋白转变成生活于冰水、无需血红蛋白（一些物种甚至不需要肌红蛋白）的进化历程，并且估计出基因丧失的时间范围，为生物进化的基本原则——自然选择和遗传变异增添了新的证据。

与南极冰鱼中基因缺失（gene loss）的故事相反，乌干达基巴莱森林中的疣猴通过基因获得（gene gain）来辨认出营养较丰富的树叶，而科学家们解开其全彩视觉和反刍消化系统"进化创新"之谜的关键还是DNA证据。所有猿类和旧大陆（非洲和亚洲）猴类的视觉都具有三元辨色力（可以看到蓝、绿、红三原色所构成的颜色光谱），而大部分哺乳类只有二元辨色力（可以分辨蓝色和黄色，但无法分辨红色和绿色）。由于热带地区一大半植物的嫩叶呈红色，因而只有这些具有三元辨色力的灵长类可以独享既柔软可口又富有营养的嫩叶。对哺乳类的视蛋白基因分析发现，人类和黑猩猩及其他猿类都有3种视蛋白基因，而其他哺乳类只有2种视蛋白基因。显然，人和上述灵长类动物的视蛋白基因数量随其进化历程而增加，基因重复（gene duplication）和功能分歧（functional divergence）则是其基本进化机制。通俗地说，上述视蛋白基因先通过制作"拷贝"来倍增DNA信息，再靠这些不同复制品接受自然选择的考验，各奔前程，最终进化出具有不同功能的"新"、"旧"基因。当然，更令人惊叹的是，这些不同功能的"同源"基因在同一个生物体中必须各司其职、和平共处才行。同样，作为反刍动物的乌干达疣猴也是采用基因重复和功能分歧的套路，在继续保持与非反刍猴类几乎完全相同的溶菌酶基因的同时，发展出另外两个具有新功能的基因，以满足疣猴对大量嫩叶的消化需求。

几乎每一本进化生物学教科书中都会列举一些研究案例和科学发现，但对发表于各类学术刊物的大量原始"素材"进行合理剪裁却并非易事。本书作者肖恩·卡罗尔教授显然是一个讲故事的高手,他将一个个涉及不同物种在不同地域和不同生境中的进化故事娓娓道来,向我们展示了令人惊异的、鲜活的进化线索及其分子证据。是的,他精心制作的这一道道赏心悦目的"大餐"(作者语)都是与开放问题答案有关的线索和证据,而非答案本身;但正是这些构成证据的故事,显示出比普通教科书大得多的威力。我想,即使是对进化理论持怀疑甚至否定态度的人也无法回避自然的证据吧。

可以说,今天的进化生物学家是如此幸运,因为我们进入了基因组时代——获取一个生物物种的全部DNA序列(称为全基因组测序)已越来越便利而经济,呈现在我们面前的海量信息中不乏新的生物进化证据。诚如书中所言:"基因组学能让我们看到进化过程的深层内涵。达尔文之后的一个多世纪内,人们只能在雀鸟或飞蛾的繁殖和生存中观察自然选择的作用。而现在,我们可以看到'适者'是如何产生的,因为DNA中包含的各种信息是达尔文无法想象或期望的,完全是新的、不同的。不过,这些信息让他的进化理论更加坚不可摧。我们现在可以识别DNA中特定的变化,了解这些变化如何让物种适应不断改变的环境,进而进化出新的生命形式。"

写到这里,我起身拉开窗帘,发现不知不觉间窗外竟大雪纷飞。前方的比日神山已披上银装,西藏巨柏依然孤傲地耸立于山间。如同卡罗尔教授在书末所担忧的一样,人类活动和全球气候变化极大地影响着生物的进化历程,而对在青藏高原这类极端环境和生态敏感地区艰难适应的生物而言更是雪上加霜。除了呼吁和祈祷之外,我们至少应当努力了解这些物种各自独特的适应机制,才能最大限度地降低威胁其生存与发展的环境扰动,以免它们走上灭绝的不归路。

钟扬(复旦大学/西藏大学教授)
2012年11月于西藏大学林芝校区

附：钟扬教授著作总目

专 业 著 作

《数量分类的方法与程序》，1990年，钟扬，陈家宽，黄德世编著，武汉大学出版社。

《分支分类的理论与方法》，1994年，钟扬，李伟，黄德世编著，科学出版社。

《简明生物信息学》，2001年，钟扬，张亮，赵琼著，高等教育出版社。

专 业 译 著

《水生植被研究的理论与方法》，1992年，李伟，钟扬编译，华中师范大学出版社。

《分子进化与系统发育》，2002年，吕宝忠，钟扬，高莉萍等译；赵寿元，张建之等校，高等教育出版社。

《生物信息学（中文版）》，2003年，钟扬，王莉，张亮主译；李亦学，钱晓茵，张晓宁校，高等教育出版社。

《计算分子进化》，2008年，钟扬，张文娟，梅旖，王莉译；杨继校，复旦大学出版社。

《延续生命》，2017年，索顿（钟扬），乔琴，张体操，臧婧泽，金燕，顾卓雅译；赵佳媛校，科学出版社。

编委/编译/审校

《药物基因组学——寻求个性化治疗》，2005年，蒋华良，钟扬，陈国强，罗小民等译；钟扬，南蓬校，科学出版社。

《科学编年史》，2011年，编委：汪品先、杨雄里、陈运泰、胡亚东、钟扬等，上海科技教育出版社。

科 普 著 作

《基因计算》，2003年，钟扬，张文娟，王莉，赵佳媛著，上海教育出版社。

科 普 译 著

《基因·女郎·伽莫夫》，2003年，钟扬，沈玮，赵琼，王旭译，上海科技教育出版社。

《林肯的DNA》，2005年，钟扬，李作峰，赵佳媛，赵晓敏译，上海科技教育出版社。

《大流感：最致命瘟疫的史诗》，2008年第一版，2013年第二版，钟扬，赵佳媛，刘念译；金力校，上海世纪出版集团。

《DNA博士》，2009年，钟扬，赵佳媛，杨桢译，上海世纪出版集团。

《基因密码》，2016年，钟扬，黄艳燕，刘天猛，陈科元，徐翌钦译，上海科学技术出版社。

科 普 审 校

《造就适者》，2012年，杨佳蓉译；钟扬校，上海科技教育出版社。

《听基因讲祖先的故事》，2005年，长谷川政美，任文伟，杨莉琴著；曹缨，钟扬审订，上海科技教育出版社。

图书在版编目(CIP)数据

钟扬文选/钟扬著.本书编委会编.—上海:复旦大学出版社,2018.6(2018.7重印)
ISBN 978-7-309-13623-4

Ⅰ.钟… Ⅱ.①钟…②本… Ⅲ.植物学-文集 Ⅳ.Q94-53

中国版本图书馆 CIP 数据核字(2018)第 071125 号

钟扬文选
钟　扬　著　本书编委会　编
责任编辑/唐　敏　胡欣轩

复旦大学出版社有限公司出版发行
上海市国权路 579 号　邮编:200433
网址:fupnet@fudanpress.com　http://www.fudanpress.com
门市零售:86-21-65642857　团体订购:86-21-65118853
外埠邮购:86-21-65109143　出版部电话:86-21-65642845
常熟市华顺印刷有限公司

开本 787×1092　1/16　印张 25.75　字数 350 千
2018 年 7 月第 1 版第 2 次印刷

ISBN 978-7-309-13623-4/Q·106
定价:78.00 元

如有印装质量问题,请向复旦大学出版社有限公司出版部调换。
版权所有　　侵权必究